WATER POLLUTION AND MANAGEMENT

WATER POLLUTION AND MANAGEMENT

Editors

Dr. D.S. MALIK

Dr. SUNIL KUMAR

UMESH BHARTI

BIOTECH

2011

BIOTECH BOOKS

Delhi

ISBN : 978-81-7622-227-3

Published by : **BIOTECH BOOKS**
4762-63/23, Ansari Road, Darya Ganj,
New Delhi - 110 002.
Phone : 23262132
email : biotechbooks@yahoo.co.in

Laser Typesetting : **Twinkle Graphics**
Delhi - 110 088

PRINTED IN INDIA

Preface

Water is a prime resource, a basic human need and a precious national asset. Today water resources have been most exploited natural system since man strode the earth. All over the world, industrial development, rapid population growth and urbanization have caused indiscriminate exploitation of natural water resources without any regards to its protection, conservation and management aspects. Pollution of water bodies is increasing steadily due to natural and anthropogenic factors along with increasing living standards and wide spectrum of consumption patterns of human's activities. The problems of water pollution are not a recent one but remain with the growth of human civilization. Therefore, research scientists, social workers, environmentalists and non-Govt. organizations are critically observing the various water pollution aspects and their effect on health of all living organisms and their associated environment. The management of water pollution is most important environmental issue of most immediate concern with survival of living organism, public health, and all economic development of all nations. Due to extensive exploitation of water resources, the availability of water could not longer be considered as plentiful and in exhaustible but has become a valuable and scare source. As the demand for water continues to increases, sustainable, holistic comprehensive water conservation, recycling, reuse and management policy should be strictly motivated in all steps of industrial processes and other stages of life support system. In the present book, the research papers related with water pollution and its management concerned with scientific knowledge based data, suggested various biological water pollution indicators, focused on water quality indices and find out a lot of bioremediation technology, nanotechnology, water treatment methods to combat the pollution load from the point and non point source of water pollution.

We express thanks to all authors for their valuable contributions for the book and hopeful that this publication will be useful to research scientists, environmentalists, academicians and other varied users, who are mainly concern with water resource management. All efforts have been made to avoid any omissions and errors but still some may have come to light after publication. It is hoped that those oversight will not detract the readers from consulting the information presented by different authors in their respective papers.

Editors

Dr. D.S. Malik, Dr. Sunil Kumar & Umesh Bharti

Contents

1

Bottom Analysis by Using the Benthos in Derelict Water Body of Aligarh Region

Habeeba A. Kabir, Saltanat Parveen, Asif Ali Khan, Shazia Ansar, Uzma Ahmad and Altaf H. Ganai

ABSTRACT

The need to assess the environmental condition of a pond encouraged the design of specific biotic indices to evaluate the response of benthic communities to human induced changes in water quality. The pond receives rain water from the surrounding area along the shore line. Physico-chemical factors namely; temperature, hardness, T.D.S., pH, D.O., CO_2, alkalinity and conductivity were analysed with the help of method given in Trivedy and Goel (1984). For benthos analysis bottom mud sample was collected with the help of mud scrapper. Separation and identification was done with the help of method given in Edmondson (1959) and APHA (1998).

Keywords

Benthos, Diversity, Eutrophication, Phytobenthos, Zoobenthos.

INTRODUCTION

The development of biological indicators used as a tool for the assessment and hence protection of biological diversity. Benthic invertebrates are often used as bio-indicators to detect and monitor the environmental changes. Benthic species are relatively long living sessile organisms unable to avoid unfavorable condition. In this way the benthic community plays a vital role to determine water and sediment quality condition and their presence and absence indicate temporal as well as spatial differences. Water is an integral constitute of life and one of the most important natural resource. The benthic zone is one of the ecological region of a body of water. It comprises the bottom such as the ocean floor or the bottom of lake the sediment surface and subsurface layers.

Fresh water present a wonderfull pictures of thousands of living organisms ranging from microscopic to macroscopic organisms including chemosynthetic bacteria, saprophytic fungi, micro and macro invertebrate algae and other higher aquatic plants. These communities are named as Plankton, Neuston, Nekton, Pleuston and Benthos etc. Benthos are sensitive to watershed condition and exhibits sufficient stability in assemblage structure over time to make them useful to monitor stream and indicator of water quality (Gauffin and Tarzwell-1952). The density of benthos in a water body is useful index of water quality although density may fluctuate widely with changes in the season and space Hynes (1960). The benthic algae productivity accounts for large percentage of total primary production in shallow esturine water bodies (Jonson 1991). Sometimes it may be even greater than that of phytoplankton (Hargrave 1983). The study of benthic communities has been found to be best indicator of pollution (Venugopal 1982).

The benthos is composed of bottom dwelling organisms and has a major categorical divison between phyto and zoobenthos. The term benthos is derived from two greek words *Ben-* meaning the organisms living in or on the sea, and *thos-* meaning the bottom of sea or lakes. Benthos as an organism associated microscopic to macroscopic fungi, micro and macro invertebrates, algae and other higher aquatic plants. The need for comprehensive studies on benthic fauna is important for better understanding of benthic community and also to evaluate the eutrophic condition of water body. Benthos has significantly contributed for better understanding of benthic community and also to evaluate the eutrophic condition of waterbody. Benthos forms an important part of food chain especially for fish. Benthos played a critical role in the natural flow of energy and nutrients.

Ecologist evaluate environmental quality using benthic sample as they are important indicator of streams, river or lake quality.

Medical college pond is a sewage fed pond, situated on the back side of the medical college residential complex at a distance of about 0.71 hectares, having a depth of 1 to 2 meter in different season. The basin is simple flat and slightly sloping. The bottom samples are black in colour consisting of sand, silt clay, mud with decayed plant matter and organic substances. These various types of deposits support various kind of benthic organisms.

METHODOLOGY

For the present work bottom mud scrapper with tow line was used to collect the sample from the bottom of the point for the quantitative and qualitative analysis. This mud scrapper as described by Michael (1984). Separation was done with the help of sieves having different mesh sizes arranged in a series according to APHA (1998). Identification was done with the help of keys given in APHA (1998) and Edmondson (1959), counting was done with the help of sadgwick rafter cell. Counts were made as number of benthos/m^2. Physico-chemical analysis was done with the help of method given in Trivedy and Goel (1984).

RESULT AND DISCUSSION

Phytobenthos is represented by algae group viz, Cholorophyceae, Myxophyceae, Bacillariophyceae, Euglenophyceae and Desmidiaceae. Cladocera, Copepods, Oligochetes, Rotifers, Ostracods and Insect constitute the major dominating group was the group show continuous occurrence. The presence of higher number of chironomous throughout the study period showed eutrophic condition of water body. In my study the mollusc are absent throughout the study period it shows the water body is highly polluted because filter feeding mechanisms of mollusc are blocked. In my study chironomous was one of the most dominating group in the pond indicate good water quality as well as favorable condition for insect community. In my study the presence of coleopteran found throughout the study period because the adult beetles are tolerant of wide changes in pH and dissolved oxygen and must rise to the surface to respire atmospheric oxygen concentration. Many adults can't use dissolved oxygen and must rise to the surface to respire atmospheric oxygen.

The diptera is one of the most dominating species and found throughout the study period and chironomous was one of the most abundant group the reason is that they stored oxygen in their in the body. This allows them to live in the area that has limited oxygen condition such as eutrophic water body. In the present study the molluscs are absent throughout the study period. Because the clams and mussels are exclusively filter feeder and in polluted water body the filter feeding mechanism of mollusc are blocked. Due to organically polluted pond and dissolved oxygen concentration become reduced or eliminated and abundance of tubificid and oligochaetes are commonly found concomitant with a precipitous reduction and exclusion of most other benthic animals. As long as anaerobic sedimentary metabolisms do not accumulate the rich food supply and freedom from competing benthic animals and predator permit rapid growth.

The total number of benthic community was dominated during winter season it indicates winter season is one of the most favorable condition for benthic community. The higher number of Rotifers during August 320 No/m^2 and lower during the month of July (211 No/m^2). The maximum number of Cladoceran 130 No/m^2 during December and minimum 71 No/m^2 during August. The higher number of Copepods 134 No/m^2 during March and minimum 101 No/m^2 during September.

The higher numbers of Oligochaetes were recorded maximum 340 No/m^2 during December and lower 207 No/m^2 during July. The maximum number of Diptera community was found during September and minimum number 345 No/m^2 given in Table-2. Dipteran is one of the most dominating groups throughout the study period it indicates the good water quality for insect community and higher number of Chironomous indicates the water body is highly eutrophic. The number of benthic community decreases during winter and spring season

Table 1 : Monthly Variation in Different Physico-chemical Parameters in Medical Pond

April	8.7	26	24	19.2	5.2	–	423	–	144	36	180	400	140	40.86	2496
May	8.3	32	35	18	1.8	absent	235	–	80	45	125	230	60.5	42.5	2168
June	8.1	35	30	17.5	1.6	–	312.8	–	60	60	120	291	98.5	16.37	2050
July	8.0	37	34	18	3.4	–	225	–	56	46	102	212	70.8	7.55	2200
Aug	9.0	32	35	19	4.7	–	245	–	72	58	130	197	80.2	11.64	2340
Sep.	9.6	33	34	19.5	2.4	–	198	–	40	64	104	202	82.5	10.28	2398
Oct	9.2	29	27	16.0	3.6	–	212	–	188	110	298	184	67.8	11..9	2458
Nov	9.2	24	22	15.0	6.8	–	240	–	78	227	305	230	91.0	4.09	2598
Dec	9.0	19	20	16.0	9.4	–	230	–	94	103	197	198	44.08	22.36	2652
Jan	9.3	15	17	15. 5	9.2	–	245	–	56	102	158	319	52.8	24.3	2698
Feb	8.5	20	27	15.4	8.7	–	162	–	140	32	172	312	60.12	42.9	2525
Mar	8.5	26	24	15.7	4.7	–	153	–	56	112	168	350	61.3	13.8	2528

because during this season the predator pressure increase and high turbidity decreases oxygen.

In Medical pond insect remained highest but maximum (45.9%) was observed in July, 2007 with a tow line population (36.4 %) in March, 2008. Highest percentage (8.9%) of Cladocera were recorded in July, 2007, where as highest percentage of Copepods were recorded (8.3%) in the month of January, 2008 and lowest (4.9%) in the month of October, 2007. Oligochaetes were found to be higher (20.3%) in September, 2007 and lower (14.4%) during January, 2008. High percentage of Ostracods (20.4%) in the month of March, 2008 and low percentage (6.5%) in the month of December, 2007 and Rotifers (21.7%) were recorded in the month of August, 2007 and minimum (15.1%) in the month of July, 2007 respectively (Table 3).

PHYSICO-CHEMICAL PARAMETERS

Water Temperature

Water temperature in ranged between 15°C to 37°C. The temperature was observed to be lowest during and higher during seasons shows in Table-1.

pH

This parameter showed variation between 8.1 to 9.6 shows in Table-1. In general lower pH values were observed during and higher during season.

Electrical Conductivity

In general this parameter ranged between 270 to 340 µS. The conductivity was observed to be lower 2050 during June and higher 2698 during January shows in Table-1.

Dissolved Oxygen

The D.O. values were higher 9.4 during December and lower 1.6 during June shows in Table-1.

Hardness

The total hardness was observed to be higher 423 during April and lower 153 during March season shows in Table-1.

The fluctuation in the values of hardness higher and lower is due to continuous washerman activity and continuous addition of surf in the pond.

Calcium

The value of calcium ranged between 44 to140 shows in Table-1.

Magnesium

This parameter showed variation between 7.5 to 42.5 shows in Table-1.

Alkalinity

The value of alkalinity ranged between 102-305 shows in Table-1.

Transparency

This parameter showed variation between 15.0 to 19.5 cm shows in Table-1.

Acknowledgment

We greatly like to thank the chairman of zoology for their help and laboratory facilities during my experimental work. I am also indebted to my supervisor because the work doesn't get a good result without her support.

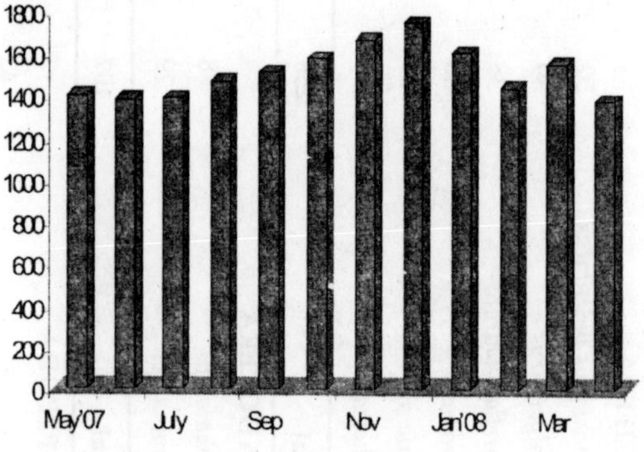

Fig. 1 : Histogram showing monthly variation in Zoobenthic community.

Table 2 : Distribution and Abundance of Zoobenthos (No/m²) in Medical Pond

Months/Genera	*May'07*	*June*	*July*	*Aug*	*Sep*	*Oct*	*Nov*	*Dec*	*Jan'08*	*Feb*	*Mar*	*Apr*
ROTIFERA												
Branchionus bidentatus	52	26	28	68	39	64	62	67	62	69	36	48
B. calyciflorus	39	56	29	62	58	26	64	28	48	52	39	28
Keratella quadrata	38	39	52	52	39	45	29	38	29	48	39	49
K. tropica	60	59	52	64	42	29	25	44	68	58	64	63
Filinia sp.	39	37	28	36	35	54	62	72	45	38	27	27
Notholca	28	29	22	38	26	29	29	34	29	39	36	29
Total	**256**	**240**	**211**	**320**	**239**	**247**	**267**	**283**	**281**	**304**	**241**	**241**
CLADOCERA												
Daphnia sp.	78	97	29	32	54	59	58	92	67	53	48	69
Bosmina sp.	29	28	54	39	29	39	42	38	29	27	42	38
Total	**107**	**125**	**83**	**71**	**83**	**98**	**100**	**130**	**96**	**80**	**90**	**107**
COPEPODA												
Cyclops sp.	42	29	62	45	54	45	52	61	69	49	55	49
Diaptomus sp.	52	49	45	45	56	34	58	34	65	45	29	39
Total	**94**	**78**	**107**	**90**	**110**	**79**	**110**	**95**	**134**	**94**	**84**	**88**

Contd...

Months/Genera	*May'07*	*June*	*July*	*Aug*	*Sep*	*Oct*	*Nov*	*Dec*	*Jan'08*	*Feb*	*Mar*	*Apr*
OSTRACODA												
Cypridopsis	32	51	62	64	39	56	42	29	44	28	27	39
Nauplius	24	25	19	22	17	32	26	21	34	18	14	49
Eggs	45	42	65	46	45	56	62	65	54	59	46	43
Total	**101**	**119**	**146**	**132**	**101**	**144**	**130**	**115**	**132**	**105**	**320**	**131**
OLIGOCHAETA												
Tubifex	52	49	54	79	87	95	72	87	58	46	45	42
Chaetogaster	42	36	29	48	70	62	72	81	46	39	42	36
Nais	62	64	30	39	61	59	68	79	46	39	42	62
Aelosma niveum	72	62	53	45	38	62	38	39	43	38	52	58
A.quaternarium	28	32	41	28	53	39	49	54	38	29	51	38
Total	**256**	**243**	**207**	**239**	**309**	**317**	**299**	**340**	**233**	**211**	**257**	**236**
EPHEMEROPTERA												
Baetis	29	29	58	57	28	62	40	52	61	39	22	54
Caenis	38	49	55	37	38	60	77	66	52	38	32	49
Total	**67**	**78**	**113**	**94**	**66**	**112**	**117**	**118**	**113**	**77**	**54**	**103**

Contd...

Months/Genera	*May'07*	*June*	*July*	*Aug*	*Sep*	*Oct*	*Nov*	*Dec*	*Jan'08*	*Feb*	*Mar*	*Apr*
PLECOPTERA												
Atoperala	40	38	39	29	21	28	19	42	38	36	21	34
Total	**40**	**38**	**39**	**29**	**21**	**28**	**19**	**42**	**38**	**36**	**21**	**34**
COLEOPTERA												
Berosus	30	32	44	38	37	52	48	56	49	38	34	23
Hydaticus	34	22	28	32	46	44	50	55	38	33	29	46
Hydracarina	52	41	42	44	64	66	85	74	63	64	59	35
Hydranchna	28	26	19	17	24	26	36	38	29	19	17	34
Total	**144**	**121**	**133**	**131**	**171**	**188**	**219**	**223**	**179**	**154**	**139**	**138**
DIPTERA												
Chironomous	239	248	249	260	292	237	258	247	266	257	247	97
Tanypus	49	39	39	37	39	45	47	52	51	48	34	86
Pentaneura	28	26	32	36	38	29	34	39	37	29	27	65
Culicoides	29	35	37	35	46	48	72	66	52	49	47	50
Total	**345**	**348**	**357**	**368**	**415**	**359**	**411**	**403**	**406**	**383**	**355**	**308**
Grand Total	**1410**	**1390**	**1396**	**1474**	**1515**	**1582**	**1671**	**1750**	**1612**	**1444**	**1561**	**1386**

Table 3 : Monthly Percent Composition of Different Groups of Zoobenthos in Medical Pond

Months/Genera	*May'07*	*June*	*July*	*Aug*	*Sep*	*Oct*	*Nov*	*Dec*	*Jan'08*	*Feb*	*Mar*	*Apr*
Rotifera	18.1	17.2	15.1	21.7	15.7	15.6	15.9	16.1	17.4	21.0	15.4	18.3
Cladocera	7.5	8.9	5.9	4.8	5.4	6.1	5.9	7.4	5.9	5.5	5.7	8.2
Copepoda	6.6	5.6	7.6	6.1	7.2	4.9	6.5	5.4	8.3	6.5	5.3	7.2
Ostracod	7.1	8.5	10.4	8.9	6.6	9.1	7.7	6.5	8.1	7.2	20.4	8.0
Oligochaets	18.1	17.4	14.8	16.2	20.3	20.0	17.8	19.4	14.4	14.6	16.4	18.2
Ephemeroptera	4.7	5.6	8.0	6.3	4.3	7.7	7.0	6.7	4.7	5.3	3.4	4.5
Coleoptera	2.8	2.7	2.7	1.9	1.3	1.7	2.5	2.4	2.3	2.4	2.5	1.5
Plecoptera	10.2	8.7	9.5	8.8	11.2	11.8	13.0	12.7	11.1	10.6	8.9	8.9
Diptera	24.4	25.0	25.5	24.9	27.3	22.6	24.5	23.0	25.1	26.5	22.7	24.7

REFERENCES

APHA (1998) : Standard Methods for Examination of Water and Wastewater. American Public Health Association, AWWA, WPCF Washington, D.C. (U.S.A.), 1193 pp.

Edmondson, W.T. (1959) : Ward and Whipple's Freshwater Biology, 2nd Ed. John Wiley and sons Inc. New York, 1248pp. Eds. J. Rodriguez and W.K.W. Li, Sci. Mar., 58 : 109-117.

Gaufin, AR, and Tarzwel, CM. (1952) : Aquatic invertebrates as indicators of stream pollution. Public Health Rep., 6757-6.

Hargrave, B.T., N.J. Prouse, G.E. Philips and P.A. Neame (1983) : Primary production and respiration in pelagic and benthic communities at 2 intertidal sites in the upper bay of Fundy. *Canadian J. Fish and Aquat Sci.*, 40, 229-243.

Hynes, H.B.N. (1960) : Biology of Polluted Water. Liverpool Univ. Press, 202 pp.

Jonsson, B. (1991) : A C14 incubation technique for measuring micro phytobenthic primary productivity in intact sediments crores. *Limnol.* Oceanogra., 36(7) : 1485-1499.

Michael, P. (1984) : Ecological methods for field and laboratory investigations. Tata McGraw-Hill Publishing Co. Ltd., New Delhi, 404 pp.

Trivedy, R.K. and Goel, P.K. (1984) : Chemical and Biological Methods for Water Pollution Studies. Environmental Publications, Karad, India, 215 pp.

Venugopal, P., S.K. Devi, K.N. Remani and R.V. Unnithan, (1982) : Trace metals in the sediments of the Cochin backwaters. Mahasagar, 15(4): 205-214.

2

Nanotoxicity : Human and Environmental Risk Assessment of Nanomaterials

K.P. Singh and Manjul Mungali

ABSTRACT

Nanotechnology has created plenty of opportunities for engineers and scientists to manufacture superior and more efficient devices and products. Nanomaterials (NMs) are now widely used in consumer products as well as for research applications. At nanosize range, the properties of materials differ substantially from bulk materials of the same composition, mostly due to the increased specific surface area and reactivity, which may lead to enhanced bioavailability and toxicity. Various biotests have been successfully used for evaluation of ecotoxic properties of pollutants to invertebrates, algae and bacteria. These tests are now increasingly applied for evaluation of hazard of nanoparticles at different levels of the aquatic food-web. Nanotechnology is poised to become a major factor in the world's economy and part of our everyday lives in the near future. Hundreds and hundreds kilogram of Nanoparticles already enter the environment annually, but still very little is known of their interactions with biological systems. Recent studies indicate that some nanoparticles are not completely benign to biological and environmental targets. The challenge for toxicologists is to identify key factors that can be used to predict toxicity, permit targeted screening, and allow material scientists to generate new, safer nanoparticles with this structure-toxicity information in mind. The aim of this chapter is to summarize some known facts about nanomaterials and discuss future perspectives, regulatory issues and tasks of the emerging branch of toxicology, that is, nanotoxicology.

Keywords

Aquatic, Bioaccumulation, Environmentally hazardous NPs, Fullerene, Nanotoxicity.

INTRODUCTION

Our environment is our surrounding. This includes living and non-living things around us. The non-living components of environment are land, water and air. The living components are germs, plants, animals and people. All plants and animals adjust to the environment in which they are born and live. A change in any component of the environment may cause discomfort and affect normal life. Any unfavorable change or degeneration in the environment is known as 'Environmental Pollution'. Among pollutants chemical and bio-pollutants causes pollution in air, water and rest of the environment. In the increasing order of pollution new materials are always a threat to human and animal life due to their unknown properties, impact and nature. Nanomaterials are the new incumbent in this series.

Nanotechnology is an emerging multidisciplinary technology that involves the synthesis of molecules in the nanoscale (0-100 nm) size range. Given the excitement associated with all of the nanotechnology applications, evaluating the potential hazards related to exposures to nanoscale materials and its products has become an emerging area in toxicology and health risk assessment. In particular, some reports indicate that inhaled ultrafine/nanoparticles, following deposition in the alveolar regions of the lung, largely escape alveolar macrophage surveillance and transmigrate to the pulmonary interstitium or the systemic circulation following deposition in the alveolar regions of the lung (Oberdorster *et al.* 2001). Alternatively, other recent studies indicate that the toxicity of some nanoparticles may be related, in large part, to the surface reactivity of the particles, indicating that the particle surface–cellular interactions may take precedence over the core particle or particle size/surface area influencing the development of inflammatory and cytotoxic responses in the lung.

Nanoparticles and Nanotoxicity

Nanoparticles (NPs) are of similar dimensions as biologically active molecules and structures involved with chemical control (proteins), information storing (DNA and RNA) and support structures (membranes, actin and microtubules). Figure-1 shows the gold nanoparticle. The potential impact of such particles on the environment has led to the study of nanotoxicity. Early *in vitro* studies demonstrated that both polymeric Dendrimers as well as metal NPs can be cytotoxic due to the disruption of– and nanopore formation in cell membranes (Hong *et al.* 2005). Such NP/cell membrane interaction is dependent on the surface charge as well as the size of the NP (Nel *et al.* 2006). Both cationic gold and positively charged Dendrimers particles exhibited this ability to disrupt the phospholipids cell membrane (Hong *et al.* 2005).

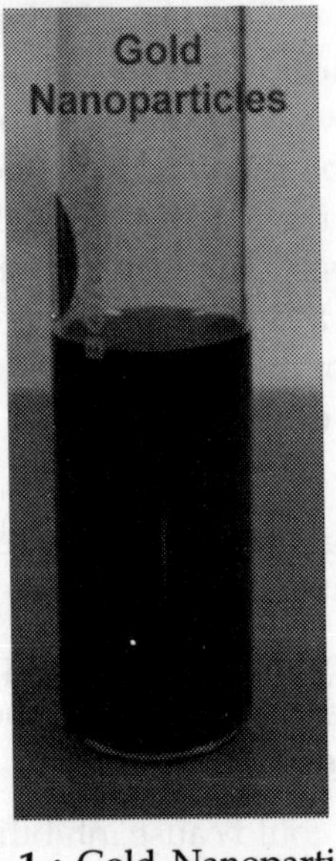

Fig. 1 : Gold Nanoparticles.

More recent publications indicate that metal and metal oxide NPs cause cytotoxicity by adsorbing to cell membranes where interaction with sulphidoproteins disrupt these membranes. In some cases ion formation causes severe cytotoxic effects. The formation of reactive oxidant species (ROS), by both organic and inorganic NPs, causes oxidative damage to cell membranes and its effect on cellular anti-oxidant defense mechanisms also plays a role in the cytotoxicity of NPs (Hong *et al.* 2005; Park *et al.* 2008). The identification of naturally occurring NPs, the improving technical ability to synthesize both organic and inorganic NPs as well as improving methods for evaluating the environmental impact of such particles have led to the realization that some of these NPs are toxic to biological systems.

Reasons Which Make Nanoparticles Toxic

The toxicity of Nanoparticles can be attributed to the features below :

1. Particle size and surface area are important material characteristics from a toxicological perspective. As the size of a particle decreases, its surface area increases and also allows a greater proportion of its atoms or molecules to be displayed on the surface rather than the interior of the material. The change in the physico-chemical and structural properties of engineered nano material with a decrease in size could be responsible for a number of material interactions that could lead to toxicological effects. Chemical composition of the particle which is responsible for its reactivity.
2. Surface charge of the particle is responsible for electrostatic interactions.
3. Hydrophobicity and sometimes lipophilic groups which may allow the particle to interact with proteins and membranes respectively.
4. Nanostructure could cause inhibition of enzyme activity either competitive or noncompetitive.

5. Accumulation of an inert particle in the body could also trigger tissue formation around the foreign entity and thus leading to formation of a scar tissue.

Pathway for Nanoparticle Entry into the Cell

Active cell uptake includes endocytosis or phagocytosis by specialised cells while NPs can also penetrate nonphagocytic mammalian cell membranes like that of red blood cells and become internalized at vesicles and mitochondria causing serious cell damage and cell death, figure-1 represented the pathway for nanoparticle entry into the cell (Moore, 2006). One characteristic common to both organic and inorganic NPs is that the finer the NP, the stronger the adverse health response seems to be. NP size (surface area), surface charge and chemical nature influence the severity of adverse health reactions (Reijnders, 2006). Although some of the effects of NP pollution on human health have similar consequences whether caused by organic or inorganic NPs. Due to the different physico-chemical surface characteristics of organic and inorganic NPs, regulations and control mechanisms for the protection of the environment may differ substantially in some respects depending on the type of NP.

Processes that control transport and removal of NPs in water and wastewater are yet to be investigated to understand the fate of NPs. Studies on the effect of NPs on plants and microbes are also rare. The fate of nanomaterials in aqueous environment is controlled by many biotic/abiotic processes such as solubility/dispersability, interactions between the nanomaterials and natural/anthropogenic chemicals in the ecosystem. In addition, ecological risk assessment is essential to understand environmental implications of nanomaterials. Before unknowingly dumping a huge amount of dangerous nanomaterials into the environment, we need to investigate the solubility and degradability of engineered NPs in soils and waters, to establish baseline information on their safety, toxicity and adaptation of soil and aquatic life.

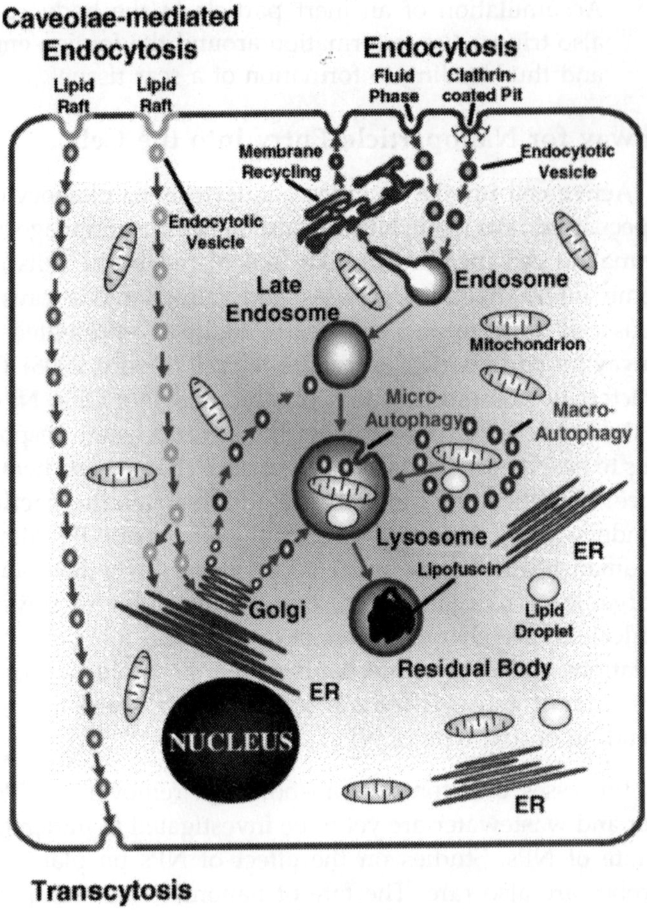

Figure 2 : Potential endocytotic pathways for nanoparticle entry into cells.

Source of Nanoparticles in our Environment

Industrial revolution is the main cause of environmental pollution that results into contamination of water, air and soil with hazardous substances. In the environment, NPs released into water/groundwater sources or into the atmosphere will tend to form aggregates, enhancing the possibility of NP elimination through sedimentation. Natural or artificially enhanced aggregation will reduce but not necessarily eliminate all free NPs. The concentration, environmental distribution, mobility and reactivity of such NP aggregates will differ significantly from that of the original free NPs. However, it cannot be assumed that the NP aggregates will be nontoxic to the environment unless proven by thorough scientific investigations. Another factor that may impact on environmental nanotoxicity is the changes that may occur once NPs is released. Loss of functionalization and chemical/structural changes that may occur due to its exposure to metabolic or other processes prior to environmental release may completely change the nature of a specific NP compared to its characteristics prior to release. Very little is known regarding the environmental fate and chemical stability of many of the NPs in use today. It is this unknown quantity and lack of information regarding the fate of NPs released into the environment, combined with the lack of regulation of the rapidly expanding nanoindustry that is causing concern with interest groups worldwide. It is therefore understandable that much of the available data on NP toxicity is aimed at its impact on human health. However, not nearly enough data is available at present and much of the relevant experimental data is generated using human cell cultures and animals (Hoyt, 2008). This makes investigations into nanotoxicity to humans difficult to quantify with precision. The potential areas/industries which lode NPs in the environment are :

- Electronic
- Pharmaceutical
- Cosmetics and sunscreens
- Energy
- Environmental
- Catalytic
- Food
- Biomedical
- Defense
- Material and Textiles

Nano-Pollutants

- A pollutant must be considered to be any substance that is released (on purpose or accidental) from a specific area of safe confinement and-or handling and has the potential to cause harm to the environment (Nowack, 2007).
- For nanomaterials to be classified as pollutants, the possibility of environmental exposure to such nanoparticles must exist (Nowack, 2007).
- Furthermore, such exposure must result in a hazard to the environment (Nowack, 2007).
- Analogous to the natural disaster precipitated release of conventional hazardous substances, NPs may be released unintentionally in large quantities. Population and industrial growth in areas where natural disasters may occur, as well as the effect natural disasters may have on storage or nano-related production facilities (Young, 2004), may well become a major environmental risk factor.

Effects of NP Pollution of Water and Aquatic Biota

Direct release or accidental spillage of industrial effluent containing NPs into waterways or aquatic systems will impact on human health. Direct contact with such water will lead to ingestion of nanopollutants through skin contact, ingestion and water aerosols. Ingestion of vegetable matter irrigated with contaminated water or aquatic organisms like fish and shellfish could also lead to humans being exposed to the effects of bioaccumulation (Moore, 2006). Research and development of technology suitable for the effective removal of various types of NPs from the aquatic environment must therefore be treated as a very high priority.

Aquatic Nanotoxicity

Outburst of nonmaterial research will certainly pump a lot of NPs to the environment, which will ultimately lead to the so-called nanoparticle pollution, by deposition of nanoparticle in ground water and soil. As the nanotechnological industries are developing, there is no doubt that nanoscale products and by-products will enter the aquatic environment. Uptake of nanoparticles into the aquatic biota is a major concern. Potential routes include direct ingestion or entry across epithelial boundaries such as gills, olfactory organs or body wall. At the cellular level, prokaryotes like bacteria may well be largely protected against the uptake of many types of nanomaterials, since they do not have mechanism for the bulk transport of supramolecular and colloidal particles across the cell wall. However, with eukaryotes (i.e., protists and metazoans) the situation is very different, since they have highly developed processes for the cellular internalization of nanoscale and microscale particles, namely endocytosis and phagocytosis, respectively.

In 2007, **Seetharam and Sridhar** studied toxicity of fullerene-C60 in two aquatic species, Daphnia and Pimephales elevated lipid peroxidation (LPO) in brain, significantly increased LPO in gill, and resulted in significant increase in expression of genes related to the inflammatory response and metabolism. In contact with water, C60 spontaneously forms a stable aggregate (nano-C60) with nanoscale dimensions (d = 25–500 nm). Prokaryotic exposure to these aggregates even at relatively low concentrations is growth-inhibitory (0.4 ppm) and decreases the rate of aerobic respiration (4 ppm).

Impact of Nanotoxicity in Human beings

Nanotoxicity can be subdivided into organic and inorganic nanotoxicity although these different types of NPs may have similar cytotoxic effects. Some human health effects occurs independent of whether the NPs are organic or inorganic and synergistic effects between the two types of NPs may lead to increased levels of toxicity (Reijnders, 2006; Hoyt, 2008). NP toxicity is mainly due to its large surface areas and ability to generate reactive oxygen species (ROS). Inflammation and fibrosis are indications of nanotoxicity noticed at organism level while oxidative stress, antioxidant activity and cytotoxicity occurs at the cellular level (Hoyt 2008). Currently, inhalation is the primary method of airborne NP ingestion by humans as the lungs are in direct contact with the environment (Yang 2008). NPs exist easily as aerosols due to their small size and are considered to be part of the respirable fraction (the amount of fine and ultra-fine particles < 5¼ m inhaled on a regular basis) (Hoyt, 2008). Figure-3 is a simple representation of bio kinetics of nanoparticles.

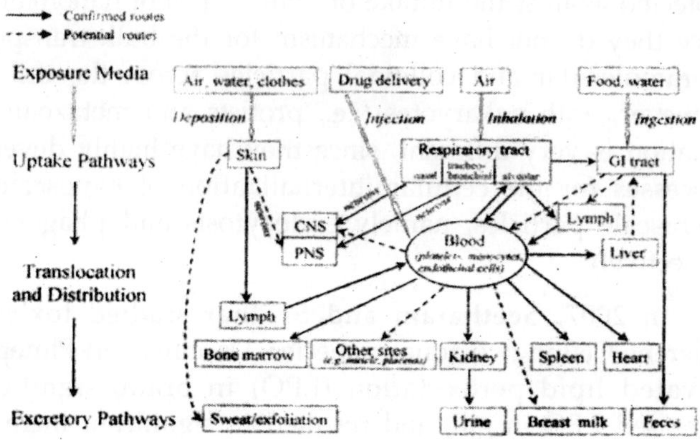

Figure 3 : Kinetics of Nanoparticles

Although small particles in the micron and sub-micron range have been implicated in human pulmonary and cardiac diseases, NPs and NP aggregates (< 100 nm) penetrates even further than these ultrafine particles and may cause even more severe symptoms (Thrall, 2006). All particles that make up the respirable fraction reach the alveoli in the lungs and due to the small size of NPs, Brownian motion will largely determine their fate. According to the particle deposition curve for particles in the lungs, 50-80% of NPs will attach to the alveoli compared to the 10–20% of larger particles. Macrophages tasked to ingest these contaminants fail to cope with these large numbers or fail to recognise them as invasive entities. Large numbers of NPs entering the lungs are therefore not removed by the body's primary defense system for inhaled small particulate matter (Hoyt, 2008). These NPs may enter the bloodstream (Nowack, 2007) where it will circulate and if toxic, may cause severe renal and cardiovascular damage as the large numbers of NPs cannot be removed efficiently by the kidneys. Oxidative stress caused in blood vessels and tissue by the circulating NPs may lead to a loss of dilation in blood vessels resulting in cardiac stress (Nowack, 2007; Hoyt, 2008).

The observed translocation of inhaled carbon NPs along the olfactory nerve to the olfactory bulb in the brain has been observed (Reijnders, 2006) and NPs capable of crossing the blood-brain barrier are associated with neurological damage (Nowack, 2007; Hoyt, 2008). Mouse microglia, specialized cells protecting the brain from external stimulae were shown to suffer damage due to oxidative stress from the NP induced formation of reactive oxygen species (ROS). Microglia engulfs chemical particles via phagocytosis and destroys them by releasing chemicals in an oxidative burst mechanism. Prolonged stimulation by ROS was implicated in severe brain cell damage, similar to that observed for degenerative diseases like Parkinson's and Alzheimer's (Thrall, 2006).

Adsorption through the skin and ingestion are also recognized as alternative routes of human exposure to NPs (Hoyt, 2008). NPs may reach the epidermis and penetrate to the blood stream through diffusion but ingestion remains a poorly understood mechanism for NP adsorption [Hoyt, 2008]. However, Reijnders (Reijnders, 2006) cites several authors indicating the possibility of NPs to cross the intestinal barrier for uptake into the lymphatic system (2-3% of ingested dose).

The surface characteristics of the NPs will determine their uptake into liver, spleen and kidneys through the blood stream and inflammation of these areas may occur. Similar toxic effects as described for the inhalation of NPs are expected even if exposure occurs through these alternative routes (Hoyt, 2008). Treatment of nanotoxicity is mainly symptomatic unless chronic illness develops and is based to great extent on experience with ultrafine particles (Hoyt, 2008). The aggregation of NPs, size dependent sedimentation onto cells or diffusion towards cells is the main parameters determining NP uptake by mammalian cells (Nowack, 2007). NP adsorption to cell membranes, determined by the surface characteristics of the NPs, also appears to play an important role.

Possible Solution for Removal of Environmentally Hazardous NPs

Current problems in dealing with nanopollution, existing waste water treatment and water purification technologies are not primarily designed to deal with nanopollution. Due to adsorption and other chemical interactions with NPs, dissolved chemicals may well escape in the treated water. Pre-coagulation and membrane filtration represent suitable treatment options but increased fouling by sub-micron particles may interfere with efficiency. End-of-pipe treatment is yet to be implemented on a significant scale and will rely on the producers of nanomaterials with significant pollution potential to commit to the cost implications of such treatment. Evaluation of source reduction mechanisms will allow the implementation of process specific treatments but special attention should be given to capturing NPs in liquid rather than gaseous medium as well as the recycling of NPs.

Containment of NPs in the process is important, therefore good filtration technology will be required. End of pipe treatment at production facilities includes filtration and electrostatic precipitation. This is mainly for aerosol NPs and not very effective, especially as NPs decrease in size. Carbonaceous NPs respond well to catalytic oxidation filters and continuously regenerating traps but actually increase the secretion of volatile droplets of less than 30 nm.

Magnetic NPs can theoretically be scavenged using magnetic filters but the efficiency is reduced with decreasing particle size < 200nm (Reijnders, 2006). Coating may render NPs much less hazardous and must be considered (Reijnders, 2006). This treatment option will of course depend on specific characteristics and applications of NPs. Regulation of NP release will have to be enforced by governmental agencies and it is doubtful that such enforcement will be possible prior to the development of adequate life-cycle and toxicity information.

Mass based risk assessment parameters may be inappropriate as NPs of smaller mass may differ radically from the bulk or larger particles of the same material. A number based system as for air-pollution quality would be more appropriate (Reijnders, 2006).

RESENT TECHNIQUES FOR THE REMOVAL OF NPs FROM THE ENVIRONMENT

Filtration and Adsorption Technology

Nanofiltration (NF) is a membrane dependent, pressure driven process that removes bacteria, viruses, cysts, organic material (Costa, 2006) and some dissolved ions like As and metals, also nitrates, dyes and specialty chemicals (Smith, 2006). NF is also effective in reducing COD loads and removing salt ions from textile waste waters for reuse, at much lower operating costs than reverse osmosis (RO). Pre-treatment with UF allows optimal permeate production rates and fouling protection for NF systems (Gozalvez-Zafrilla, 2008). Coatings made from nanomaterials can also be used to minimize bio-fouling of existing membrane technologies (Smith 2006). Zhang (Zhang, 2006) showed that composite Al_2O_3 membranes with $Mg(OH)_2$ nanotubes as active adsorbent could be used to remove almost 90% of Ni_2^+ ions from wastewater and repeated use of these membranes were possible prior to regeneration (Zhang, 2006).

The high surface area of clays, nanofibres and ceramic NPs allows efficient adsorption of pollutants, (Smith, 2006). Nanoporous zeolites, polymers and attapulgite clays are useful in removing pollutants through adsorption. Magnetic nanoparticles are also being investigated for use in water treatment (Smith, 2006). Organic carbon NPs and organic colloids are exceptionally strong adsorbents for organic pollutants and may provide a future method for the reduction of steroid hormones and herbicides in the environment (Nowack, 2007). Nanofibre technology used in the filtration industry offers high specific surface area, interconnectivity of pores with the added potential to add functional or active groups on the surface at the nanoscale (Ramakrishna 2006,

Barhate, 2008). The large surface to volume ratio enables effective adsorption of organic materials, bacteria, viruses and protozoan cysts although these nanofibers also can be structurally modified to remove inorganic species as well. Nanofibers' surface to volume ration can be further increased by synthesis of porous nanofibers. Nanofiber based filtration experiments reports minimal fouling and flow restriction compared to membrane dependent processes which reduces the energy demand of related technologies (Smith, 2006).

Silica NPs are used in slurries for chemical-mechanical metal polishing and chemical-mechanical planarization (CMP) (Yang, 2007; Chin, 2006). Such waste water contains high solid loads and conventional coagulation is difficult due to changes in turbidity and the production of high sludge volumes (Chin, 2006). Composite TiO_2/Al_2O_3 membranes can also be used successfully to remove SiO_2 NPs from oxide CMP waste via lectrofiltration (Yang, 2007). Magnetic seeding aggregation with magnetite NPs proved to be very effective and provide results similar to membrane filtration. Recycling and reuse of the magnetite particles proved possible but still requires optimization (Chin, 2006).

Al (III) is used for the preparation of polyaluminium chloride (PAC) used in flocculation and coagulation processes for water and waste water treatment. $Al_{1,3}$ (a nanosized particle) is believed to be the most effective form of Al (III) available. However, the separation and purification of this "forced hyd-olysis" PAC product presents difficulties. Fractionation of the different nano-sized particles using UF membranes with a MWCO of alternatively 1 and 3 kDa can be used to purify $Al_{1,3}$ from a mixture of Al (III) hydrolysation products (Huang *et al.* 2006).

An industrial park in Taiwan produces effluent in which the levels of COD are closely related to the number of NPs present in the effluent. This NP fraction comprises mainly of CaF_2 and silica. Poor coagulation performance of NPs using polyaluminium chloride, high sludge volumes and the poor settling characteristics of these NPs do not allow for adequate treatment using biological methods. Although increased dosage of PAC improved the removal of the NPs it was established that charge neutralization was not the mechanism of choice for destabilisation in this system. Si particles in filtered, chemically treated water tended to agglomeration and possibly associate with compounds responsible for the high COD load. Rapid sand filtration of this fraction with particles > 0.45¼ m was suggested for the adequate removal of the Si NPs (Chang, 2007).

CONCLUSION

The importance of nanotechnologies to our well being is beyond debate, but its potential adverse impacts need to be studied all the more. Nanotoxicology as a new discipline should make an important contribution to the development of a sustainable and safe nanotechnology. An improved understanding of the risk factors related to nanomaterials in the human body and the ecosystem will aid future development and exploitation of a variety of nanomaterials. Awareness is growing about the need to develop an infrastructure for characterizing and measuring nanomaterials in complex matrices and for developing reference materials, both for calibration of instruments used for assessing exposure and dosimetry, and for benchmarking toxicity tests.

REFERENCES

Barhate, R.S., Sundarrajan, S., Pliszka, D., Ramakrishna, S., (2008) : The potential of nanofibres in filtration, Filtration+Separation, 32-35.

Costa, A.R., De Pinho, M.N., (2006): Comparison of the performance of ultrafiltration and nanofiltration in surface water treatment, Desalination, 199: 73-75.

Chang, M.R., Lee, D.J., Lai, J.Y., (2007): Nanoparticles in wastewater from a science-based industrial park – Coagulation using polyaluminium chloride, Journal of Environmental Management, 85: 1009-1014.

Gozalvez-Zafrilla, J.M., Sanz-Escribano, D., Lora-Garcia, J., Hidalgo, M.C.L., (2008) : Nanofiltration of secondary effluent for wastewater reuse in the textile industry, Desalination, 222: 272-279.

Hong, S., Hessler, J.A., M.M.B., Holl, Leroueil, P., Mecke, A., Orr, B.G., (2005) : Physical interactions of nanoparticles with biological membranes: The observation of nanoscale whole formation, American Chemical Society, doi:10.1016/j.chs.09.004.

Hoyt, V.W., (2008) : Mason, E., Nanotechnology, Emerging health issues, Journal of Chemical Health and Safety, 10-15, doi:10.1016/j.jchas.2007.07.015.

Huang, L., Wang, D., Tang, H., Wang, S., (2006): Separation and purification of nano-$Al_{1,3}$ by UF method, Colloids and Surfaces A: Physicochemical Engineering Aspects, 275: 200-208.

Moore, M.N. (2006): Do nanoparticles present ecotoxicological risks for the health of the aquatic environment? *Environ. Int.*, 32 : 967-976.

Nowack, B., Bucheli, T.D., (2007): Occurrence, behaviour and effects of nanoparticles in the environment, Environmental Pollution, 150 : 5-22.

Nel, A., Xia, T., Madler, L., Li, N., (2006): Toxic potential of materials at the nanolevel, Science, 311, 622-627.

Oberdorster, G., Maynard, A., Donaldson, K., Castranova, V., Fitzpatrick, J., Ausman, K., Carter, J., Karn, B., Kreyling, W., Lai, D., Olin, S., Monteiro-Riviere, N., Warheit, D., Yang, H., (2005): Principles for characterising the potential human health effects from exposure to nanomaterials: Elements of a screening strategy, Particle Fibre Toxicology, 2.

Park, E.J., Yi, J., Chung, K.H., Ryu, D.Y., Choi, J. and Park, K., (2008) : Oxidative stress and apoptosis induced by titanium dioxide nanoparticles in cultured BEAS-2B cells, Toxicology Letters, doi: 10.1016/j.toxlet.2008.06.869.

Ramakrishna, S., Fujihara, K., Teo, W.-E., Yong, T., Ma, Z. and Ramaseshan, R., (2006) : Electrospun nanofibers : solving global issues, Materialstoday, 9(3), 40-50.

Seetharam R. J. and Sridhar K. R.; (2007): Nanotoxicity: Threat posed by nanoparticles; Current Science 93:6, 769-770.

Smith, A., (2006) : Nanotech – the way forward for clean water, Filtration and Separation, 32-33.

Thrall, L., (2006) : Are red blood cells defenceless against smaller nanoparticles? Environmental Science & Technology, 4327.

Thrall, L., (2006) : Study links TiO_2 nanoparticles with potential for brain-cell damage, Environmental Science & Technology, 4326.

Reijnders, L., (2006) : Cleaner nanotechnology and hazard reduction of manufactured nanoparticles, Journal of Cleaner Production, 14, 124-133.

Yang, W., Peters, J.I., Williams, R.O., (2008) : Inhaled nanoparticles – A current review, International Journal of Pharmaceutics, 356, 239-247.

Young, S., Balluz, L., Malilay, J., (2004) : Natural and technological hazardous material releases during and after natural disasters: a review, Science of the Total Environment, 322, 3-20.

Zhang, S., Cheng, F., Tao, Z., Gao, F, and Chen, J., (2006): Removal of nickel ions from wastewater by $Mg(OH)_2/MgO$ nanostructures embedded in Al_2O_3 membranes, Journal of Alloys and Compounds, 426, 281-285.

3

Nutrient Accumulation Pattern of *Eichhornia crassipes* Mart. (Solms.) in Natural Wetlands with Different Trophic Condition

D. Balasubramanian, K. Arunachalam, A. Arunachalam and A.K. Das

ABSTRACT

The potential of *Eichhornia crassipes* to take up nutrients from nutrient-rich waters, including organically polluted environments, were proposed as being advantageous for the improvement of water and sediment quality. In the present study, *Eichhornia crassipes* allocated more nitrogen (N), phosphorus (P) and potassium (K) to the aboveground parts (leaves and petioles) when compared to roots. Further, nutrient allocation to leaves and petioles were showed a declining trend from May'06 to December'06 or January'07 and increased markedly after February'07. Conversely, nutrient allocation ratio to root parts gradually inclined towards winter (until January'07). Nutrient allocation pattern significantly varied among sites, plant parts and sampling months. Nutrient assimilation rates (NAR) were higher in Dulikoto and Pichola Nishi wetlands compared to Nirjuli wetland. Among nutrients, potassium assimilation rate was relatively much higher than the rate of assimilation of nitrogen and phosphorus. Both tissue and water N : P ratio fluctuated reasonably throughout the sampling months. However, N.P ratio for plant tissue and water in Dulikoto wetland tended to increase towards winter season.

Keywords

Accumulation; nutrients; trophic condition; water hyacinth (Eichhornia crassipes); water quality; wetlands.

INTRODUCTION

Nutrient concentrations in plant tissues are an approximate measure of the consumption of resources, a basic feature to predict the ecological strategy of a species (McJannet *et al.* 1995). Aquatic plants play an important role in maintaining the purification capability of water and the entire aquatic ecosystem. They can significantly impact the assimilation of pollutants such as nitrogen and phosphorus from water by biological absorption or physicochemical adsorption (Eriksson *et al.* 1999; Haseeb *et al.* 2004). Aquatic macrophytes often seem to be more important in controlling ecosystem fertility than are the system's abiotic factors (Hobbie, 1992; Bastian and Hammer, 1993). Water hyacinth is one of the most rapid growing aquatic macrophytes and is ranked eighth among the world's top 10 weeds (Holm *et al.* 1977). *Eichhornia crassipes* is occasionally used in sewage treatment systems because the plants take up nutrients in surplus of growth needs directly from its substrates (Gossett and Norris, 1971; Haller and Sutton, 1973; Ornes and Sutton, 1975; Wooten and Dodd, 1976; Gopal and Sharma, 1981; Wilson *et al.* 2005). Natural wetland systems colonized by water hyacinth could serve as "nature's kidneys" for proper effluent treatment to preserve the earth's precious water resources from getting polluted. It has been used for decontaminating inorganic nutrients, toxic metals as well as persistent organic pollutants. Water hyacinth has exceptionally high affinity and accumulation capacity for several metals (Zaranyika *et al.* 1994; Polprasert and Katiwada, 1998; Zhu *et al.* 1999). Therefore, it is used as a biomarker and is introduced in wetlands for water phytoremediation (Satyakala and Jamil, 1992). Apart from the utilization of *Eichhornia crassipes* for the treatment of domestic wastewaters, in recent years there is significant amount of effort also has been made in bioremediation of heavy metals and radionuclides (Prasad and De Oliveira Freitas, 2003; Malik,

2004). Shawky *et al.* 2005 reported that maximum sorption of 64000 U^{6+} µg/g by the non-living water hyacinth roots.

The water hyacinth has an excellent ability to take up nutrients and other chemicals from its environment, and the chemical composition of the water hyacinth depends strongly on its environment (Musil and Breen, 1977; Poddar *et al.* 1991). Further, Poddar *et al.* 1991 reported a nitrogen content of 1.78% (db) in water hyacinths growing in marshy land where the nitrogen level in the water was only 2.40 mg/L. Studies in which water nutrient levels were controlled have shown positive relationships between water and plant nitrogen levels (Rogers and Davis, 1972; Wilson *et al.*, 2005). Roger and Davis reported that a single water hyacinth has the capability of removing more than 3 mg of phosphorus (P) and over 20 mg of nitrogen (N) per day. Increases in water nitrogen may also lead to enhanced phosphorus uptake (Reddy *et al.* 1989) or have no effect (Sato and Kondo, 1981), and can be used as a supplementation of water phosphorus (Ornes and Sutton, 1975; Reddy *et al.* 1990). The efficiency of nutrient accumulation of water hyacinth and other aquatic macrophytes is mainly due to their high biomass production. It is therefore particularly necessary to study the *Eichhornia crassipes* in wetlands with different trophic levels (eutrophic, mesotrophic and oligotrophic), because they accumulate large amount of nutrients from their substrates (water).

The potential of *Eichhornia crassipes* to contain high-nutrient standing stocks, and its capacity to take up nutrients from nutrient-rich waters, including organically polluted environments, were proposed as being advantageous for the improvement of water and sediment quality. Many investigators reported that nutrient accumulation capacity of water hyacinth, mostly under mesocosm and artificially fertilised experimental ponds, (Haller *et al.* 1970; Gossett and Norris, 1971; Rogers and Davis, 1972; Dunigan *et al.* 1975; Ornes and Sutton, 1975; Wolverton *et al.* 1976; Shiralipour *et al.* 1981; Center and Spencer, 1981; Richards, 1982; Reddy and Tucker, 1983; Reddy and DeBusk, 1984; Reddy and DeBusk, 1987; Reddy *et al.* 1989;

Tchobanoglous *et al.* 1989; Reddy *et al.* 1990) but, studies on nutrient assimilation rates and allocation to different organs (leaves, petioles and roots) of the single water hyacinth plant along different nutrient regimes under natural conditions are scanty. Few studies have examined water-plant nutrient associations in water hyacinth growing at field sites where nutrients are heterogeneously distributed both in space and time (Gross *et al.* 1995; Jackson and Caldwell, 1993). Hence, in this study, we examined the capability of *Eichhornia crassipes* to take up nutrients from wetlands with different trophic condition *viz.*, eutrophic, mesotrophic and oligotrophic. The objectives of the present study were :

1. To estimate the amount of nutrients that can be accumulated by water hyacinth in different nutrient regime, and
2. To determine the nutrient allocation pattern to particular organs (leaves, petioles and roots) of *Eichhornia crassipes*.

MATERIALS AND METHODS

Plant Samples Collection

Fresh water hyacinth samples in six replicates collected from each wetland stations (Dulikoto, Pichola Nishi and Nirjuli) were brought to laboratory and drained for half of an hour. In laboratory, all the samples were rinsed of adhering macroinvertebrates and sediment particles through running tap water and then separated into leaves, roots and petioles. The plant samples from three wetlands were separately oven-dried at 85°C to constant weight for 48 hours and dry weights recorded.

Plant Nutrient Analysis

For all the chemical analyses, the oven-dried plant samples were ground in a laboratory Wiley mill and sieved through 0.5 mm mesh screen. The ground plant material was redried at 65°C to a constant weight and then analysed for N, P and K.

Nitrogen (N) was determined by Kjeldhal method (Allen *et al.* 1974). After a tri-acid wet oxidation in $HNO_3 + H_2SO_4 + HClO_4$ (with selenium catalyst) was performed, phosphorus (P) was determined according to molybdenum blue method (Jackson, 1973) and potassium (K) was determined following flame photometric procedure (Allen *et al.* 1974).

Nutrient Allocation

The monthly nutrient (N, P and K) assimilation rates of water hyacinth were calculated by multiplying N, P and K contents of individual ramet with the total biomass in each wetlands (Dulikoto, Pichola Nishi and Nirjuli). The nutrient allocation ratio was defined as the ratio of a certain tissue nutrient concentration (N, P or K) to total plant nutrient (total plant N, P or K). For example, LNR (Leaf Nitrogen Ratio) = leaf N/total plant N, PNR (Petiole Nitrogen Ratio) = petiole N/ total plant N and RNR (Root Nitrogen Ratio) = root N/total plant N. Similarly, LPR (Leaf Phosphorus Ratio) = leaf P/total plant P, PPR (Petiole Phosphorus Ratio) = petiole P/total plant P, RPR (Root Phosphorus Ratio) = root P/total plant P, LKR (Leaf Potassium Ratio) = leaf K/total plant K, PKR (Petiole Potassium Ratio) = petiole K/total plant K and RKR (Root Potassium Ratio) = root K/total plant K.

Data Analysis

All data were analysed statistically using Microsoft Excel, STATISTICA 6.0, and ORIGIN 7.0. The differences between nutrient accumulations along sites, plant parts and sampling months were analysed through Post-Hoc least significant mean deviation (LSMEANS). Two-way analysis of variance (ANOVA) was used to compare the variations in substrate nutrients (N, P and K) and nutrient assimilation rates along three wetlands and sampling months. The level of significance (P) in all the cases was held at 0.05. In order to understand the seasonal variations, the monthly data were clubbed into 4 seasons *viz.*, rainy (May-September), autumn (October-November), winter (December-February) and spring (March-April).

RESULTS

Nutrient Accumulation Pattern

Seasonal variations in nitrogen phosphorus and potassium allocation to leaves, petioles and roots of *Eichhornia crassipes* in Dulikoto, Pichola Nishi and Nirjuli wetlands are presented in **Figures 1, 2** and **3** respectively that showed the water hyacinth allocated more nitrogen contents to its leaves than the petioles and roots. Among the sites, nitrogen accumulation was highest in Dulikoto (eutrophic wetland), followed by Pichola Nishi (mesotrophic wetland) and Nirjuli (oligotrophic wetland). Nitrogen contents were lower during summer months and in general greater during winter months in all three wetlands (**Figures 1, 2, 3**). Three-way analysis of variance (ANOVA) showed that nitrogen contents varied significantly among the sites ($F = 2432.89$, $P < 0.05$), various plant parts ($F = 4666.60$, $P < 0.05$) and sampling months ($F = 317.27$, $P < 0.05$) (**Table 5**). Moreover, the interaction among sites*plant parts*month was also quite significant. Tissue nitrogen contents established a significant positive correlation with the substrate nitrogen (**Table 7**). The highest level of tissue nitrogen were 8.32% (Jan'07) for leaves, 4.49% (Dec'06) for petioles and 3.56% (Jan'07) for roots in Dulikoto wetland; 4.51% (Dec'06) for leaves, 2.27% (Dec'06) for petioles and 1.86% (Dec'06) for roots in Pichola Nishi wetland and in the case of Nirjuli wetland, 4.32% (Dec'06) for leaves, 1.41% (Dec'06) for petioles and 1.13% (Dec'06) for roots.

Seasonal fluctuations in phosphorus accumulation followed the same trend as that of nitrogen. However, phosphorus contents among various plant parts were significantly different. Tissue phosphorus concentration was 1.38% (Dec'06) in leaves, 2.03% (Dec'06) in petioles and 0.77% (Jan'07) in roots in Dulikoto wetland; 0.94% (Jan'07) for leaves, 1.29% (Jan'07) for petioles and 0.97% (Jan'07) for roots in Pichola Nishi wetland and 0.64% (Jan'07) for leaves, 0.56% (Jan'07) for petioles and 0.43% (Jan'07) for roots in Nirjuli wetland. In Dulikoto and Pichola Nishi, large quantities of phosphorus was accumulated in petiole whereas in Nirjuli leaves served as storage organ for phosphorus. In general, phosphorus contents were lower during summer months that increased markedly during winter months. As for nitrogen, the amount of phosphorus accumulation was highest in Dulikoto wetland which was followed by Pichola Nishi and Nirjuli wetlands. Further, three-way ANOVA (**Table 5**) showed that phosphorus contents were significantly varied among sites ($F = 3060.75$, $P < 0.05$), plant parts ($F = 2623.80$, $P < 0.05$), sampling months ($F = 776.51$, $P < 0.05$) and interaction between sites*plant parts*month ($F = 40.75$, $P < 0.05$). **Table 7** showed that tissue phosphorus (except for leaves and roots in Dulikoto wetland) contents were positively correlated with phosphorus content in water.

Among the nutrients, potassium was accumulated in large amounts by different plant parts in all three wetlands (**Figures 1, 2 3**). As in phosphorus, petioles were allocated with large amount of potassium in Dulikoto and Pichola Nishi, while in Nirjuli leaves accumulated more potassium. The highest concentration of tissue potassium recorded were 5.19% (Oct'06) for leaves, 10.26% (Nov'06) for petioles and 1.86% (Oct'06) for roots in Dulikoto wetland; 4.35% (Dec'06) for leaves, 6.44% (Dec'06) for petioles and 3.47% (Nov'06) for roots in Pichola Nishi wetland and in the case of Nirjuli wetland, 3.72% (Nov'06) for leaves, 2.26% (Nov'06) for petioles and 1.51% (Nov'06) for roots. Among the sites, Dulikoto accounted for large amount of potassium, followed by Pichola Nishi and Nirjuli. Potassium accumulation was greater during winter and lower during summer months. Three-way ANOVA for variations in potassium allocation pattern are given in **Table 5**. It showed that there were highly significant differences in potassium allocation pattern among sites ($F = 23777.11$, $P < 0.05$), plant parts ($F = 39534.65$, $P < 0.05$) and sampling months ($F = 707.72$, $P < 0.05$). Further, interaction among sites, plant parts and sampling months were also found to be significant ($F = 135.68$, $P < 0.05$). Like nitrogen and phosphorus, tissue potassium.

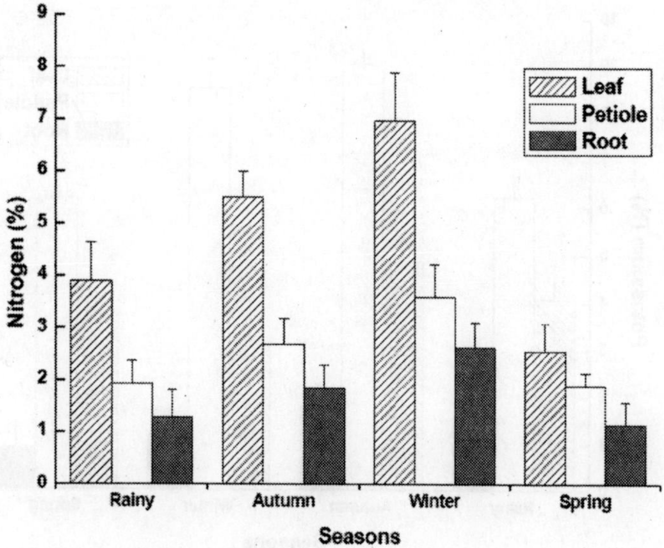
Nitrogen (%)
Leaf
Petiole
Root
Rainy
Autumn
Winter
Spring
Seasons

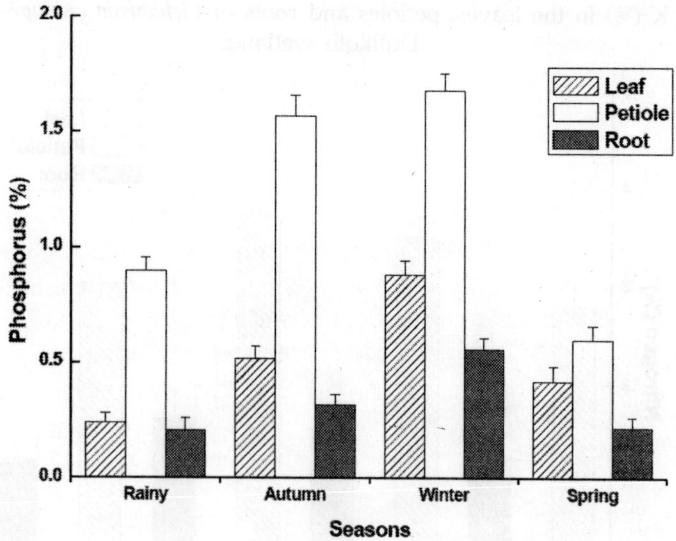
Phosphorus (%)
Leaf
Petiole
Root
Rainy
Autumn
Winter
Spring
Seasons

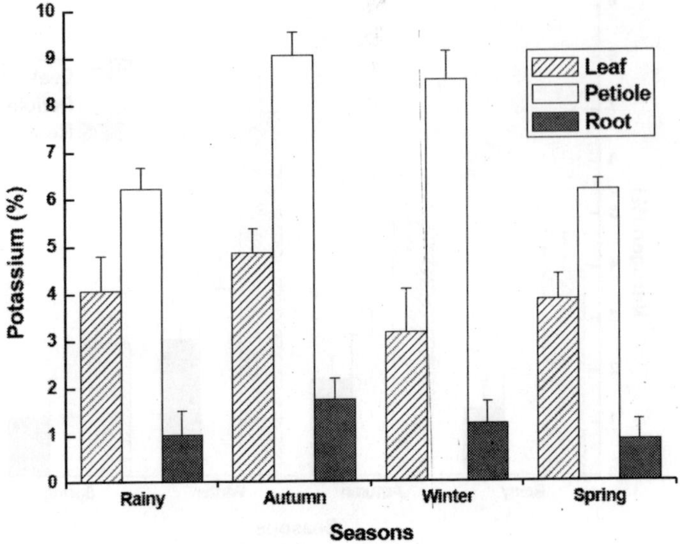

Figure 1 : Seasonal (May 2006 to Apr'07) variations in N, P and K (%) in the leaves, petioles and roots of *Eichhornia crassipes* in Dulikoto wetland.

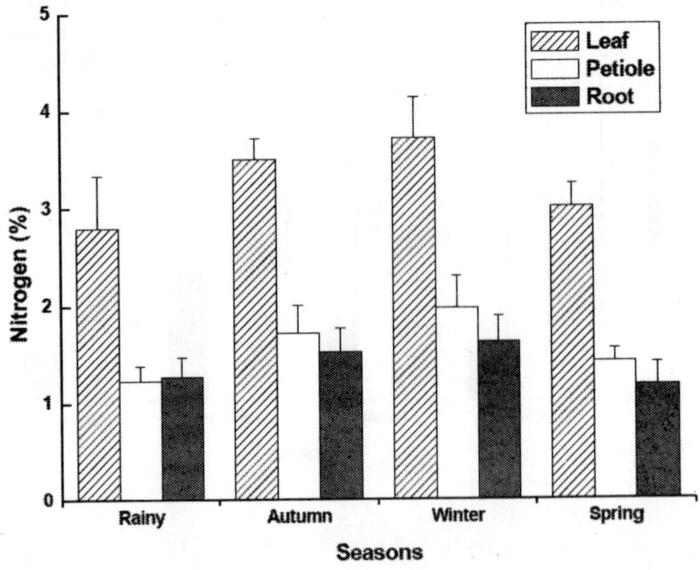

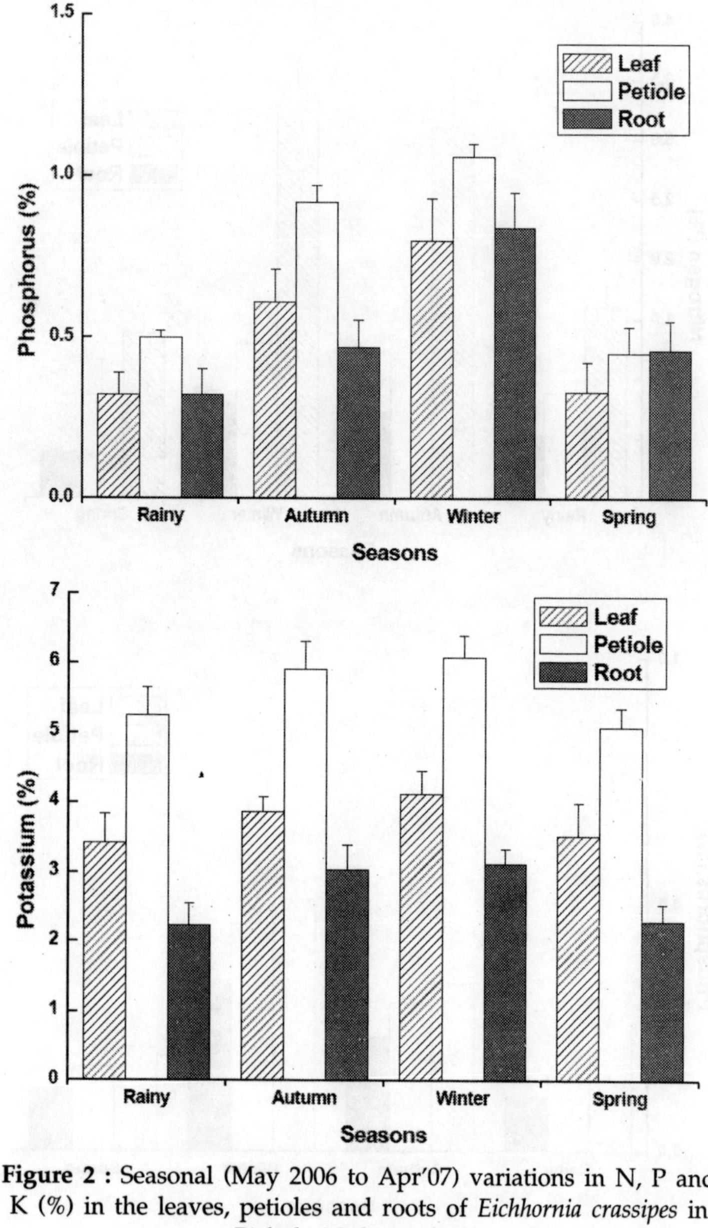

Figure 2 : Seasonal (May 2006 to Apr'07) variations in N, P and K (%) in the leaves, petioles and roots of *Eichhornia crassipes* in Pichola Nishi wetland.

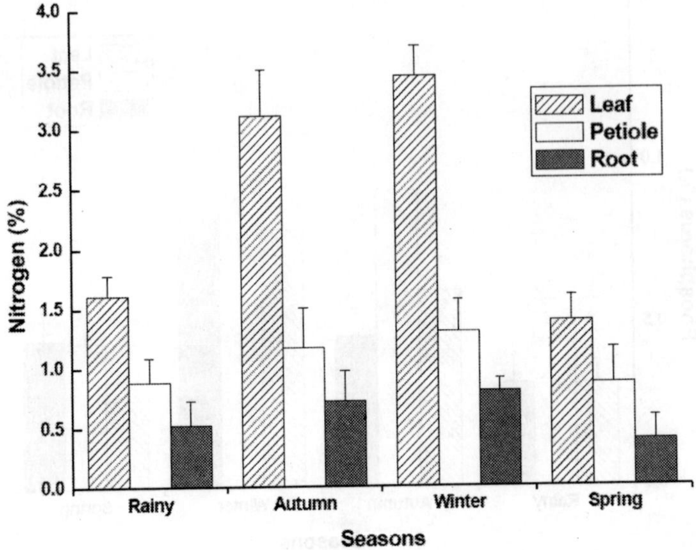

4.0
3.5
3.0
2.5
2.0
1.5
1.0
0.5
0.0
Nitrogen (%)
Leaf
Petiole
Root
Rainy
Autumn
Winter
Spring
Seasons

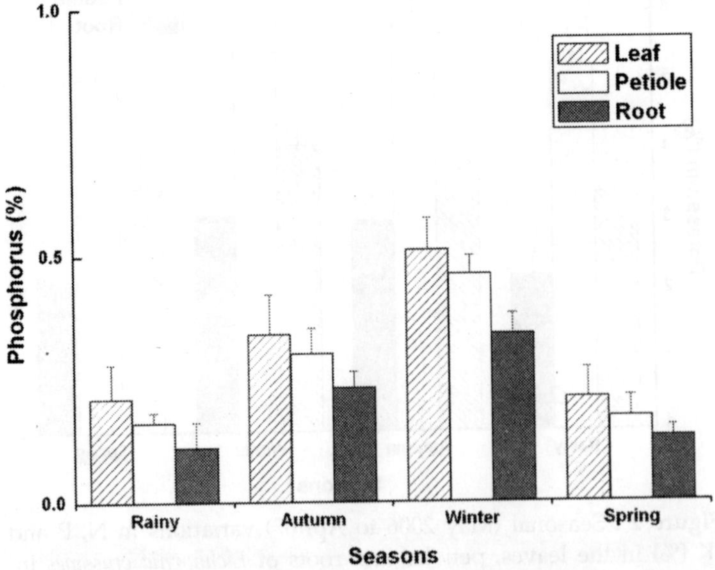

1.0
0.5
0.0
Phosphorus (%)
Leaf
Petiole
Root
Rainy
Autumn
Winter
Spring
Seasons

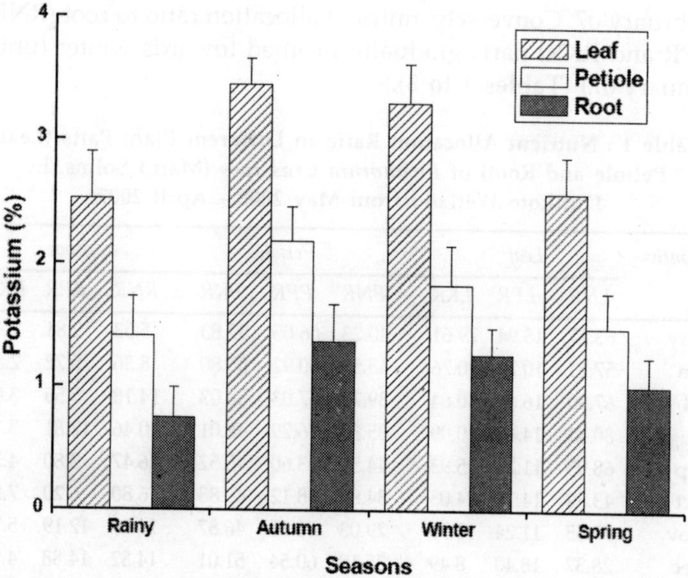

Figure 3 : Seasonal (May 2006 to Apr'07) variations in N, P and K (%) in the leaves, petioles and roots of *Eichhornia crassipes* in Nirjuli wetland.

(Except for leaves in Dulikoto and Pichola Nishi wetland) also showed a significant positive correlation with the substrate potassium contents (**Table 7**).

Nutrient Allocation Ratio

Nutrient allocation ratio among various plant parts of water hyacinth in Dulikoto, Pichola Nishi and Nirjuli wetlands have been summarised in **Tables 1, 2** and **3** respectively. They explained that *Eichhornia crassipes* allocated more nitrogen (N), phosphorus (P) and potassium (K) to the above ground part (leaves and petioles) when compared to belowground part root. Nutrient allocation ratio to plant parts mirrored and concordantly occurred along with the biomass allocation ratio to individual organs by *Eichhornia crassipes*. Further, nutrient allocation to leaves (LNR, LPR and LKR) and petioles (PNR, PPR and PKR) showed a declining trend from May'06 to December'06 or January'07 and increased markedly after

February'07. Conversely, nutrient allocation ratio to root (RNR, RPR and RKR) parts gradually inclined towards winter (until January'07) (**Tables 1 to 3**).

Table 1 : Nutrient Allocation Ratio in Different Plant Parts (Leaf, Petiole and Root) of *Eichhornia Crassipes* (Mart.) Solms. in Dulikoto Wetland from May 2006 – April 2007.

Months	*Leaf*			*Petiole*			*Root*		
	LNR	*LPR*	*LKR*	*PNR*	*PPR*	*PKR*	*RNR*	*RPR*	*RKR*
May	63.17	18.94	29.61	40.23	66.05	52.83	5.53	2.81	2.19
Jun	57.21	10.77	30.76	45.58	50.92	55.80	8.36	5.72	2.82
Jul	67.90	16.06	30.64	39.75	67.03	58.03	14.18	5.55	3.06
Aug	80.41	14.44	20.09	35.33	76.29	48.01	10.46	12.81	3.15
Sep	68.95	11.39	25.93	44.51	63.60	45.52	16.47	3.80	4.38
Oct	43.89	14.14	24.05	34.82	58.12	46.88	16.80	6.20	7.52
Nov	39.55	11.24	10.75	29.03	64.32	46.87	14.19	12.19	5.78
Dec	28.37	18.40	8.49	35.10	60.54	51.01	14.82	14.88	4.92
Jan	21.91	12.69	7.27	30.61	90.28	58.27	16.11	16.08	4.16
Feb	27.66	11.47	8.27	38.33	69.77	57.33	14.15	12.95	5.68
Mar	25.63	20.16	14.83	31.91	49.78	48.50	18.44	16.78	5.63
Apr	36.59	31.55	29.42	35.36	62.77	53.31	10.31	11.19	4.02

Table 2 : Nutrient Allocation Ratio in Different Plant Parts (Leaf, Petiole and Root) of *Eichhornia Crassipes* (Mart.) Solms. in Pichola Nishi Wetland from May 2006 – April 2007

Months	*Leaf*			*Petiole*			*Root*		
	LNR	*LPR*	*LKR*	*PNR*	*PPR*	*PKR*	*RNR*	*RPR*	*RKR*
May	52.99	21.56	28.57	45.52	73.11	75.80	10.86	11.34	8.17
Jun	33.92	10.06	29.98	20.95	57.34	72.99	6.37	12.49	9.29
Jul	40.02	21.68	20.79	31.10	72.98	64.68	11.63	20.23	8.55
Aug	47.69	31.62	21.33	24.09	58.15	54.99	17.28	22.45	14.77
Sep	44.30	30.47	22.17	39.99	53.21	44.66	25.41	22.12	16.32
Oct	42.15	20.07	20.98	23.38	49.47	50.24	20.96	20.71	18.45
Nov	24.55	16.79	18.52	31.27	43.82	45.66	20.64	20.79	24.19
Dec	21.51	18.76	14.81	27.21	44.15	63.66	16.83	26.82	19.06
Jan	22.04	10.85	14.67	37.79	50.23	65.42	15.42	22.75	15.99
Feb	29.52	27.99	19.44	29.87	85.66	53.14	14.80	18.05	13.84
Mar	51.05	22.10	23.95	38.71	39.31	45.63	16.56	23.19	19.02
Apr	72.51	15.97	31.68	32.82	28.67	58.19	19.99	33.13	13.24

Table 3 : Nutrient Allocation Ratio in Different Plant Parts (Leaf, Petiole and Root) of *Eichhornia Crassipes* (Mart.) Solms. in Nirjuli Wetland from May 2006 – April 2007.

Months	*Leaf*			*Petiole*			*Root*		
	LNR	*LPR*	*LKR*	*PNR*	*PPR*	*PKR*	*RNR*	*RPR*	*RKR*
May	36.18	49.12	46.29	36.07	34.02	41.66	11.51	14.89	9.61
Jun	40.39	29.22	39.31	38.16	40.09	31.33	8.27	12.82	9.34
Jul	38.81	23.78	39.27	37.88	33.75	30.44	12.46	13.39	9.45
Aug	46.25	31.52	31.67	27.60	40.21	32.22	15.47	13.96	15.39
Sep	26.04	32.24	34.32	25.76	44.25	43.18	10.16	17.30	10.95
Oct	32.64	17.35	32.02	30.14	28.01	34.23	10.75	18.53	19.73
Nov	38.47	22.16	27.96	30.74	53.86	41.77	12.71	23.66	16.12
Dec	27.93	21.49	25.98	30.51	62.86	44.86	11.54	18.35	14.08
Jan	24.45	13.52	25.58	30.35	46.75	53.16	7.31	17.70	12.26
Feb	13.33	24.94	32.27	29.49	44.36	39.68	8.16	15.62	19.72
Mar	42.29	39.19	53.89	38.19	40.49	37.67	13.13	31.25	20.71
Apr	114.94	49.05	59.75	33.70	41.54	40.54	7.26	16.94	13.10

Values are mean of 6 replicates.

LNR – Leaf Nitrogen Ratio; LPR – Leaf Phosphorus Ratio; LKR – Leaf Potassium Ratio; PNR – Petiole Nitrogen Ratio; PPR – Petiole Phosphorus Ratio; PKR – Petiole Potassium Ratio; RNR – Root Nitrogen Ratio; RPR – Root Phosphorus Ratio; RKR – Root Potassium Ratio.

Nutrient Assimilation Rate

Monthly variations in nutrient assimilation rates (g/m^2/day) for water hyacinth in Dulikoto, Pichola Nishi and Nirjuli wetlands are summarised in **Table 4**. It showed that among three wetland sites, nutrient assimilation rates were higher in Dulikoto and Pichola Nishi wetlands compared to Nirjuli wetland. The nitrogen, phosphorus and potassium assimilation rates of *Eichhornia crassipes* were higher during October'06 to January'07. Among nutrients, potassium assimilation rates were relatively much higher than the rate of assimilation of nitrogen and phosphorus. Two-way ANOVA showed that nitrogen (N), phosphorus (P) and potassium (K) assimilation rates were significantly different among sites and sampling months (**Table 6**). Further, it was found that the interaction among sites and sampling months also significantly influenced the nutrient assimilation rates ($P < 0.05$).

Table 4 : Monthly Variations in Nutrient Assimilation Rate (g/m²/day) of *Eichhornia crassipes* (Mart.) Solms at Three Study Sites from May 2006 – April 2007.

Months	*Dulikoto*			*Pichola Nishi*			*Nirjuli*		
	N	*P*	*K*	*N*	*P*	*K*	*N*	*P*	*K*
May	11.20	3.01	26.62	6.99	1.12	15.40	1.24	0.18	1.85
Jun	17.34	5.78	36.02	21.08	2.88	28.34	2.42	0.50	4.05
Jul	18.54	7.14	57.23	19.75	4.05	41.87	2.57	0.48	4.77
Aug	34.66	6.54	88.12	19.50	3.86	50.91	4.98	0.69	7.75
Sep	36.17	11.51	90.45	20.03	7.31	58.41	5.80	0.83	7.25
Oct	74.66	18.62	131.47	38.14	9.98	64.48	13.49	2.49	17.12
Nov	87.01	32.70	260.20	77.12	23.15	138.50	19.26	3.43	26.85
Dec	135.52	35.67	190.98	59.85	17.06	90.48	17.42	3.43	15.52
Jan	136.39	20.66	162.99	43.26	18.10	77.32	9.62	2.83	9.83
Feb	51.94	15.29	110.69	32.16	8.25	60.06	5.83	1.07	6.12
Mar	37.90	9.56	101.30	14.91	5.23	38.95	3.76	0.60	4.61
Apr	47.69	7.28	85.99	13.30	4.01	33.34	1.54	0.25	2.76
Mean	**57.41**	**14.48**	**111.84**	**30.51**	**8.75**	**58.17**	**7.38**	**1.40**	**9.04**

Values are mean of 6 replicates.

Table 5 : Three-Way ANOVA for Variations in Plant Nutrient Accumulation Pattern Among Sites, Plant Parts and Sampling Months

Source of Variations	*Factors/F-Value*						
Source of Variations	*Sites (df = 2)*	*Plant Parts (df = 2)*	*Month (df = 11)*	*Sites* Plant Parts (df = 4)*	*Sites Month (df=22)*	*Plant Parts* Month (df=22)*	*Sites* Plant Parts* Month (df=44)*
Nitrogen (N)	2432.89*	4666.60*	317.27*	175.13*	90.66*	57.43*	24.93*
Phosphorus (P)	3060.75*	2623.80*	776.51*	1212.86*	72.12*	93.44*	40.75*
Potassium (K)	23777.11*	39534.65*	707.72*	11399.84*	80.23*	126.81*	135.68*

**Values are significant at $P < 0.0500$ level; df – degree of freedom.*

Table 6 : Summary of Two–Way Factorial ANOVA for Variations in Nutrient Assimilation Rates Among Three Wetland Sites and Sampling Months

Factors	*Source of Variations/F-Values*		
	Sites (df = 2)	*Month (df = 11)*	*Sites * Month (df = 22)*
Nitrogen (N)	4768.72*	922.49*	271.97*
Phosphorus (P)	4082.83*	913.85*	192.62*
Potassium (K)	22630.56*	2659.90*	667.41*

**Values are significant at P < 0.0500 level; df – degree of freedom.*

Table 7 : Correlation Between Water Nutrients (N, P and K) and Nutrient Contents in Different Parts (Leaves, Petioles and Roots) and Whole Ramet of *Eichhornia Crassipes*

Wetlands	*Plant Parts/Ramet*	*Nutrient Contents*		
		N	*P*	*K*
Dulikoto	Leaves	**0.683***	0.046	–0.067
	Petioles	**0.411***	**0.436***	**0.455***
	Roots	**0.602***	0.172	**0.347***
	Ramet	**0.588***	**0.389***	**0.464***
Pichola Nishi	Leaves	**0.591***	**0.843***	0.188
	Petioles	**0.613***	**0.683***	**0.406***
	Roots	**0.600***	**0.764***	**0.400***
	Ramet	**0.326***	**0.934***	**0.314***
Nirjuli	Leaves	**0.591***	**0.588***	**0.407***
	Petioles	**0.638***	**0.685***	**0.317***
	Roots	**0.467***	**0.558***	**0.692***
	Ramet	**0.664***	**0.551***	**0.377***

N – Nitrogen; P – Phosphorus; K – Potassium; n = 84; Case-wise deletion of missing data)

Values are significant at 0.0500 p – level.

DISCUSSION

The N and P concentrations at which water hyacinth biomass accumulation was maximum are classified as hypertrophic in South Africa (Holmes, 1996), reinforcing the fear that water bodies with excessive nutrient loading are at risk to water hyacinth invasion. Although, resource allocation pattern is closely linked to the survival strategies of plants (Bastian and Hammer, 1993), we can choose biomass, energy or nutrient status related to structural function as the index of resource allocation. After analysing the interrelationships between biomass, energy and nutrients, Abrahamson 1982 earlier reported that nutrients are more important resources for evaluating resource allocation than biomass, and that the allocation pattern of critical resources reflects the ecological strategy of the plant.

Nutrient concentration in plants may reflect the bioavailability more accurately than that in their environments. Plant nutrient concentrations of the limiting nutrients are even determinants for plant community structure (Tilman, 1990). Abdelhamid and Gabr 1991 state the nutrient content is lower in the stem and root compared with the nutrient content in the leaves. Olsen 1953 and Sutcliffe 1962 reported that shoots are the main organs of nutrient absorption. However, the results of the present study showed that above-water parts (leaves and petioles) allocated with the more nutrients than the under-water roots in all the three wetlands (**Figures 1, 2, 3**). In fact, nutrient allocation to particular plant parts closely followed the biomass allocation strategy in *Eichhornia crassipes*. Similarly, Sato and Kondo 1981 found that high level of nitrogen in petiole and lamina of *Eichhornia crassipes* cultivated in mesocosm. The present study further revealed that elongated petioles of water hyacinth in Dulikoto and Pichola Nishi wetlands were exploited for accumulation large amount of phosphorus and potassium. Earlier, Boyd and Scarsbrook 1975 also reported that levels of tissue nitrogen, phosphorus and potassium increased with petiole length (plant height). However, Agami and Reddy 1990 reported that P accumulation was evenly distributed between roots and shoots, although N concentration in water hyacinth petioles was ~2 fold higher than the roots.

Nutrient accumulation by the individual organs increased as level of eutrophication increased in all the wetlands. In this study too Dulikoto (eutrophic) wetland showed a highest nutrient accumulation when compared to Pichola Nishi (mesotrophic) and Nirjuli (oligotrophic) wetlands. Among the various plant parts, overall nutrient (N, P and K) levels in the leaves and petioles of *Eichhornia crassipes* deduced the extent of eutrophication very clearly than the roots. Further, relative allocation of N, P and K to different tissues of water hyacinth differed significantly between sites. Evidently, three-way ANOVA for tissue nutrient contents (**Table 5**) showed that they were significantly varied among sites, plant parts, sampling months and interaction between sites*plant parts*month. It is worth highlighting here that nutrient allocation ratio in leaves and petioles reasonably declined during winter when the root nutrient allocation ratio generally increased (**Tables 1, 2, 3**). It may be due to the active petiole and leaf growth during summer months to acquire more photoassimilates. Similarly, increasing root nutrient allocation ratio could be attributed to the production of colossal roots in order to enhance nutrient acquisition. In fact, nutrient allocation ratio in individual plant parts mirrored and concordantly occurred along with the biomass allocation to individual organs by *Eichhornia crassipes*. The present study results were comparable to the observations of Bernard and Lauve (1995), who proposed that in spring and summer shoots grow rapidly, partly through translocation of reserve food and nutrient elements from rhizomes, estimated to contribute to almost 100% of the total spring above-ground growth. Also, reallocation of biomass between compartments is essential for surviving water level changes.

In the present investigation, tissue nitrogen contents were lowest during summer months and gradually increased to maximum during winter months (November' 06 – January'07). A similar trend had been earlier reported by other researchers as well (Tucker, 1981; Reddy and Tucker, 1983; Moorhead *et al.* 1988). Plant's nitrogen absorption and accumulation is often affected by the environmental conditions, nutrient concentrations, hydrodynamic characteristics etc. (Xiao *et al.* 2007) and studies by Cary and Weerts 1983b, 1984; Reddy and Debusk 1987 and Fisher and Reddy 1987 revealed that temperature influences the nutrient absorption capacity of water hyacinth. For instance, in tropical where the temperature regime is higher, registered better efficiency in the absorption of nitrogen and phosphorus. Shiralipour *et al.* 1981 reported that nitrogen concentration affects phosphorus uptake in water hyacinth. It is observed that in P-rich growth media, a small increase in N concentration could enhance P uptake, and *vice-versa*. In the present study, nitrogen uptake was higher than the phosphorus accumulation. Hence, it could be said that higher concentration of nitrogen in substrate reduced the tissue phosphorus accumulation. Phosphorus uptake by floating macrophytes could be affected by the N/P ratio of the substrate (Reddy *et al.* 1989, 1990, Jayaweera and Kasturiarachchi, 2004). Further, when P is deficient, other nutrients (such as N) are taken up by the plant in greater amounts than normal (Gerloff, 1969). Haller *et al.* 1970 and Haller and Sutton 1973 reported that increasing P concentration in growth media resulted in increasing P uptake by water hyacinth. Hadad and Maine 2007 also reported that when the concentration of P in water increases, P accumulation takes place in leaves of aquatic macrophytes. Another factor that may influence the absorption capacity is the N: P ratio in the water. Reddy and Tucker 1983 reported that the best rates of nitrogen and phosphorus absorption were observed when the N : P ratio in the water is around 2.3 and 5, when highest production of *Eichhornia crassipes* biomass was recorded. Sato and Kondo 1981 referred that greatest productivity of *Eichhornia crassipes*, occurs when the N:P ratio in the water was close to 3.6. The present study results showed that N:P ratios varied greatly from 4.08 to 13.57 in Dulikoto, 2.54 to 12.99 in Pichola Nishi and in Nirjuli wetland N:P ratio varied from 1.68 to 10.35. However, the highest biomass and productivity occurred when water N:P ratios were close to the values reported by Reddy and Tucker (1983). The tissue N:P ratios were lower than the substrate N:P ratio except in Nirjuli wetland where N:P ratios were higher than that of water N:P ratio. The present results were in accordance with earlier studies Sharma and Oshode, 1991; Pinto-Coelho and Greco, 1999 suggested that in natural habitat phosphorus could be a limiting factor for the growth of *Eichhornia crassipes*.

The growth and nitrogen (N) and phosphorus (P) content of water hyacinth are positively related to the N and P levels of the water (Sastroutomo *et al.*, 1978; Taheruzzaman and Kushari, 1989). In the present study also, substrate nutrients (N, P and K) availability showed a significant positive correlation with tissue nutrient contents and nutrient assimilation rates (Table 7). The annual mean of nutrient assimilation rates given in Table 4 revealed that nutrient assimilation rates were higher in Dulikoto (eutrophic) wetland, followed by the Pichola Nishi (mesotrophic) and Nirjuli (oligotrophic) wetlands. The nutrient assimilation rates of water hyacinth in the present study were comparable with or higher than the values reported earlier by Rogers and Davis (1972); Sato and Kondo (1981). However, N assimilation rates were considerably much higher than the values (3.3 g.N/m^2/day) reported by Reddy and DeBusk (1985). Lower assimilation rates could be attributed to loss of plant detritus from the stands and by surface leaching from water hyacinth ramet. Many researchers implied that plant detritus is an integral part of the water hyacinth mat and nutrient release from that is obvious (Reddy and Sacco, 1981; Ogwada *et al.* 1984).

It has been reported that water hyacinth has high rates of potassium (K) accumulation when compared to other aquatic macrophytes (Lawrence, 1968; Boyd, 1969a, 1969b; Boyd, 1970; Knipling *et al.* 1970; Boyd and Vickers, 1971; Scarsbrook and Davis, 1971; Parra and Hortenstine, 1974; JAAS, 1982; Sutton and Portier, 1983; Zhou and Yang, 1984; Madsen and Adams, 1988; Reddy *et al.*, 1991; Zhou *et al.*, 2004). In this study also potassium was accumulation was greater than nitrogen and phosphorus. The reason why a plant requires a large amount of potassium is that the element has a very low affinity for protein, so that a high concentration is required to maintain the potassium-protein complex which is essential for optimal enzyme activity (Sutcliffe and Baker, 1978). The K content in water hyacinth can be 5%, or more of dry weight basis. It was found that the mean K content was 4.3% for dry stem material and 6.6% for dry leaf material, but only 1.4% for dry root (Yu, 1989). Stem and leaf material represented to about 86% of the total dry mass of the plant. Hence, the accumulation of K in petioles and leaves are higher than the root in all three wetlands (Figures 1, 2, 3). Percentages of K in water hyacinth are generally considerably higher than in terrestrial plants, *e.g.*, threefold higher than in straw of rice, corn and Chinese sorghum (Ma, 2001). In the present investigation, Dulikoto wetland accounted for large amount of potassium followed by Pichola Nishi and Nirjuli. Further, nutrient assimilation rate for potassium found to be higher than that of the nitrogen (N) and phosphorus (P) (Table 4). Potassium assimilation rates were ranged from 26.62 $g/m^2/day$ (May'06) to 260.20 $g/m^2/day$ (Nov'06) in Dulikoto wetland, from 15.40 $g/m^2/day$ (May'06) to 138.50 $g/m^2/day$ (Nov'06) in Pichola Nishi wetland and in Nirjuli wetland it varied from 1.85 $g/m^2/day$ (May'06) to 26.85 $g/m^2/day$ (Nov'06). Potassium accumulation gradually increased with active vegetative growth. Similar observations were also made by Dunlop and Bowling (1971).

Over all, study revealed that nutrient assimilation rates as well as nutrient allocation to different tissues of *Eichhornia crassipes* were increased with eutrophication level. Evidently, three-way ANOVA showed that (Table 6) nutrient assimilation rates differed significantly between wetlands (eutrophic, mesotrophic and oligotrophic). Amongst different parts of water hyacinth, petioles and leaves accounted for large amount of nutrient accumulation. Hence, it may be reasonable to use above-water parts (leaves and roots) of water hyacinth as index to designate the level of aquatic environmental conditions, in fact they clearly indicating the level of eutrophication. While comparing different nutrients, potassium consumptions were luxurious in all three wetlands. Thus, the local *Eichhornia crassipes* having higher capacity to remove potassium (K) followed by nitrogen (N) and phosphorus (P) in an aquatic environment.

REFERENCES

Abrahamson, W. Q. (1982) : On the comparative allocation of biomass, energy, and nutrients in plant. *Ecology*, 63: 982–991.

Agami, M. and Reddy, K. R. (1990) : Competition for space between *Eichhornia crassipes* (Mart.) Solms. and *Pistia stratiotes* L. cultured in nutrient-enriched water. *Aquatic Botany*, 38 : 195–208.

Bastian, R.K. and Hammer, D.A. (1993) : The use of constructed wetlands for wastewater treatment and recycling. In: 'Constructed wetlands for water quality improvement' ed. By Moshiri, G. A.; Lewis Publ., Boca Raton, FL, pp: 59–68.

Bernard, J. M. and Lauve, T. E. (1995) : A comparison of growth and nutrient uptake in *Phalaris arundinacea* L. growing in a wetland and a constructed bed receiving landfill leachate. *Wetlands*, 15 (2) : 176–182.

Boyd, C.E. and Scarsbrook, E. (1975) : Chemical composition of aquatic weeds. In : Proceedings of the symposium on water quality and management through biological control Gainesville, University of Florida. pp : 144-150.

Boyd, C.E. (1969a) : The nutritive value of tree species of water weeds. *Economic Botany*, 23 : 123–127.

Boyd, C.E. (1969b) : Production, mineral nutrient absorption, and biochemical assimilation by *Justicia americana* and *Alternanthera philoxeroides*. *Archiv. für. Hydrobiologia*, 66 : 139–160.

Boyd, C.E., and Vickers, D.H. (1971) : Variation in the elemental content of *Eichhornia crassipes*. *Hydrobiologia*, 38 : 409–414.

Cary, P. R. and Weerts, P. G. J. (1983b): Growth of *Salvinia molesta* as affected by water temperature and nutrition. II. Effects of phosphorus level. *Aquatic Botany*, 17 : 61-70.

Cary, P. R. and Weerts, P. G. J. (1984) : Growth of *Salvinia molesta* as affected by water temperature and nutrition. III. Nitrogen-phosphorus interactions and effect of pH. *Aquatic Botany*, 19 : 171-182.

Center, T. D. and Spencer, N.R. (1981) : The phenology and growth of water hyacinth (*Eichhornia crassipes* [Mart.] Solms) in a eutrophic north-central Florida lake. *Aquatic Botany*, 10 : 1–32.

Dunigan, T. A.; Phelan, R. A. and Shamsuddin, Z. H. (1975): Use of water hyacinth to remove nitrogen and phosphorus from eutrophic waters. *Journal of Hyacinth Control*, 13 : 59–61.

Dunlop, J. and Bowling, D. J. F. (1971) : The movement of ions to the xylem exudates of maize roots. *Journal of Experimental Botany*, 22 : 434–444.

Eriksson, P.G. and Weisner, S. W. B. (1999) : An experimental study of effects of submersed macrophytes on nitrification and denitrification in ammonia-rich aquatic systems. *Limnol. Oceanogr.*, 44 : 1993-1999.

Fisher, M.M. and Reddy, K.R. (1987) : Water hyacinth (*Eichhornia crassipes* (Mart.) Solms.) for improving eutrophic lake water: Water quality and mass balance. In : 'Aquatic Plants for Water Treatment and Resource Recovery' ed. by Reddy, K. R. and Smith, W. H.; Magnolia Publ., Inc., Orlando, FL, pp : 969-976.

Gerloff, G.C. (1969) : Evalulating nutrient supplies for the growth of aquatic plants in natural waters. In: 'Eutrophication: Cause, Consequences, Correlations' ed. By Likens, G.E.; National Academy of Science. Washington, DC, pp : 537–555.

Gopal, B. and Sharma, K.P. (1981) : Water-hyacinth (*Eichhornia crassipes*), the most troublesome weed in the world. Hindasia, Delhi. pp : 128.

Gossett, D.R. and Norris, W.E. (1971) : Relationship between nutrient availability and content of nitrogen and phosphorus in tissue of the aquatic macrophyte, *Eichhornia crassipes* (Mart.) Solms. *Hydrobiologia*, 38 : 15.

Gross, K. L.; Pregitzer, K. S. and Burton, A. J. (1995) : Spatial variation in nitrogen availability in three successional plant communities. *Journal of Ecology*, 83 : 357–367.

Hadad, H. R. and Maine, M.A. (2007) : Phosphorus amount in floating and rooted macrophytes growing in wetlands from the Middle Paran´a River floodplain (Argentina). *Ecological Engineering*, 31 : 251–258.

Haller, W.T. and Sutton, D.L. (1973) : Effect of pH and high phosphorus concentrations on growth of water hyacinth. *Journal of Hyacinth Control*, 11 : 59-61.

Haller, W.T. and Sutton, D.L. (1973) : Effect of pH and high phosphorus concentrations on growth of water hyacinth. *Journal of Hyacinth Control*, 11 : 59-61.

Haller, W.T.; Knipling, E.B. and West, S.H. (1970): Phosphorus absorption by and distribution in water hyacinths. *Proceedings of Soil and Crop Science Society of Florida*, 30 : 64-68.

Haseeb, Md. and Irfanullah, B.M. (2004) : Factors influencing the return of submerged plants to a clear-water, shallow temperate lake. *Aquatic Botany*, 80 : 177-191.

Hobbie, S.E. (1992) : Effects of plant species on nutrient cycling. *Trends in Ecology and Evolution*, 7 : 336–339.

Holm, L.G., Plucknett, D.L., Pancho, J.V. and Herberger, J. P. (1977) : The World's Worst Weeds. Distribution and Biology, University Press of Hawaii, Honolulu, Hawaii, USA.

Holmes, S. (1996) : South African Water Quality Guidelines: Aquatic Ecosystems, vol. 7. Department of Water Affairs and Forestry, South Africa.

JAAS (Jiangsu Academy of Agricultural Science) Laboratory of Agricultural Land Ecology, Modernization Research Institute, (1982) : Potential crop productivity and ability of potassium recovery of three aquatic plants. *Jiangsu Agricultural Science,* 4 : 46.

Jackson, R.B. and Caldwell, M.M. (1993) : Geostatistical patterns of soil heterogeneity around individual perennial plants. *Journal of Ecology,* 81 : 683–692.

Jayaweera, M.W. and Kasturiarachchi, J.C. (2004) : Removal of nitrogen and phosphorus from industrial wastewaters by phytoremediation using water hyacinth (*Eichhornia crassipes* (Mart.) Solms). *Water, Science and Technology,* 50 : 217–25.

Lawrence, J. M. (1968) : Dynamics of chemical and physical characteristics of water, bottom muds and aquatic life in a large impoundment on a river. Auburn University Agriculture Experiment station, Zoology-Entomology Department, Series of Fisheries., No.6, p. 216.

Madsen, J. D. and Adams, M. S. (1988) : The nutrient dynamics of a submerged macrophyte community in a stream ecosystem by *Potamogeton pectinatus. L. J. Fresh. Ecol.,* 4 : 541-550.

Malik, A. (2004) : Metal bioremediation through growing cells. *Environ. Int.,* 30 : 261–78.

McJannet, C.L., Keddy, P. A. and Pick, F. R. (1995) : Nitrogen and phosphorus tissue concentration in 41 wetland plants: a comparison across habitats and functional groups. *Functional Ecology,* 9: 231–238.

Musil, C. F. and Breen, C.M. (1977) : The Application of Growth Kinetics to the Control of *Eichhornia crassipes* (Mart) Solms, through Nutrient Removal by Mechanical Harvesting, University of Natal. *Hydrobiologia,* 53 (2) : 165-171.

Ogwada, R.A.; Reddy, K.R. and Graetz, D. A. (1984) : Effects of aeration and temperature on nutrient regeneration from selected aquatic macrophytes. *Journal of Environmental Quality,* 13 : 239–243.

Olsen, C., (1953) : The significance of concentration for the rate of ion absorption by higher plants in water culture. IV. The influence of hydrogen ion concentration. *Physiol. Plantarum,* 6 : 848-858.

Ornes, W.H. and Sutton, D.I. (1975) : Removal of phosphorous from static sewage effluent by water hyacinth. *Journal of Hyacinth Control,* 13: 56-61.

Pinto-Coelho, R.M. and Greco, M.K.B. (1999) : The contribution of water hyacinth (*Eichhornia crassipes*) and zooplankton to the internal cycling of phosphorus in the eutrophic Pampulha Reservoir, Brazil. *Hydrobiologia,* 411 : 115–127.

Poddar, K.; Mandal, L. and Banerjee, G.C. (1991) : Studies on water hyacinth (*Eichhornia crassipes*) – Chemical composition of the plant and water from different habitats. *Indian journal of Vetinary*, 68 : 833–837.

Polprasert, C. and Katiwada, N.R. (1998) : An integrated kinetic model for water hyacinth ponds used for wastewater treatment. *Water Researc*, 32 : 179–85.

Prasad, M.N.V. and De Oliveira Freitas, H.M. (2003) : Metal hyperaccumulation in plants-biodiversity prospecting for phytoremediation technology. *Electronic Journal of Biotechnology*, 6 : 110–46.

Reddy, K.R. and DeBusk, W.F. (1984) : Growth characteristics of aquatic macrophytes cultured in nutrient-enriched water. Part I. Water hyacinth, water lettuce and pennywort. *Econommic Botany*, 38 : 229–239.

Reddy, K.R. and DeBusk, W.F. (1985) : Nutrient removal potential of selected aquatic macrophytes. *Journal of Environmental Quality*, 14 : 459-462.

Reddy, K.R. and DeBusk, W.F. (1987) : Nutrient storage capabilities of aquatic and wetland plants. In : 'Aquatic Plants for Water Treatment and Resource Recovery' ed. by Reddy, K.R. and Smith, W.H.; Magnolia Publ., Inc., Orlando, Fl, pp. 337-357.

Reddy, K.R. and Sacco, P. D. (1981): Decomposition of water hyacinth in agricultural drainage water. *Journal of Environmental Quality*, 10 : 255-266.

Reddy, K.R. and Tucker, J.C. (1983) : Productivity and nutrient uptake of water hyacinth: *Eichhornia crassipes*. Part I. Effects of nitrogen source. *Economic Botany*, 37 : 237–247.

Reddy, K.R.; Agami, M. and Tucker, J.C. (1989) : Inuence of nitrogen supply rates on growth and nutrient storage by water hyacinth (*Eichhornia crassipes*) plants. *Aquatic Botany*, 36 : 33-43.

Reddy, K.R.; Agami, M. and Tucker, J. C. (1990): Inuence of phosphorous on growth and nutrient storage by water hyacinth [*Eichhornia crassipes* (Mart.) Solms.] plants. *Aquatic Botany*, 37 : 355-265.

Richards, J.H. (1982) : Developmental potential of axillary buds of water hyacinth, *Eichhornia crassipes* Solms. (Pontederiaceae). *American journal of Botany*, 69 : 615–622.

Rodríguez Gallego, L. R.; Mazzeo, N.; Gorga, J.; Meerhoff, M.; Clemente, Kruk, C.; Scasso, F.; Lacerot, G.; García, J. and Quintans, F. (2004): The effects of an artificial wetland dominated by free-floating plants on the restoration of a subtropical, hypertrophic lake. *Lakes Research and Management*, 9 : 203 – 215.

Rogers, N.H. and Davis, D.E. (1972) : Nutrient removal by water hyacinth. *Weed Science*, 20: 423-428.

Sato, H. and Kondo, T. (1981) : Biomass production of water hyacinth, and its ability to remove inorganic minerals from water. I. Effect of the concentration of culture solution on the rates of plant growth and nutrient uptake. *Japanese Journal of Ecology*, 31 : 257-267.

Satyakala, G. and Jamil, K. (1992) : Chromium-induced biochemical changes in *Eichhornia crassipes* (Mart.) Solms and *Pistia stratiotes* L. *Bulletin of Environmental Contamination and Toxicology*, 48 (6) : 921-928.

Scarsbrook, E. and Davis, D. E. (1971): Effect of sewage effluent on growth of five vascular aquatic species. *Journal of Hyacinth control*, 9 : 26-30.

Sharma, B. M. and Oshode, O. O. (1991) : Effect of nutrients on the biomass of water hyacinth (*Eichhornia crassipes* (Mart.) Solms.). *Hydrobiologia*, 38 : 401–408.

Shawky, S., Abdel-Geleel, and Aly, A. (2005) : Sorption of uranium by non-living water hyacinth roots. *Journal of Radio analysts and Nuclear Chemistry*, 265 (1) : 81-84.

Shiralipour, A., Garrard, L.A. and Haller, W.T. (1981) : Effect of Nitrogen Source, Biomass Production and Phosphorus Uptake in Water hyacinth. *Journal of Aquatic Plant Management*, 19 : 40 – 43.

Sutcliffe, J. F. and Baker, D.A. (1978) : Plants and mineral salts. Studies in biology no.48, Edward Arnold publ., London, pp : 60.

Sutcliffe, J.F. (1962) : Mineral salts absorption in plants, Pergamon. pp : 194.

Sutton, D.L. and Portier, K.H. (1983) : Variation of nitrogen, phosphorus and potassium contents of *Hydrilla* in south Florida. *Journal of Aquatic Plant Management*, 21 : 87-92.

Tchobanoglous, G.; Maitski, F.K.; Thomson, K. and Chadwick, T.H. (1989): Evolution and performance of city of San Diego pilot scale aquatic wastewater treatment system using water hyacinth. *Journal of Water Pollution Control Federation*, 61 : 11 – 2.

Tchobanoglous, G.; Maitski, F.K.; Thomson, K. and Chadwick, T.H. (1989) : Evolution and performance of city of San Diego pilot scale aquatic wastewater treatment system using water hyacinth. *Journal of Water Pollution Control Federation*, 61 : 11–12.

Tilman, D. (1990) : Mechanisms of plant competition for nutrients: the elements of predictive theory for competition. In : 'Perspective on Plant Competition' ed. by Grace, J. and Tilman, D.; Academic Press, New York, pp : 117–141.

Tucker, C.S. (1981) : The effect of ionic form and level of nitrogen on the growth and composition of *Eichhornia crassipes* (Mart.) Solms. *Hydrobiologia*, 83: 517–522.

Wilson, J. R.; Holst, N. and Rees, M. (2005): Determinants and patterns of population growth in water hyacinth. *Aquatic Botany*, 81: 51-67.

Wolverton, B.C., Barlow, R.M., McDonald, R.C. and Pierson, R.W. (1976) : Application of vascular aquatic plants for pollution removal, energy, and food production in a biological system. In: 'Biological control of water pollution' ed. by Tourbier, J. and Pierson Jr., R.W.; University of Pennsylvania Press, Philadelphia, pp : 141-149.

Wooten, J.W. and Dodd, J.D. (1976): Growth of water hyacinths in treated sewage efûuent. *Economic Botany*, 30 : 29-37.

Xiao, K.; Yu, D. and Wu, Z. (2007) : Differential effects of water depth and sediment type on clonal growth of the submersed macrophyte *Vallisneria natans*. *Hydrobiologia*, 589 : 265-272.

Yount, J.L. and Crossman Jr., R.C. (1970) : Eutrophication control by plant harbesting. *Journal of Water Pollution Control Federation*, 42 : 174-183.

Zaranyika, M.F.; Mutoko, F. and Murahwa, H. (1994): Uptake of Zn, Co Fe and Cr by water hyacinth (*Eichhornia crassipes*) in Lake Chivero, Zimbabwe. *Science of Total Environment*, 153 : 117–21.

Zhu, Y.L.; Zayed, A.M.; Qian, J. H.; de Souza, M. and Terry, M. (1999): Phytoaccumulation of trace elements by wetland plants: II. Water hyacinth. *Journal of Environmental Quality*, 28: 339–44.

4

Effect of Solar Ultraviolet Radiaton on Lake Ecosystem : A Review

D.S. Malik, S. Kumar and Swati Nagar

ABSTRACT

The role of stratospheric ozone layer in absorbing biologically harmful UV radiation is well known. With the downward trend of ozone layer, penetration of UV-B increases. Hence aquatic ecosystems become vulnerable to damage due to exposure to harmful UV-B radiation. UV-B radiation not only affects the distribution of aquatic organisms but also affect the productivity of the ecosystem by damaging the photosynthetic components. Although, plethora studies on phytoplankton, zooplankton, fishes etc has been conducted on the responses of organisms to UVR like mat formation, Diel Vertical Movement, pigmentation, etc. But these show species specific responses. Since in nature, everything is inter-related and sustenance of ecosystems depends on how physical and biological factors interact with each other. Thus, it should be emphasized to study the interactions of aquatic organisms with other important factors like DOC, temperature, condition of clouds, zenith angle of solar radiation etc and how changes in any of the factor further affect the responses of aquatic organisms.

Keywords

Ultruviolet radiation, Lake ecosystem, Aquatic organism.

INTRODUCTION

The role of stratospheric ozone layer in absorbing biologically harmful UV radiation is well known. However, there has been a downward trend of total column ozone of -61 % per decade for April and -3% per decade for June between 1979 and 1999 while UV-B is predicted to increase at the rate of 8% per decade (Hader et a1. 2003).UV radiation especially UV-B (280-320 nm) is found to be harmful for the aquatic organisms. The energy level of photons of UV-B is high which makes its photochemically active form of radiation. More importantly, it is absorbed by biologically important molecules such as DNA, proteins, chromophores etc. This leads to wide ranging effects on aquatic organisms (Bhandari *et al.* 2007). Lake ecosystem constitute 0.26% of the total freshwater which generates nearly 3% of its net primary production. UV-B radiation affect the growth and productivity by a number of mechanisms involving several molecular targets within the exposed cell (Hader *et al.* 1998). Extent of damage depends on the interaction of UV-B with other physical, biochemical and biological factors etc.

ULTRAVIOLET RADIATION AND DOC

Many of the optical properties of the natural waters may be modified by exposure to UVR. DOC (dissolved organic carbon) is identified as the most important factor controlling the attenuation of ultraviolet radiation in lakes (Scully and Lean 1994; Morris *et al.*, 1995; Morris and Hagreaves, 1997). DOC and POC (particulate organic carbon) are degradation products of living organisms. These substances are of great importance in the cycling of carbon in aquatic ecosystems, UV-B radiation has been found to break down high-molecular weight substances and make them available to bacterial degradation. In addition, DOC is responsible for short-wavelength absorption in the water column. In fresh water ecosystems, penetration of solar radiation is limited by high concentration of dissolved and particulate matter (Naganumalo *et al.* 1996). The study in North American lakes showed photochemical degradation of DOC caused decrease in UV absorbance (Blumthaler, 1993). Acidification also decreases DOC concentrations (Anesio *et al.*, 2003 and Porcal *et al.* 1996) depending on its concentration; DOC can have positive or negative effects on phytoplankton growth. Low concentration contributes to nutrient recycling (N and P) (Qualls *et al.* 2003) and availability while high concentrations negatively affect phytoplankton growth by shading (Klug, 2002).

The impact of exponential increase in UV penetration may be particularly pronounced in freshwater ecosystems with low DOC concentration below 2 mg L^{-1} (Schindler *et al.* 1996). Thus, variable levels of DOC and UV exposure may be important factor in determining the distribution and abundance of planktonic and shallow benthic organisms as well as influence the spawning depth of vertebrates such as amphibians and fish that lay their eggs in shallow surface waters. DOC level is relatively low in high mountain lakes, thus aquatic organism in such lakes are more prone to damage due to UV-B exposure. However, it has been observed that organism in such lakes develops various adaptations to UVR. In high mountain lake in spain, UV-A exerted the main pressure (Carrillio *et al.* 2002).

UV EXPOSURE AND TEMPERATURE

When UV-B exposure is combined with other aspects of global climate change such as increasing temperature, this may pose additional stress on aquatic organisms. There is a complex interaction of the effects of temperature and UV-B exposure on abundance, reproduction (Persaud & Williamson, 2005) and survival (Gillespie, 2008) of the zooplanktons like Daphnia etc. With the increase in temperature, reproduction decreases while survivability, first decreases then increases to certain extent due to increase in NER mechanism. With the warming of climate, this effect may be more pronounced in nature. Williamson (2002) has demonstrated that elevated temperatures increased UV tolerance in *Daphnia catawba* and *Leptodiaptomus minutus*, species that depend heavily on photoenzymatic repair (PER), but decreased UV tolerance in *A. girodi*, a species that has less PER. Ultraviolet inhibition of growth (percentage reduction of μM_{max} at each temperature) increased linearly with decreasing temperature, consistent with the hypothesis that net inhibition represents the balance between temperature independent photochemical damage and temperature-dependent biosynthetic repair.

UV AND ALTITUDE

The thickness of the atmosphere decreases as a function of altitude due to reduced atmospheric pressure. This results in increase in solar irradiance (UV-B penetration) at high altitude lakes. Blumthaler *et al.*, 1993 have found an altitude effect of 19 & 11% per 1000m by comparing annual totals of UV-A and UV-B at an altitude of 3576 msl and 577 msl respectively. Another study shows the observed increases in irradiance with altitude (altitude effect) of the daily totals of global irradiance as 8%±2% per 1000 m (total irradiance), 9%±2% per 1000 m (UV-A irradiance) and 18%±2% per 1000 m (erythemal effective irradiance) during the summer. The altitude effect of the simultaneously measured erythemal effective irradiance between Innsbruck (577 m a.s.1.) and Hafelekar (2300 m a.s.1.), horizontally separated by 2.5km, shows a slight dependence on the solar elevation: 15.1%±1.8% per 1000 m at 60° solar elevation and 18.6% ± 2.9% per 1000 m at 20° solar elevation (Blumthaler *et al.* 1997). Thus, exposure to ultraviolet radiation increases with the altitude.

UV AND OTHER PHYSICAL ATTRIBUTES

The incidence of solar energy to the earth's surface within critical UV region depends strongly on the factors such as the zenith angle of the sun, the thickness of the ozone layer, cloud cover, altitude above sea level, local air pollution etc. Ozone layer absorbs the UV-C radiation wholly while its depletion results in the increase in the UV-B radiations into the earth's surface. Clouds have a strong effect on UV radiation but the response depends on the type, height and opacity and it is quite complex (Gautbeir, 1994 and Li *et al.* 1995). For example, when sky is overcast, reductions of upto 80% have been observed, but during partly cloudy condition, UV levels can be enhanced by upto 20% due to increased scattering (Mc Keinzie *et al.* 1994).

ULTRAVIOLET RADIATION AND BIOLOGICAL FACTORS

Sometimes, SUVR are harmful as well as beneficial. UV-A radiation (320-400 nm) may cause damage or stimulates beneficial photorepair of UV-B damage. The nature of both direct and indirect effects of UVR in nature depends on both the photon flux density and the spectral composition of the radiation incident on aquatic organisms across environmental UVR gradients in space (depth, transparency, elevation) and time (diel, seasonal, interannual) (Williamson *et al.*, 2001) Laboratory and mesocosm studies show that UV-B along with other stressors can have detrimental effects on primary producers, consumers and decomposers in both freshwater as well as marine ecosystems.

Phytoplankton

These organisms inhabit the top layers of the habitats (euphotic zone) where they receive sufficient solar radiation (photosynthetic available radiation, PAR) for photosynthetic processes. However, algae from different habitats show a wide range of sensitivities to UV radiation (Jokiel and York 1984, Xiong *et al.* 1996) with certain species producing photoprotective pigments (Carreto *et al.* 1990, Garcia-Pichel and Castenholz 1991). Economically important cyanobacterium, *Arthrospira platensis* shows decrease in yield on exposure with UVR (Kunshan *et al.* 2008). UV-B radiation exposure also results in a wide range of responses at the cellular level, including motility, protein biosynthesis, photosynthesis, nitrogen fixation and survival in cyanobacteria (Dillion *et al.*, 2003 and Kumar *et. al.* 2003). Protective strategy include mat or crust formation (with UV-B absorbing substances like MAA at the top), migration of individuals within mat or change in morphology as observed in *Anthrospira platensis* (Gao *et al.* 2008). In Canadian Rocky Mountain lakes solar UV-A and UV-B were found to decrease algal density and alter community composition. (Tank *et al.*, 2004) However, some studies indicated that after long-term exposure to solar UV, phytoplankton can adapt to the radiation (Lesser *et al.* 2002).

UV radiation also affects the photosynthetic pigments of the photosystem thus affecting its production in *Nostoc* sp., UVR exposure causes photo bleaching of phycobiliproteins-phycobilisomes (phycobilisomes absorb strongly in the UV region 250-300nm).This explains the faster bleaching effect than on chlorophyll *a*. However, the clonogenic activity is not significantly impaired by UV-R. (Romula *et al.* 1997)

In shallow lakes, physical avoidance of UV radiation by phytoplankton via vertical migration into deeper waters was not possible in these shallow littoral enclosures. However, migrations by some epilimnetic phytoplankton into sediments (Hansson 1996) may have allowed some taxa to gain physical refuge from UV radiation. (Vinebrooke & Leavitt, 1999). Photorepair is limited at low temperatures. While at 6°C solar UV radiation significantly inhibited growth in natural phytoplankton samples from a mountain lake in the U.S.A no such inhibition was observed at 14°C, indicating that the repair processes compensate the UV inhibition at the elevated temperature (Doyle, 2005). Mixing is an important factor in plankton survival. In contrast to marine habitats with high mixing, lakes often show stable thermal stratification. Consequently, lake plankton communities show vertical distribution (Vuorio *et al.* 2003) and populate certain horizontal bands of optimal light conditions (Perez, 2002) using buoyancy and active motility for niche selection. Besides direct effects on cellular targets, UV-B also operates via the production of reactive oxygen species (ROS) (Lessar, 2006). Phytoplanktons defend themselves by activating antioxidant systems. However, UV-B decreases the activity of antioxidant enzymes and ROS scavengers (Zang, 2005).

Zooplankton

Ultraviolet radiation (UV) has recently also received a lot of attention since it may have negative effects on many aquatic taxa (Bancroft *et al.* 2007). UV is known to affect both behaviour (Rhode *et al.* 2001; Boeing *et al.* 2004; Hansson and Hylander,

2009) and mortality (Hairston, 1979; Ringelberg *et al.* 1984) in zooplankton communities and is predicted to be an environmental issue of increasing prominence for future aquatic ecosystems (Bronmark and Hansson, 2002; Rautio and Korhola, 2002). Despite these predictions, little is known about the quantitative effects of long term exposure to UV radiation on aquatic organisms, especially in relation to other processes occurring in natural systems (Siebeck *et al.* 1994). Several previous studies investigating zooplankton vertical migration have noted that its causes may differ both temporally (within a system) and spatially (among lakes) (Leech and Williamson, 2000; Winder *et al.*, 2004; Leech *et al.* 2005). Other studies have shown that the susceptibility of zooplankton to environmental factors such as UV radiation may be species-specific (Rhode *et al.* 2001; Hansson *et al.* 2007). Some species of Daphnia, for example, are more vulnerable to UV than are several copepod species (Leech and Williamson, 2000; Hansson and Rylander, 2009b). This differential susceptibility has led some to argue that UV radiation could be structuring zooplankton communities (Williamson *et al.* 2001). However, since the UV threat co-varies with other parameters such as temperature, food foliage and fish predation, it is difficult to single out the UV effect (Marinone *et al.* 2006).

The responses of zooplankton towards the UV-R dicladocerons are the least UV-tolerant gp. Thus generally shows surface avoidance and vertical migration from surface during peak time when UV rays are highest. While, other groups of rotifers and copepods are less tolerant (Leech *et al.* 2005). Some organisms have evolved adaptations to reduce radiation damage, such as the various types of protective pigmentation of freshwater zooplankton. However, strong pigmentation also increases vulnerability to visually hunting predators. Hence, where both UV radiation and predation are intense, zooplankton may be sandwiched between conflicting selective pressures : to be pigmented and to be transparent at the same time.

Ficher *et al.* (2006) suggested that UV-R, along with temperature, algal resources; and predators, may affect zooplankton Dicl Vertical Migration (DVM) in aquatic ecosystems. *Daphnia catawba* shows much response to UV-R by DVM (deeper during daytime) as compare to *Leptodiaptomus minutus* which was found to be at the bottom of the pond during day as well night. Temperature plays a significant role in the reponses of the zooplanktons to the exposure of UV-R. Species heavily dependent on photoenzymatic repair (PER) shows increased UV tolerance *(D. catawba* and *L. minutus)* with the increase in temperature (Williamson *et al.* 2002).

Fishes

The response of individual species, especially during early development may help to understand the vulnerability of that species and its potentiality in natural fishery. Damage to fishes include the destruction of microridges in the epidermis, exposed neuromast cells of the skin (Sharma *et al.* 2005), in *Plecoglossus altivelius;* damaged both gill filaments and lamellae in *Catla catla* (Sharma *et al.* 2006); UV-B sensitivity of *Pagrus major* developed during ontogenic development. So, care should be taken at early stages (Sharma *et al.* 2007). The exposure of UV-B radiation affects the survival rate of surface feeder *Catla* larvae (Sharma *et al.* 2008). It also affects digestive physiology and immune system of fish by decreasing enzymatic activities (Sharma *et al.* 2010). *In Plecoglossus platessa* UV-B exposure increases mortality and induced loss of positive buoyancy. In larvae and juvenile, long term exposure during and after metamorphosis affected ventilation rate at normoxia and ventilator regulation during hypoxia incubations (Steeger *et al.* 2001).

Moreover, the effect of UV can also be found on the *Salmo gairdneri Richardsoni* at high altitude lakes during 3-month UV exposure which results in the sequential pathogensis of the UV-induced dorsal fin lesions (Bullock & R. R. Coutts, 2006) while in *Salmo trutta* exposure of UV causes the disappearance of mucous cells disappeared from the irradiated epidermis (Noceda *et al.* 1997). Markkula *et al.* (1981) during an experiment on common carp *(Cyprinus carpio)* showed the potentially harmful impact of increased solar UV-B radiation on fish immune functions. Also the short term and long term effects on the exposure of different UV-B doses are different from each other. Thus UV can be considered as an environmental stress factor.

UV AND PRIMARY PRODUCTION

At weak light intensities the primary production per unit of time (photosynthetic rate) increases linearly with PAR up to a saturation value. This maximum photosynthetic rate depends on genetic factors, temperature, and on the adaptation state of the phytoplankton species present in the water body. If the light intensity increases further, the excess light can be converted to heat or otherwise destroyed resulting in a constant photosynthetic rate. At very high light intensities the photosynthetic rate decreases, a phenomenon called photo inhibition. It is due to mainly reversible damages of the photosynthetic apparatus and is caused by intense PAR, but most pronounced by wavelengths below 400 nm (UV-light) (Pasini *et al.* 1998).

In situ experiments were performed by many researchers to evaluate the role of UV-radiation on the primary production. Depth at which the primary production is inhibited by SUVR was observed to be different in different lakes depending on the physiography of the aquatic ecosystem. At some places it is 6 m while at other it is found to be 2m or so. Gala and Giesy (1991) have demonstrated the inhibitory effect of solar ultraviolet radiation on the primary production of offshore Lake Michigan phytoplanktons assemblage. *In situ* incubation of the natural phytoplankton assemblage in Antarctica waters indicate under normal condition UV-B radiation is responsible for a loss of approx 4.9% of primary production in the euphotic zone, whereas UVR with wavelengths between 320-360 nm causes a loss of approx 6.2%.

Relative sensitivity to UVR also varies seasonally like during spring erythemal dose was found to be ED_{50} = 17.6 kJ/m^2 UV-B while during the fall (ED_{50} = 30.5 kJ/m^2 UV-B) > summer (ED_{50} = 131.6 kJ/m^2 UV-B). The extent of inhibition was found to be directly related to irradiance and to increases with time of exposure (Belay, 1981). Altitude also indirectly affects the primary production by altering the UV-R intensity and other factors. Primary production in two high mountain lakes: Lake Scuro (245 1m a.s.l.) and Lake Cadagno (1923 m a.s.l.) were studied. Being difference in their UV-B intensity and attenuation coefficients for UVB-light (*L. cadagno*, Kd=2.24 m^{-1}; *L. scuro* 0.25 m^{-1}) which is due to their concentrations of biomass (chlorophyll-*a* in *L. cadagno* 2 mg 1^{-1}, in *L. scuro* 0.7 mg 1^{-1}) and dissolved organic carbon (DOC in *L. cadagno* 2 mg 1^{-1}, in *L. scuro* 0.5 mg 1^{-1}). Photoinhibition due to UV-radiation was detected in both lakes, although down to different depths. In *L. cadagno* it was not detectable below 1 meter, in *L. scuro* there was still a significant photoinhibition at 2 meters. Thus, the role of UV-radiation in high mountain water systems should take into account the fluctuations in radiation intensity as well as the fluctuation in the concentration of DOC.

Surface photo-inhibition reached 72-78% in the samples exposed to PAR and UV-R as compared to the control treatment from which UV-R was excluded, however, areal primary production showed only an 8-14% decrease in the presence of UV-R. Although the filters applied absorb ca. 15% PAR as well, comparison of the treatments pointed out that it was UV-AR (320-400 nm), and not UV-BR (280-320nm) that predominantly inhibited photosynthesis (Palffy *et al.* 2003). The copepod Boeckella, living in Lake Titicaca with very high solar UV levels, counters the detrimental effect by incorporating photoprotective MAAs. (Helbling *et al.*, 2002) Copepods cannot synthesize these substances but acquire them from their algal diet (*e.g.*, dinoflagellates) : (Moeller *et al.* 2005).

CONCLUSION

After recognizing the importance of ozone layer, there have been many of studies showing the harmful effects of Ultraviolet radiation. In some studies, it has been observed that only UV-B is harmful but sometimes UV-A also shows harmful effects. In aquatic ecosystem, the extent of damage largely depends upon the physical factors such as temperature, transparency, depth of spectral distribution, level of DOC and other particulate matter etc. Altitude and latitude also plays a very important role as it has found that the level of temperature, DOC level etc., varies with the altitude. Many new studies strengthen the evidences showing the effect of solar radiation on the phytoplankton and zooplanktons. Being surface dwellers in the lake, they are more prone to SUVR exposure.

In lakes, phytoplankton abundance may vary by orders of magnitude depending upon future climate-DOM-UV intervention (Lewitt *et al.* 2003). Lakes often shows thermal stratification and as a consequences plankton communities show vertical distributions where the UV regime can strongly control species composition (Cozza *et al.* 2004). There are many studies on the effect of SUVR on phytoplankton, zooplanktons, fishes etc., but these are species specific. Although such studies are very useful and important but uncertainty remains about its interaction with other factors while assessing the effect of SUVR in an ecosystem. Though some mesocosm studies including both plankton and zooplankton and their grazers suggests that species composition and population structure may be more influenced by UV-B than overall algal biomass (Roy *et al.* 2006). This suggest that there is a need for more mesocosm studies that shows interactions of physical & biological components of ecosystems thus help in better understanding of their interaction.

REFERENCES

Bancroft. B.A., Baker, N.J. and Blaustem, A.R. (2007) : Effects of UV-B radiation on marine and freshwater organisms: a synthesis through meta-analysis. Ecology Letters, 10 : 332–345.

Belay, Amha. (1981) : An experimental investigation of inhibition of phytoplankton photosynthesis at lake surfaces. *New Phytol.* 89 : 61-74.

Blumthaler, M, Ambach, W. and Huber, M. (1993) : Altitude effect of solar UV radiation dependent on albedo, turbidity and solar elevation. *Meteorol.* Z. : 2,116-120.

Blumthaleri M., Ambach, W. and Ellinger, R. (1997) : Increase in solar UV radiation with altitude, Journal of Photochemistry and Photobiology B: Biology, Volume 39, Issue 2, June 1997, Pages 130-134.

Boeing. W.J. Leech, D.M. Williamson. C E *et al.* (2004): Damaging UV radiation and invertebrate predation : conflicting selective pressures for zooplankton vertical distribution in the water column of low DOC lakes Oecologia. 138 : 603-612.

Bronmark. C and Hansson, L.A. (2002) : Environmental issues in lakes and ponds current state and perspectives. *Environ. Conserv.* 29 : 290-307.

Bullock, A.M. and Coutts, R.R., (2000) : Impact of solar ultraviolet radiation upon the skin of rainbow trout, *Salmo gairdneri Richardson*, farmed at high altitude in Bolivia. *Journal of Fish Diseases.* Vol. 8 Issue 3 : 263 – 272.

Carreto, L.I., M.O. Carignan, G. Daleo, and S.G. de Marco. (1990): Occurrence of mycosporine-like amino acids in the red tide dinoflagellate Alexandrium excavatum: UV photoprotective compounds? *Journal of Plankton Research,* 12:909-92.

Carrillo, P., Medina-Sanchez, J.M. and Villar-Argaiz, M. (2002) : The interaction of phytoplankton and bacteria in a high mountain lake : importance of the spectral composition of solar radiation. *Limnol. Oceanogr,* 47 : 1294-1306.

Cozza, R., Chiappetta, A., Petrarulo, M., Salimonti, A., Rende, F., Bitonti, M.B. and Innocenti, A.M. (2004): Cytophysiological features of Posidonia oceanica as putative markers of environmental conditions, *Chem. Ecol.,* 20, 215–223.

Dillon, J.G., Miller, J.R. and Castenholz, R.W. (2003) : UV-acclimation responses in natural populations of cyanobacteria (*Calothrix* sp.), *Environ. Microbiol.,* 5 : 473-483.

Doyle, S.A., Saros, J.E., and Williamson, C.E., (2005): Interactive effects of temperature and nutrient limitation on the response of alpine phytoplankton growth to ultraviolet radiation, *Limnol. Oceanogr.,* 50, 1362–1367.

Fischer, Janet M., Nicolai, Jaclyn L., Williamson, Craig E., Persaud, Anurani D. and Lockwood, Ryan S. (2006) : Effects of Ultraviolet Radiation on Diel Vertical Migration of Crustacean Zooplankton : An *in situ* Mesocosm Experiment, Hydrobiologia. 563 : 217-224.

Gala, William R. and Giesy, John P. (1991) : Effects of ultraviolet radiation on the primary production of natural phytoplankton assemblages in Lake Michigan. Ecotoxicology and Environmental Safety Volume 22, Issue 3, 345-361.

Gao W.H., Villafane, V.E., Watanabe, T, Helbing E.W. (2005): Effects of solar UV radiation on morphology & photosynthesis of filamentous cyanobacterium Arthrospira platensis Appl Environ Microbiol. 71(9): 5004–5013.

Garcia-Pichel, F., and R.W. Castenholz. (1991): Characterization and biological implications of scytonemin, a cyanobacterial sheath pigment. *Journal of Phycology* 27 : 395-409.

Gauthier C., He and S. Yang (1994) : Role of clouds and ozone on spectral ultraviolet-B radiation and biologically active UV dose over Antarctica. In Ultraviolet radiation in Antarctica: Measurement & Biological effect. Vol.62 (Edited by S. Weiler and P.A. Penhale), 83-91.

Gillespie, Leigh-Anne B. (2008) : The effect of temperature on the survival and reproduction of Daphnia magna following exposure to Ultraviolet- B radiation. Thesis for Honour course, The University of Winnipeg.

Hader,D.-P., Kumar, H.D., Smith, R.C. and Worrest R.C. (1998). Effects on aquatic ecosystems. Journal of Photochemistry and Photobiology B : Biology 46 : 53–68.

Hader, D.P., H.D. Kumar *et al.* (2003): Aquatic ecosystems: effects of solar ultraviolet radiation and interactions with other climatic change factors. Photochem. Photobiol. Sci., 2, 39–50.

Hairston, N.G. (1979) Adaptive significance of color polymorphism in 2 species of Diaptomus (Copepoda). *Limnol., Ocenogr.*, 24 : 15-37.

Hansson (1996) : Algal recruitment from lake sediment in relation to grasing, sinking, and dominance pattern in the phytoplankton community. *Limnol. Oceanogr.* 41 : 1312-1323.

Hansson, L.A., Hvlunder, and Sand Sammaruga, R (2007): Escape from UV threats in zooplankton: a cocktail of behavior and protective pigmentation Ecology. 88. 1932-1939.

Hansson, L.A. and Hylander, S. (2009) : Effects of ultraviolet radiation on pigmentation, photoenzymatic repair behavior and community ecology of zooplankton Photochem. Photobiol Sci. 8, 1266-1275.

Hansson, L-A and Hylander, S. (2009) : Size- structured risk assessments govern Daphnia migration. Proc R Soc. B, 276: 331-336.

Helbling, E.W., Zaratti,F., L Sala, O., Palenque, E.R., Menchi, C.F. and Villafane, V.E. (2002) : Mycosporine-like amino acids protect the copepod Boeckella titicacae (Harding) against high levels of solar UVR. *J. Plankton Res.*, 24, 1191–1216.

Jokiel, P.L., and R.H. York. (1984) : Importance of ultraviolet radiation in photo inhibition of microalgal growth. Limnology and Oceanography 29 : 192-199.

Klug, J.L., (2002) : Positive and Negative effects of allochothonous dissolved organic matter and inorganic nutrients on phytoplankton growth. Can J. Fish. *Aquat. Sci.* 59 : 85-95.

Kumar, A., M.B. Tyagi, P.N. Jha, G. Srinivas and A. Singh. (2003): Inactivation of cyanobacterial nitrogenase after exposure to ultraviolet-B radiation, Curr. MicrobioL, 46 : 380-384.

Kunshan Gao, Zengling Maa. (2008) : Photosynthesis and growth of Arthrospira (Spirulina) platensis (Cyanophyta) in response to solar UV radiation with special reference to its minor variant. Environmental and Experimental Botany 63 : 123-129.

Leavitt, P.R., Cumming, B.F., Smol, J.P., Reasoner M., Pienitz, R. and D.A. Hodgson (2003) : Climatic control of ultraviolet radiation effects on lakes, *Limnol. Oceanogr*, 48, 2062–2069.

Leech, D.M., Padeletti, A. and Williamson. C.E. (2005): Zooplankton behavioral responses to solar UV radiation vary within and among lakes. *J. Plankton Res.* 27 : 461-471.

Leech, D.M. and Williamson, C.E. (2000) : Is tolerance to UV radiation in zooplankton related to body Size, taxon, or lake transparency. *Ecol Appl.*, 10: 1530-1540.

Lesser, M.P. 2006 : Oxidative stress in marine environments: biochemistry and physiological ecology, Annu. Rev. Physiol., 68 : 253–278.

Lesser, M.P., Barry, T.M. and Banaszak, A.T. (2002) : Effects of UV radiation on a chlorophyte alga (*Scenedesmus* sp.) isolated from the fumaroles fields of Mt., Erebus. Antarctica, *J. Phycol.*, 38 : 473-481.

Li, Z., Baker, H.W. and Moreau L (1995) : The variable effect of clouds on atmospheric absorption of solar radiation. *Nature*, 376 : 486-490.

Marinone. M.C. Marque. S.M., Suarez, D.A. *et al.* (2006): UV radiation as a potential driving force for zooplankton community structure in Palagonian lakes. *Photochem Photobiol*.. 82 : 962-971.

Mc Kenzie, R.L. (1994) : UV radiation monitoring in New Zealand. In stratospheric ozone Depletion/ UV-B radiation in the Biosphere (Edited by R.H. Biggs and HEB Joyner). Springer Verlag. Berlin. 239-246.

Morris, D.P., Zagarese, H., Williamson, C.E., Balseiro, E.G., Hargreaves, B.R., Modenutti, B., Modeller, R. and Queimalinos, C. (1995) : The attenuation of Solar UV radiation in lakes and the role of dissolved organic carbon. *Limnol. Oceanogr.* 40 : 1381-1391.

Moser, K.A., SmoL I.P., MacDonald, G.M. and Larsen, C.P.S. (2002) : 19th century eutrophication of a remote boreal lake: a consequence of climate warming? *J. Paleolimnol.* 28. 269-281.

Naganuma, T., Konishi, S., Inoue, T., Nakane, T. and Sukizaki, S. (1996): Photodegradation or photoalteration? Microbial assay of the effect of UV-B on dissolved organic matter, Marine Ecol. Progr. Ser. 135 :

309-310.

Noceda, Carlos, Sierra, Santiago Gonzalez, Juan Luis Martinez (1997) : Histopathology of UV-B irradiated brown trout Salmo trutta skin. DISEASES OF AQUATIC ORGANISMS Vol. 31 : 103-108.

Palfy, Karoly and Voros, Lajos (2003) : Effect of ultraviolet radiation on phytoplankton primary production in Lake Balaton, *Hydrobiologia*. Volume 506-509, Numbers 1-3, 289-296.

Pasini, Piero and Schanz, Ferdinand. (1998) : Influence of UV-radiation on the primary production of two high mountain lakes in the Piora Region. Edited by Peduzzi, R., R. Bachofen and M. Tonolla (Eds) Lake Cadagno: a meromictic Alpine lake Documenta Ist. ital. Idrobiol., 63 : 65-70.

Perez, A.P., Diaz M, M, Ferraro M.A., Cusminsky G.C. and Zagarese H.E. (2003): Replicated mesocosm study on the role of natural ultraviolet radiation in high CDOM, shallow lakes. Photochem. *Photobiol. Sci. Feb*; 2(2) : 118-123.

Persaud, A.D., and Williamson, C.E. (2005) : Ultraviolet and temperature effects on planktonic rotifers and crustaceans in northern temperate lakes. Freshwater Biology.Volume 50, Issue 3, p. 467-476.

Qualls, R.G., Richardson, C.J. (2003) : Factors controlling oncentration export and decomposition of dissolved organic nutrients in the Everglades of Florida, Biogeochemistry. 62, 197-229.

R.E. Moeller, S. Gilroy, C.E. Williamson, G. Grad and R. Sommaruga, (2005): Dietary acquisition of photoprotective compounds (mycosporine-like amino acids, carotenoids) and acclimation to ultraviolet radiation in a freshwater copepod, *Limnol.Oceanogr.*, 50 : 427–439.

Rautio, M and Korhola, A (2002) : Effects of ultraviolet-radiation and dissolved organic carbon on the survival of subarctic zooplankton. *Polar Biol.*, 25 : 160-468.

Rhode, S.C., PawlowskI. M and Tollrian, R (2001) : The impact of ultraviolet radiation on the vertical distribution of zooplankton of the genus Daphnia. *Nature*, -112, 69-72.

Ringelberg, J., Keyser, A.L. and Flik. B.J.G. (1984) : The mortality effect of ultraviolet-radiation in a translucent and in a red morph of Acanthodiaptomus-denticornis (Crustacea, Copepoda) and its possible ecological relevance. *Hydrobiology*, 112 : 217-222.

Romula Araoz. and Hader, Donat-P (1997) : Ultraviolet radiation induces both degradation and synthesis of phycobilisomes in *Nostoc* sp : a spectroscopic and biochemical approach. FEMS Micrology, *Ecology*. 23 : 301-313.

Roy, S., Mohovic, B., Gianesella, S.M.F., Schloss, I., Ferrario, M. and Demers, S. (2006): Effects of enhanced UV-B on pigment-based phytoplankton biomass and composition of mesocosm-enclosed natural marine communities from three latitudes, *Photochem. Photobiol.* 82, 909–922.

Samuel Hylander and Lars-anders Hansson (2010) : Vertical migration mitigates UV effects on zooplankton community composition. *Journal of Plankton Research*. 32 : 971-980.

Scully, N.M., and Lean, D.R.S. (1994) : The attenuation of Ultraviolet radiation in temperate lakes. Ergebn. *Limnol.*, 43 : 135-144.

Sharma. J. G., Masuda. R. and Tanaka M. (2007) : Orientation behaviour of Pargus major larvae exposed to UV-B radiation in laboratory conditions. International Journal of radiation Biolog., 83 : 40-42.

Sharma, J.C., Mittal, P. and Chakrabarti. R. (2008): Development of survivorship model for UV-B irradiated *Catla catla* larvae. Aquatic Ecology Springer. 2 : 17-23.

Sharma, Jaigopal, Rao, Y Vasudeva, Kumar, S., and Chakrabarti, Rina (2010): Impact of UV-B radiation on digestive enzymes and immune system of larvae of Indian major carp, *Catla catla* International Journal of Radiation Biology, Vo1. 86, No. 3, P. 181-186.

Sharma, J.G. and Chrakrabart, R. (2006): Effects of UV-B radiation on the gills of *Catla catla* during early development. Toxicological & Environmental Chemistry (Taylor & Francis) 82 : 367-371.

Sharma, J.G., Masuda, R. and Tanaka, M. (2005) : Ultra structural study of skin and eye of UV-B irradiated of Plecoglossus altivelus. *Journal of Fish Biology*, (The Fisheries society of the British Isles), 167 : 1-7.

Siebeck, O, Vail, T L., Williamson, C.E. *et al.* (1994) : Impact of UV-B radiation on zooplankton and fish in pelagic freshwater ecosystems Arch, *Hydrobiol Beih.*, 43 : 101-114.

Steeger, H.U., Freitag, .I.F, Wiemer, M and Paul, R.J. (2001): Effects of UV-B radiation on embryonic, larval and juvenile stages of North Sea plaice (*Pleuronectes platessa*) under simulated ozone-hole conditions. Helgol Mar. res. 55 : 56-66.

Tank, S.E. and Schindler, D.W. (2004) : The role of ultraviolet radiation in structuring epilithic alga communities in Rocky Mountain montane lakes: evidence from pigments and taxonomy, Can. *J. Fish. Aquat. Sci.* 61 : 1461-1474.

Vinebrooke, Rolf D., and Peter R. Leavitt, (1999): Differential Responses of Littoral Communities to Ultraviolet Radiation in an Alpine Lake. *Ecology*, 80:223–237.

Vourio, K., Nuottajarvi, M., Sarvoli, J (2003) : Spatial distribution of phytoplankton and picocyanobacteria in lake Tanganyika in March and April, 1998. *Aquat. Ecosyst.* Health Manag., 6 : 263-278.

Williamson, C.E., Grad, G., Lange, Hendrika J. de, Gilroy, Shawna and Karapelou, Danielle M. (2002) : Temperature-Dependent Ultraviolet Responses in Zooplankton: Implications of Climate Change. *Limnol., Oceanogr.* 47 (6) : 1844-1848.

Williamson, C.E., Patrick J. Neale, Gabriella Grad, Hendrika J. De Lange, Bruce R. Hargreaves (2001): Beneficial and Detrimental Effects of UV on Aquatic Organisms: Implications of Spectral Variation. Ecological Applications: 11 (6) : 1843-1857.

Williamson. C.E. Olson, O.G. Loti. S.E. *et al.* (2001): Ultraviolet radiation and zooplankton community structure following deglaciation in Glacier Sa" Alaska Ecology. 82 : 1748-1760.

Winder, M., Spaak, P. and Mooji, W.M. (2004) : Trade offs in Daphnia habitat selection. Ecology, 85 : 2027-2036.

Xiong, F., F. Lederer, Lukavsky, and L. Nedbal. (1996): Screening of freshwater algae (Chlorophyta, Chromophyta) for ultraviolet-B sensitivity of the photosynthetic apparatus. *Journal of Plant Physiology* 148 : 42-48.

5

Heavy Metal Correlation and Remodeling Using Multivariate Analysis Software

Padma S Vankar and Ashish Kr. Dwivedi

ABSTRACT

To analyze the source of water pollution and their intercorrelation between sources and pollutant as well, Principal component analysis or Multivariate statistical technique (MSA) is applied. Principal component analysis or Multivariate statistical technique (MSA) includes various conjugate analyses through specific statistical software which not only present appropriate data reduction but also their correlation in term of matrix and dendrogram.

Keywords

Groundwater pollution, Heavy metals, Principal component analysis, SPSS analytical software,

INTRODUCTION

Multivariate statistical techniques such as multivariate analysis of variance (MANOVA) and discriminant analysis (DA) can be applied for analyzing the data obtained from soil, water sources like river, ocean and many other sedimentary sources which may be contributed by the heavy metals. It provides an important data reduction as it used number of parameters in terms of heavy metal affording 100% correct assignations. Principal component analysis, widely used to reduce data and to extract a smaller number of independent factors (principal components) for analyzing relationships among observed variables, starts with the correlation matrix describing the dispersion of the original variables and extracting the eigenvalues and eigenvectors. An eigenvector is a list of coefficients multiplying the original correlated variables to obtain new uncorrelated (orthogonal) principal components, which are weighted linear combinations of the original variables. Principal component analysis can reduce the number of correlated variables to a smaller set of orthogonal factors, making it easier to interpret a given multidimensional system by displaying the correlations among the original variables. Principal components are those whose eigenvalues exceed 1 (Lu, Wang *et al.* 2010) and Principal component analysis and derivative methods have been widely applied to various environmental media, such as sediments to identify pollution sources and to apportion natural versus anthropogenic contributions. Correlation analysis classifies a set of observations into two or more mutually exclusive unknown groups based on a combination of internal variables (Yongming, Peixuan *et al.* 2006). Correlation analysis is often coupled with Principal component analysis to check results and to group individual parameters and variables. The purpose of Correlation analysis is to discover a system of organizing observations where a number of groups/variables share observed properties. A dendrogram is the most commonly used method of summarizing hierarchical clustering. Principal component analysis was applied to identify sources of heavy metals in street dust by applying varimax rotation with Kaiser Normalization. By extracting the eigenvalues and eigenvectors from the correlation matrix, the number of significant factors and the percent of variance explained by each of them were calculated.

METHODS

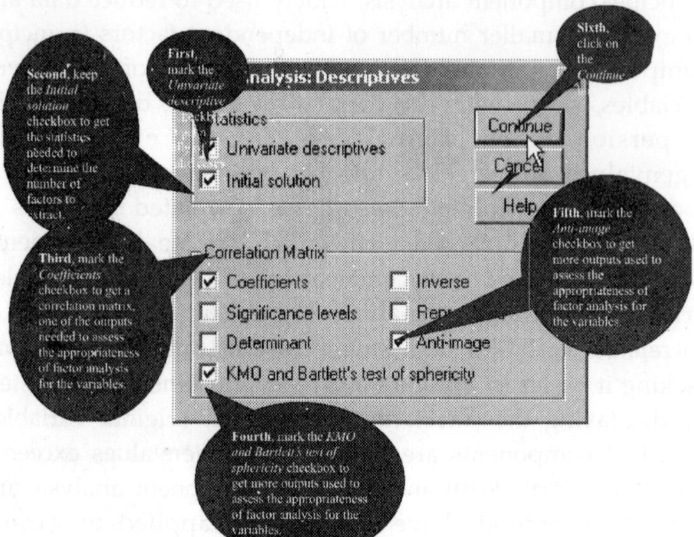
First, mark the Univariate descriptive
Second, keep the Initial solution checkbox to get the statistics needed to determine the number of factors to extract.
Sixth, click on the Continue
Analysis: Descriptives
Statistics
Univariate descriptives
Initial solution
Continue
Cancel
Help
Fifth, mark the Anti-image checkbox to get more outputs used to assess the appropriateness of factor analysis for the variables.
Third, mark the Coefficients checkbox to get a correlation matrix, one of the outputs needed to assess the appropriateness of factor analysis for the variables.
Correlation Matrix
Coefficients
Inverse
Significance levels
Determinant
Anti-image
KMO and Bartlett's test of sphericity
Fourth, mark the KMO and Bartlett's test of sphericity checkbox to get more outputs used to assess the appropriateness of factor analysis for the variables.

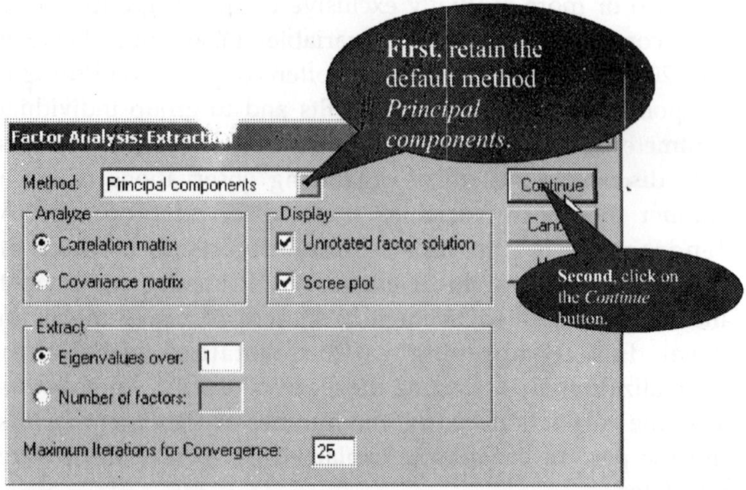
First, retain the default method Principal components.
Factor Analysis: Extraction
Method: Principal components
Continue
Analyze
Correlation matrix
Covariance matrix
Display
Unrotated factor solution
Scree plot
Second, click on the Continue button.
Extract
Eigenvalues over: 1
Number of factors:
Maximum Iterations for Convergence: 25

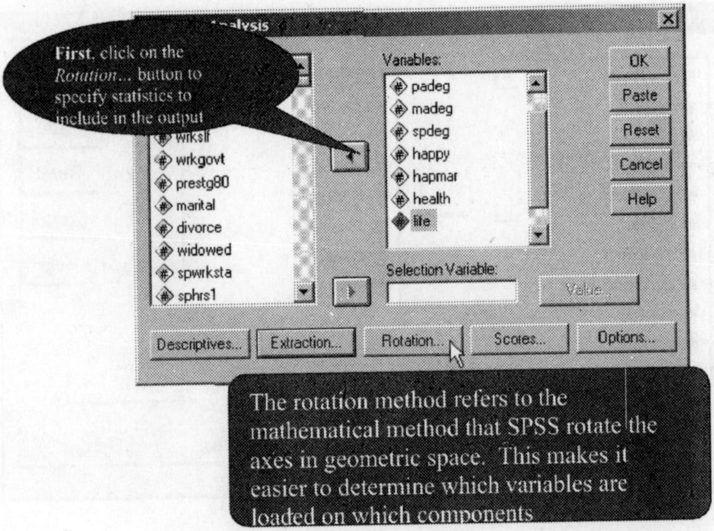
First, click on the Rotation... button to specify statistics to include in the output
Variables:
padeg
madeg
spdeg
happy
hapmar
health
life
wrkslf
wrkgovt
prestg80
marital
divorce
widowed
spwrksta
sphrs1
Selection Variable:
Value...
OK
Paste
Reset
Cancel
Help
Descriptives...
Extraction...
Rotation...
Scores...
Options...
The rotation method refers to the mathematical method that SPSS rotate the axes in geometric space. This makes it easier to determine which variables are loaded on which components

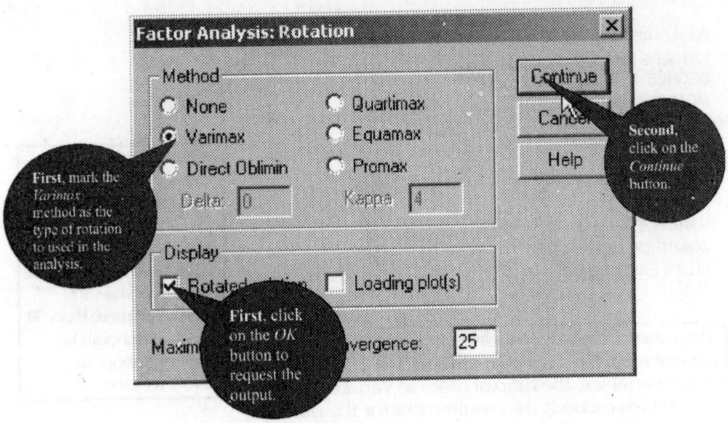
Factor Analysis: Rotation
Method
None
Quartimax
Varimax
Equamax
Direct Oblimin
Promax
Delta: 0
Kappa 4
Display
Loading plot(s)
25
Continue
Help
Second, click on the Continue button.
First, mark the Varimax method as the type of rotation to used in the analysis.
First, click on the OK button to request the output.

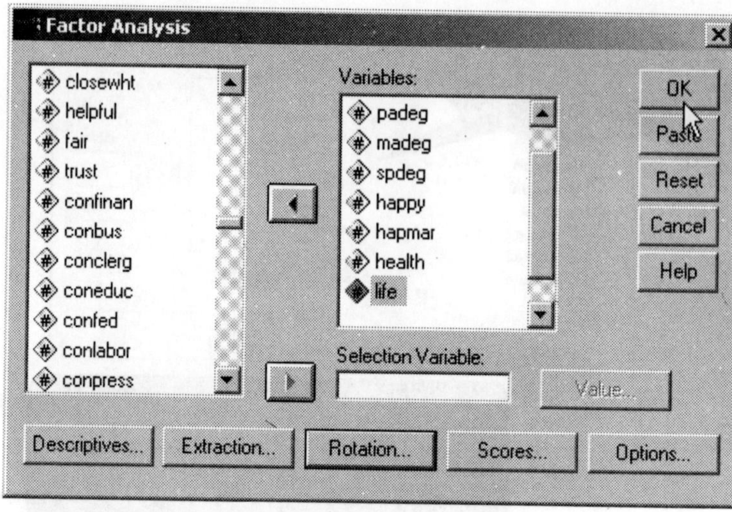

Descriptive Statistics

	Mean	Std. Deviation	Analysis N
RS HIGHEST DEGREE	1.68	1.085	68
FATHERS HIGHEST DEGREE	.96	.984	68
MOTHERS HIGHEST DEGREE	.85	.797	68
SPOUSES HIGHEST DEGREE	1.97	1.233	68
GENERAL HAPPINESS	1.65	.617	68
HAPPINESS OF MARRIAGE	1.47	.532	68
CONDITION OF HEALTH	1.76	.848	68
IS LIFE EXCITING OR DULL	1.53	.532	68

The number of valid cases for this set of variables is 68. While principal component analysis can be conducted on a sample that has fewer than 100 cases, but more than 50 cases, we should be cautious about its interpretation.

The ratio of cases to variables in a principal component analysis should be at least 5 to 1. With 68 and 8 variables, the ratio of cases to variables is 8.5 to 1, which exceeds the requirement for the ratio of cases to variables.

KMO and Bartlett's Test

Kaiser-Meyer-Olkin Measure of Sampling Adequacy.		.640
Bartlett's Test of Sphericity	Approx. Chi-Square	137.823
	df	28
	Sig.	.000

KMO and Bartlett's Test

Kaiser-Meyer-Olkin Measure of Sampling Adequacy.		.640
Bartlett's Test of Sphericity	Approx. Chi-Square	137.823
	df	28
	Sig.	.000

Communalities

	Initial	Extraction
RS HIGHEST DEGREE	1.000	.717
FATHERS HIGHEST DEGREE	1.000	.768
MOTHERS HIGHEST DEGREE	1.000	.815
SPOUSES HIGHEST DEGREE	1.000	.715
GENERAL HAPPINESS	1.000	.763
HAPPINESS OF MARRIAGE	1.000	.711
CONDITION OF HEALTH	1.000	.548
IS LIFE EXCITING OR DULL	1.000	.415

Extraction Method: Principal Component Analysis.

Communalities represent the proportion of the variance in the original variables that is accounted for by the factor solution.

The factor solution should explain at least half of each original variable's variance, so the communality value for each variable should be 0.50 or higher.

Communalities

	Initial	Extraction
RS HIGHEST DEGREE	1.000	.717
FATHERS HIGHEST DEGREE	1.000	.768
MOTHERS HIGHEST DEGREE	1.000	.815
SPOUSES HIGHEST DEGREE	1.000	.715
GENERAL HAPPINESS	1.000	.763
HAPPINESS OF MARRIAGE	1.000	.711
CONDITION OF HEALTH	1.000	.548
IS LIFE EXCITING OR DULL	1.000	.415

Extraction Method: Principal Component Analysis.

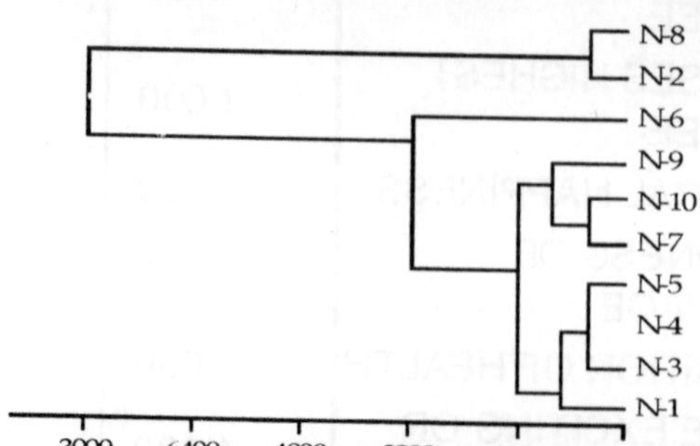

Euclidean distance between groups and subgroups has inverse relation with their correlation

Communalities

	Initial	Extraction
RS HIGHEST DEGREE	1.000	.717
FATHERS HIGHEST DEGREE	1.000	.768
MOTHERS HIGHEST DEGREE	1.000	.815
SPOUSES HIGHEST DEGREE	1.000	.715
GENERAL HAPPINESS	1.000	.763
HAPPINESS OF MARRIAGE	1.000	.711
CONDITION OF HEALTH	1.000	.548
IS LIFE EXCITING OR DULL	1.000	.415

Extraction Method: Principal Component Analysis.

On iteration 1, the communality for the variable "attitude toward life" [life] was 0.415. Since this is less than 0.50, the variable should be removed from the next iteration of the principal component analysis. The variable was removed and the principal component analysis was computed again.

DISCUSSION

A multi-dimensional hyper-space is often difficult to visualize. The main objectives of unsupervised learning methods are to reduce dimensionality, scoring all observations based on a composite index and clustering similar observations together based on multivariate attributes. Summarizing multivariate attributes by two or three variables that can be displayed graphically with minimal loss of information is useful in knowledge discovery. Because it is hard to visualize a multi-dimensional space, PCA is mainly used to reduce the dimensionality of multivariate attributes into two or three dimensions.

Correlation Matrix

		RS HIGHEST DEGREE	FATHERS HIGHEST DEGREE	MOTHERS HIGHEST DEGREE	SPOUSES HIGHEST DEGREE	GENERAL HAPPINESS	HAPPINESS OF MARRIAGE	CONDITION OF HEALTH
Correlation	RS HIGHEST DEGREE	1.000	.490	.410	.595	-.017	-.172	-.246
	FATHERS HIGHEST DEGREE	.490	1.000	.677	.319	-.100	-.131	-.174
	MOTHERS HIGHEST DEGREE	.410	.677	1.000	.208	.105	-.046	-.008
	SPOUSES HIGHEST DEGREE	.595	.319	.208	1.000	-.053	-.138	-.392
	GENERAL HAPPINESS	-.017	-.100	.105	-.053	1.000	.514	.267
	HAPPINESS OF MARRIAGE	-.172	-.131	-.046	-.138	.514	1.000	.282
	CONDITION OF HEALT	-.246	-.174	-.008	-.392	.267	.282	1.000
	IS LIFE EXCITING OR DULL	-.138	-.012	.151	-.090	.214	.161	.214

Anti-image Matrices

		RS HIGHEST DEGREE	FATHERS HIGHEST DEGREE	MOTHERS HIGHEST DEGREE	SPOUSES HIGHEST DEGREE	GENERAL HAPPINESS	HAPPINESS OF MARRIAGE	CONDITION OF HEALTH	IS LIFE EXCITING OR DULL
Anti-image Covariance	RS HIGHEST DEGREE	.511	-.101	-.079	-.274	-.058	.067	-.008	.108
	FATHERS HIGHEST DEGREE	-.101	.455	-.290	-.024	.103	-.028	.050	.028
	MOTHERS HIGHEST DEGREE	-.079	-.290	.476	.028	-.102	.043	-.052	-.121
	SPOUSES HIGHEST DEGREE	-.274	-.024	.028	.578	-.014	-.012	.203	-.039
	GENERAL HAPPINESS	-.058	.103	-.102	-.014	.666	-.325	-.085	-.085
	HAPPINESS OF MARRIAGE	.067	-.028	.043	-.012	-.325	.692	-.099	-.024
	CONDITION OF HEALTH	-.008	.050	-.052	.203	-.085	-.099	.749	-.102
	IS LIFE EXCITING OR DULL	.108	.028	-.121	-.039	-.085	-.024	-.102	.876
Anti-image Correlation	RS HIGHEST DEGREE	.701[a]	-.210	-.161	-.503	-.099	.113	-.012	.162
	FATHERS HIGHEST DEGREE	-.210	.640[a]	-.623	-.048	.187	-.049	.086	.044
	MOTHERS HIGHEST DEGREE	-.161	-.623	.586[a]	.053	-.181	.076	-.087	-.188
	SPOUSES HIGHEST DEGREE	-.503	-.048	.053	.656[a]	-.023	-.018	.309	-.055
	GENERAL HAPPINESS	-.099	.187	-.181	-.023	.549[a]	-.478	-.120	-.111
	HAPPINESS OF MARRIAGE	.113	-.049	.076	-.018	-.478	.619[a]	-.137	-.030
	CONDITION OF HEALTH	-.012	.086	-.087	.309	-.120	-.137	.734[a]	-.126
	IS LIFE EXCITING OR DULL	.162	.044	-.188	-.055	-.111	-.030	-.126	.638[a]

a. Measures of Sampling Adequacy(MSA)

Total Variance Explained

Component	Initial Eigenvalues			Extraction Sums of Squared Loadings			Sums of Squared Loadings		
	Total	*% of Variance*	*Cumulative %*	*Total*	*% of Variance*	*Cumulative %*	*Total*	*% of Variance*	*Cumulative %*
1	2.600	32.502	32.502	2.600	32.502	32.502	2.070	25.877	25.877
2	1.772	22.149	54.651	1.772	22.149	54.651	1.770	22.119	47.996
3	1.079	13.486	68.137	1.079	13.486	68.131	1.611	20.141	68.137
4	.827	10.332	78.469						
5	.631	7.888	86.358						
6	.487	6.087	92.445						
7	.333	4.161	96.606						
8	.272	3.394	100.000						

CONCLUSION

In summarized manner Principal Component Analysis is a well-established technique for dimensionality reduction and multivariate analysis. Examples of its many applications include data compression, image processing, visualization, exploratory data analysis, pattern recognition, and time series prediction. The popularity of PCA comes from three important properties. First, it is the optimal (in terms of mean squared error) linear scheme for compressing a set of high dimensional vectors into a set of lower dimensional vectors and then reconstructing the original set. Second, the model parameters can be computed directly from the data - for example by diagonalizing the sample covariance matrix. Third, compression and decompression are easy operations to perform given the model parameters - they require only matrix multiplication.

REFERENCES

Lu, X., L. Wang, *et al.* (2010): "Multivariate statistical analysis of heavy metals in street dust of Baoji, NW China." *Journal of Hazardous Materials* **173**(1-3): 744-749.

Yongming, H., D. Peixuan, *et al.* (2006): "Multivariate analysis of heavy metal contamination in urban dusts of Xi'an, Central China." *Science of The Total Environment* **355**(1-3) : 176-186.

6

Statistical Regression Analysis of Physico-chemical Parameters Estimated at Hasanpur, J.P. Nagar : An Approach to Water Quality Management

D.K. Sinha, Ritesh Saxena and Gaurav Kumar Rastogi

ABSTRACT

Correlation study was carried out between water quality and physico-chemical parameters. The statistical regression analysis for all the data points obtained for all the three major seasons of the year namely summer, rains and winter at Hasanpur, J.P. Nagar was performed. Conductivity of drinking water is found to be significantly correlated with almost all the physico-chemical parameters studied. It may be suggested that the underground drinking water quality at Hasanpur can be controlled effectively by controlling the conductivity of water. Significant correlation of conductivity with calculated values of water quality index also established the importance of conductivity in controlling water quality.

Keywords

Correlation, physico-chemical parameter, statistical regression analysis, water quality index, water quality management.

INTRODUCTION

With a whooping 17 crore population set to touch the 20

crore mark in the next two decades, Uttar Pradesh is headed for water wars in the future. J.P. Nagar comes under the dark block. The demand for water is continuously rising with the growth of population, industry and agriculture and these factors are adversely affecting the water quality also (Gupta, 2004, Gokulakrishn *et al.*, 2004, Sinha *et al.*, 1995). The total amount of available water remains more or less constant and this is bound to lead to scarcity of water which can be tackled by the management of water and more effectively with the management of water quality. The present study is aimed to suggest some effective measures for the drinking water quality management with the help of statistical regression analysis of ten data points for all the three major seasons of the year–summer, rains and winter at Hasanpur, J.P. Nagar.

The regression analysis attempts to establish the nature of the relationship between the variables and thereby provides a mechanism for prediction or forecasting (Mahajan *et al.*, 2005, Sinha, 2002, Adak *et al.*, 2001, Draper *et al.*, 1966). Water quality index (WQI) has been regarded as one of the most effective way to communicate water quality (Horton, 1965, Mishra *et al.*, 1985). The data obtained through quantitative analysis and W.H.O. water standards (W.H.O., 1971) are used for calculating water quality indices. J.P. Nagar is the 'C' class city of Uttar Pradesh. Hasanpur is an important tehsil of district J.P. Nagar. Hasanpur along with J.P. Nagar has gone through rapid industrialization, population increase and tremendous exploitation of water resources during last few decades.

MATERIALS AND METHODS

Water quality physico-chemical parameters at ten different drinking water sites at Hasanpur, J.P. Nagar for summer, rains and winter were estimated following standard methods and procedures of sampling and estimation (APHA, 1995, Merck, 1974). For the determination of pH, turbidity, conductivity, and fluoride, Century CP 901 pH meter, Century nephelometer, RI 215R conductivity meter, and Hach spectrophotometer 2010 were used respectively. All the chemicals of Anal R grade were used for quantitative analysis. The regression analysis for all the ten data points for all the three major seasons of the year, to study the correlation between different water quality parameters was performed using computer software SPSS, version-7.5.

Sampling sites with their number and names are :

(*i*) IM2 Hand Pump at Kayasthan Chamunda Road.

(*ii*) IM2 Hand Pump at Lal Masjid.

(*iii*) Ordinary Hand Pump at Khewan Street.

(*iv*) IM2 Hand Pump at Kot Kala-Shaheed.

(*v*) IM2 Hand Pump at Hasanpur-Rehra Road.

(*vi*) IM2 Hand Pump at Sambhal-Kaneta Road.

(*vii*) IM2 Hand Pump at Sambhal Bye pass.

(*viii*) Ordinary Hand Pump at Lal Bagh Street.

(*ix*) Ordinary Hand Pump at Main Market.

(*x*) IM2 Hand Pump at Shiv Temple.

RESULTS AND DISCUSSION

Site-wise and season-wise estimated values of water quality physico-chemical parameters are listed in Table-1. Site-wise and season-wise calculated values of water quality index are presented in Table-2. Following regression equations were obtained by the Statistical Regression Analysis taking conductivity as dependent variable of all the ten data points for all the three seasons namely summer, rains and winter for underground drinking water at Hasanpur, J.P. Nagar :

For Summer Season

Conductivity = + 0.26 *Temperature – 5.7

$N = 10, r^2 = 0.92, F = 87.1, S = 0.13$ ....(1S)

Where, N = number of data points

r = correlation coefficient

F = variance ratio

S = standard error of estimate

Conductivity = – 1.9*pH +15.09

N = 10, r^2 = 0.79, F = 29.27, S = 0.21 ....(2S)

Conductivity = + 0.42*Turbidity – 0.54

N = 10, r^2 = 0.75, F = 23.31, S = 0.23 ....(3S)

Conductivity = + 0.006*Alkalinity – 0.46

N = 10, r^2 = 0.92, F = 88.69, S = 0.13 ....(4S)

Conductivity = + 0.001*TS + 0.029

N = 10, r^2 = 0.95, F = 161.49, S = 0.09 ....(5S)

Conductivity = + 0.001*TDS + 0.11

N = 10, r^2 = 0.91, F = 85.05, S = 0.13 ....(6S)

Conductivity = + 0.006*TSS – 0.038

N = 10, r^2 = 0.85, F = 44.83, S = 0.18 ....(7S)

Conductivity = –0.31*DO + 1.99

N = 10, r^2 = 0.52, F = 8.72, S = 0.32 ....(8S)

Conductivity = – 9.51*BOD + 1.47

N = 10, r^2 = 0.62, F = 12.78, S = 0.29 ....(9S)

Conductivity = + 0.002*Total Hardness + 0.098

N = 10, r^2 = 0.95, F = 144.36, S = 0.11 ....(10S)

Conductivity = + 0.009*Calcium + 0.15

N = 10, r^2 = 0.95, F = 137.96, S = 0.11 ....(11S)

Conductivity = + 0.026*Magnesium + 0.11

N = 10, r^2 = 0.84, F = 42.64, S = 0.18 ....(12S)

Conductivity = + 0.008*Chloride + 0.33

N = 10, r^2 = 0.92, F = 91.82, S = 0.13 ....(13S)

Conductivity = + 1.09*Fluoride + 0.12

N = 10, r^2 = 0.64, F = 14.18, S = 0.28 ....(14S)

Conductivity = + 0.036*CO_2 – 0.16

N = 10, r^2 = 0.89, F = 69.63, S = 0.15 ....(15S)

Conductivity = + 0.004*SO_4 + 0.19

N = 10, r^2 = 0.90, F = 73.35, S = 0.14 ....(16S)

For Rainy Season

Conductivity = + 0.25*Temperature – 5.08

N = 10, r^2 = 0.91, F = 80.69, S = 0.13 ....(1R)

Conductivity = + 1.79*pH – 13.06

N = 10, r^2 = 0.61, F = 12.28, S = 0.28 ....(2R)

Conductivity = + 0.35*Turbidity – 0.38

N = 10, r^2 = 0.81, F = 34.77, S = 0.19 ....(3R)

Conductivity = + 0.006*Alkalinity – 0.45

N = 10, r^2 = 0.88, F = 61.01, S = 0.15 ....(4R)

Conductivity = + 0.001*TS – 7.26

N = 10, r^2 = 0.92, F = 96.65, S = 0.12 ....(5R)

Conductivity = + 0.002*TDS – 1.89

N = 10, r^2 = 0.89, F = 63.24, S = 0.15 ....(6R)

Conductivity = + 0.005*TSS – 0.045

N = 10, r^2 = 0.84, F = 41.66, S = 0.18 ....(7R)

Conductivity = – 0.31*DO + 2.14

N = 10, r^2 = 0.85, F = 44.75, S = 0.17 ....(8R)

Conductivity = – 0.14*BOD + 2.11

N = 10, r^2 = 0.73, F = 21.13, S = 0.23 ....(9R)

Conductivity = + 0.002*Total Hardness + 0.16

$N = 10, r^2 = 0.94, F = 123.70, S = 0.11$ ....(10R)

Conductivity = + 0.011*Calcium + 0.11

$N = 10, r^2 = 0.90, F = 74.22, S = 0.14$ ....(11R)

Conductivity = + 0.022*Magnesium + 0.23

$N = 10, r^2 = 0.90, F = 70.11, S = 0.14$ ....(12R)

Conductivity = + 0.006*Chloride + 0.41

$N = 10, r^2 = 0.94, F = 129.59, S = 0.11$ ....(13R)

Conductivity = + 0.89*Fluoride + 0.11

$N = 10, r^2 = 0.69, F = 18.08, S = 0.25$ ....(14R)

Conductivity = + 0.026*CO_2 – 0.006

$N = 10, r^2 = 0.86, F = 47.30, S = 0.17$ ....(15R)

Conductivity = + 0.005*SO_4 – 2.0

$N = 10, r^2 = 0.95, F = 139.77, S = 0.10$ ....(16R)

For Winter Season

Conductivity = + 0.20*Temperature – 3.75

$N = 10, r^2 = 0.83, F = 39.11, S = 0.17$ ....(1W)

Conductivity = – 2.02*pH + 16.12

$N = 10, r^2 = 0.85, F = 44.30, S = 0.16$ ....(2W)

Conductivity = + 0.31*Turbidity – 0.23

$N = 10, r^2 = 0.77, F = 26.14, S = 0.19$ ....(3W)

Conductivity = + 0.006*Alkalinity – 0.43

$N = 10, r^2 = 0.85, F = 46.07, S = 0.15$ ....(4W)

Conductivity = + 0.001*TS + 0.057

$N = 10, r^2 = 0.93, F = 100.46, S = 0.11$ ....(5W)

Conductivity = + 0.001*TDS + 0.11

$N = 10, r^2 = 0.87, F = 53.33, S = 0.15$ ....(6W)

Conductivity = + 0.005*TSS + 0.11

N = 10, r^2 = 0.85, F = 44.52, S = 0.16 ....(7W)

Conductivity = – 0.28*DO + 2.04

N = 10, r^2 = 0.62, F = 13.15, S = 0.25 ....(8W)

Conductivity = + 0.037*BOD + 0.23

N = 10, r^2 = 0.54, F = 9.22, S = 0.27 ....(9W)

Conductivity = + 0.002*Total Hardness + 0.15

N = 10, r^2 = 0.92, F = 88.55, S = 0.12 ....(10W)

Conductivity = + 0.009*Calcium + 0.17

N = 10, r^2 = 0.96, F = 214.78, S = 0.08 ....(11W)

Conductivity = + 0.024*Magnesium + 0.71

N = 10, r^2 = 0.87, F = 53.74, S = 0.14 ....(12W)

Conductivity = + 0.006*Chloride + 0.35

N = 10, r^2 = 0.90, F = 75.24, S = 0.12 ....(13W)

Conductivity = + 0.83*Fluoride + 0.19

N = 10, r^2 = 0.53, F = 9.13, S = 0.27 ....(14W)

Conductivity = + 0.027*CO_2 + 0.001

N = 10, r^2 = 0.83, F = 38.59, S = 0.17 ....(15W)

Conductivity = + 0.003*SO_4 + 0.12

N = 10, r^2 = 0.92, F = 97.68, S = 0.11 ....(16W)

Critical analysis of above equations for all the three seasons reveals following facts :

The conductivity shows highly significant correlation with eleven physico-chemical parameters namely temperature, total alkalinity, total solids, total dissolved solids, total suspended solids, total hardness, calcium, magnesium, free CO_2, chloride and sulphate for all the three seasons with the value of regression coefficient, (r) more than 0.90 *i.e.* there is more than

90% association in data. The values of variance ratio, (F) is high and the standard error of estimate is also low which are required for significant correlation.

The conductivity shows significant correlation with turbidity and dissolved oxygen in rains while they are moderately correlated with the conductivity in winter and summer seasons. Conductivity is found to be significantly correlated with pH in summer and winter while it is moderately correlated in rains. Biological oxygen demand and fluoride are found to be moderately correlated with conductivity in all the three seasons.

Following three regression equations were obtained by regression analysis of WQI values presented in Table-2 taking conductivity as dependent variable for ten data points for three different seasons :

For Summer Seasons

Conductivity = + 0.006*WQI(S) – 0.198

$N = 10, r^2 = 0.97, F = 3.22, S = 0.07$ ....(1wS)

For Rainy Season

Conductivity = + 0.005*WQI(R) – 0.133

$N = 10, r^2 = 0.98, F = 364.30, S = 0.06$ ....(2wR)

For Winter Seasons

Conductivity = + 0.005*WQI(W) – 0.15

$N = 10, r^2 = 0.77, F = 31.80, S = 0.18$ ....(2wR)

On the basis of above three equations it may be suggested that the conductivity is significantly correlated with calculated values of WQI in summer and rainy season and it is moderately correlated with WQI in winter seasons. The value of regression coefficient is more than 0.95 *i.e.* there is more than 95% association in data in summer and rainy season. The values of variance ratio are fairly high and standard error of estimate is very low.

CONCLUSION

This study on the basis of the statistical regression analysis reveals that the conductivity is a very important physico-chemical drinking water quality parameter as this is significantly correlated with sixteen water quality parameters for all the three seasons during the course of study more specifically in the catchment area of our study. The parameters are temperature, pH, turbidity, total alkalinity, total solids, total dissolved solids, total suspended solids, dissolved oxygen, biological oxygen demand, total hardness, calcium, magnesium, free CO_2, chloride, fluoride, and sulphate. These results are supported by the values of correlation coefficient, r which is near to unity, fairly high values of variance ratio, F and low values of standard error of estimate, S.

Temperature, total alkalinity, solids, total hardness, calcium, magnesium, free CO_2, chloride and sulphate show significant correlation with conductivity in all the three major seasons while turbidity and dissolved oxygen in rains only. Conductivity is significantly correlated with pH in summer and winter. pH value in rains, turbidity and dissolved oxygen in winter and summer, biological oxygen demand and fluoride in all the three seasons show moderate correlation with conductivity.

The most important part of any pollution study is to suggest some fruitful and effective measures for the water quality management and on the basis of this study, we may suggest more confidently that the underground drinking water quality especially at Hasanpur, J.P. Nagar can be checked effectively with reference to above mentioned sixteen water quality parameters by controlling the conductivity of drinking water. Significant correlation of conductivity with calculated values of water quality index also established the importance of conductivity in controlling the drinking water quality.

REFERENCES

Adak and Purohit, K.M. (2001) : Assessment of drinking water quality of river Brahmani, *Indian J. Env. Prot.*, 21(4) : 295.

APHA, (1995) : *Standard methods for examination of water and waste water*, 19th ed., AWWA, WPCF, Washington. D.C., U.S.A.

Draper, N.R. and Smith, H. (1996) : *Applied Regression Analysis*, Wiley, New York.

Gokulakrishn N. and Pandurangan A. (2004) : Contamination of ground water by tannery, *Indian J. Env. Prot.*, 24(1) : 18.

Gupta, G.K. and Shukla, R. (2004) : Physico-chemical and bacteriological quality in various sources of drinking water from industrial area, Pollution *Research*, 23(1) : 205.

Horton, R.K. (1965) : An index number system for rating water quality, *J. Water Poll. Cont. Fed.*, 37-300.

Mahajan, S.V.; Khare, S. and Srivastava, V.S. (2005) : A correlation and regression study, *Indian J. Env. Prot.*, 25(3) : 254.

Merck, E. (1974) : *The testing of water*, Darmsatdt, Federal Republic of Germany.

Sinha, D.K. and Srivastava, A.K. (1995): Physico-chemical characteristics of river Sai at Rai Bareli, *Indian J. Environ. Hlth.*, 37(3), 205.

Sinha, D.K. (2002): Statistical analysis of physico-chemical parameters of the water of river Sai at Rae Bareli, *Indian J. Env. Prot.*, 22(2), 184

Tiwari, T.N. and Mishra, M. (1985): A preliminary assignment of water quality index of major Indian rivers, *Indian J. Env. Prot.*, 5(4) : 206-209.

W.H.O. (1971) : International standards for drinking water, *World Health Organization*, Geneva.

7

A Study About Perception of Coastal Zone Management Issues Related to Climate Change in Mauritius

P. Unauth, R.T. Ramessur and K. Boodhoo

ABSTRACT

The aim of this study was to investigate the perceptions about impacts and causes of climate change on the coastal zone of Mauritius, the value of coastal resources and management decisions about coastal zone management in Mauritius. 70% of respondents were aware that ICZM can help create an enabling environment to inform policy-makers and other key stakeholders of the issues and benefits of Integrated Coastal Zone Management (ICZM), and help promote changes in the behaviour of both policy-makers, key stakeholders and civil society. 18% of the respondents believed that the most important coastal resources is the coral reef system, followed by beach systems together with lagoon and estuary systems and seafood resources (17%), then mangrove system (15%) and finally seagrass system (8%). The mostly chosen option listed by respondents as origins of impact on the coastal resources, were waste disposal (15%), followed by tourism (14%), urbanisation and agriculture (11%). 89% of the respondents were aware of the importance of mangroves in Mauritius and the majority of the respondents quoted coral bleaching as the origins of destruction of coral reefs around Mauritius (17%), followed by sewage (15%). 50% of the respondents said that there was no proper coordination and cooperation among various stakeholders involved in the coastal development in Mauritius.

Keywords

Climate Change, Coastal Zone Management, Indicators, Ecosystems, Coral Reefs, Mangroves.

INTRODUCTION

The principle of interdisciplinary and an integrative approach of all the problems encountered in the coastal zone are being implemented in Mauritius (1850 km^2, 20 °S and 57 °E, Western Indian ocean) as the coastal zone is being viewed as one complex multidisciplinary entity with the various conflicting uses and activities taking place within the zone (Ramessur 2002; Ramessur *et al.* 2010). Projects on lagoonal health characteristics and coastal pollution monitoring surface and submarine groundwater discharge have been implemented to provide baseline data in the field of marine ecology, marine pollution and physical oceanography (Burnett *et al.* 2006; Jowaheer 2010; Ramessur 2010; Swayne *et al.* 2007). There are indirect incentives in the form of tax credits for importation of pollution abatement facilities to encourage industry to invest in new production equipment and clean technology.

Studies in Mauritius are now aimed towards strategies to better protect manage and conserve marine ecosystems and the requirements of an Environmental Impact Assessment (EIA) is mandatory for any coastal development project as scheduled under the EPA Act and coastal water quality standards for recreational activities have been prepared. Waste treatment plants by polluting industries and major hotel complex have also become mandatory and assistance is being sought for the implementation of the provisions of the provisions of MARPOL in Mauritius and necessary equipment have been procured recently to implement an oil contingency plan for the harbour in case of oil spill. The use of GIS has gained more importance in planning processes for coral reef management and has provided information for sensible policy-making for the south east coast. A considerable amount of improvement has been noted over the recent decade following the enactment of the Environmental Protection Act (EPA) in 1991. The reef monitoring network is a sub-regional node of the Global Coral Reef Monitoring Network and supported under a GEF mid-size project. The European Commission considered the ICZM programme as a pilot one among the many programmes it supported with African/Caribbean/Pacific (ACP) countries and island states (Ramessur 2010). There is an urgent need to protect sensitive marine and coastal areas besides coral reefs such as wetlands islets, seagrass beds and mangroves and encourage the use of environmentally cleaner technology in industries including the recycling of wastes. The aim of this study was to investigate the perceptions about impacts and causes of climate change on the coastal zone of Mauritius, the value of coastal resources and management decisions about coastal zone management in Mauritius.

METHODOLOGY

The study was conducted via a survey questionnaire with 105 respondents. The targeted respondents for the coastal impact survey were students of the University of Mauritius (UOM) at Reduit, preferentially students having knowledge in the field of environmental sciences (BSc Biology, BSc Chemistry, BSc Marine Sciences, B. Eng Chemical and Environmental Sciences, MSc Chemistry with specialisation in Analytical/Environmental Chemistry and MSc sustainable Built environment). Targeted stakeholders about management issues were scientific officers, directors, technical officers and managers from Ministry of Environment and National Development Unit, Mauritius Oceanography Institute and also from the Tourism sector. The data of the survey was analysed statistically using the 12.0 Statistical Package for the Social Sciences (SPSS) version. The results were presented as pie charts or bar charts.

RESULTS AND DISCUSSION

Respondents were asked about sectors concerned with climate change in Mauritius as shown in Figure 1. It is observed that 13% of the respondents quoted that coastal zone, followed by the tourism sector with 11 % responses and human health with 10% responses. It should be noted that this is a multiple-response question. Mauritius has an extremely rich coastal zone consisting of near shore wetlands and mangroves, lagoon coral, fringing coral reef and all their associated marine life. Mauritius being a small island state depends largely on this coastal ecosystem for its tourism industry. Most of the hotels are coastal based. Moreover, the fishermen population residing in the coastal area earns their daily livelihood from fishing activities.

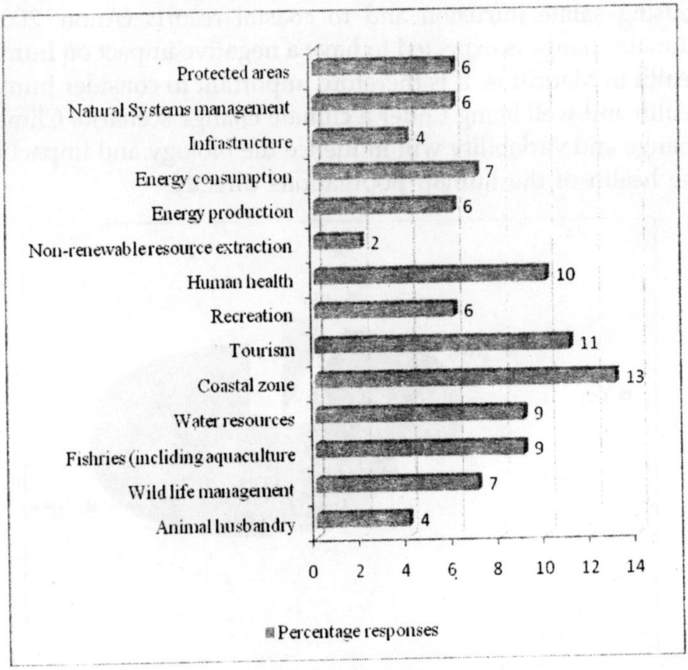

Figure 1 : Perception of respondents about sectors concerned with climate change in Mauritius.

Higher sea-level, storm surges and higher tides will cause extensive coastal flooding and damage to vital coastal infrastructure upon which tourism, transport and communications depend such as sea ports, bridges and roads. The coastal zone and in particular, the coastline of Mauritius is undergoing morphological changes as a result of natural causes and anthropogenic activities. The natural phenomena include tropical cyclones, tidal waves, and tsunamis. The anthropogenic activities include haphazard construction of ill-designed jetties and groynes, past lagoonal sand removal, increasing infrastructural development, pollution from point and non point sources, and degradation of coral reefs. The major impacts of sea-level rise in Mauritius are land loss, erosion of beaches, damage to coastal infrastructure, damage to coastal ecosystems including degradation of coral reefs and loss of wetlands. Rising sea levels may also be especially threatening to coastal aquifers causing saline intrusion and to coastal resorts (Anon. 2003). Climate change is expected to have a negative impact on human health in Mauritius. It is therefore important to consider human health and well being under a climate change scenario. Climate change and variability will influence the biology and impact on the health of the human populations directly.

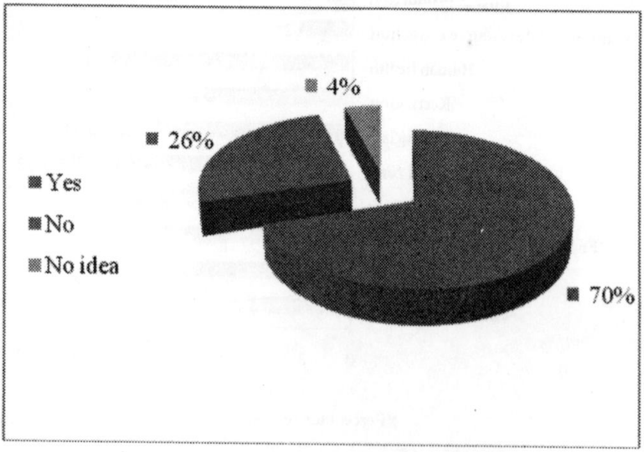

Figure 2 : Awareness of respondents about ICZM in Mauritius.

Respondents were asked if they have heard about ICZM. From Figure 2 it can be seen that 70% are aware that ICZM can generally help create an enabling environment to inform policy-makers and other key stakeholders of the issues and benefits of Integrated Coastal Zone Management (ICZM), and help promote changes in the behaviour of both policy-makers, key stakeholders and civil society. This shows that respondents have a responsibility toward their coastal environment. ReCoMaP is a regional Integrated Coastal Zone Management (ICZM) programme for the sustainable management of the coastal zones of the countries of the Indian Ocean. It is an initiative of the Indian Ocean Commission, funded by the European Union for a total of .18 million.The Programme started in August 2006 and is scheduled to end in 2011. It involves seven focus countries of the region, namely the Comoros, Madagascar, Mauritius, Kenya, the Seychelles, Somalia and Tanzania. ReCoMap's overall objective is to contribute to the reduction of poverty amongst the most deprived coastal populations in the WIO. ReCoMaP's approach to achieve this objective is to strengthen the capacities of coastal communities; local and national government officers and civil society to sustainably manage natural resource and local development. ReCoMaP has also initiated activities related to reducing the direct physical impacts of urban and tourism development on the coastal environment. More specifically, the programme is supporting activities to ameliorate the worst impacts of coastal erosion and poor waste management (Recomap 2007).

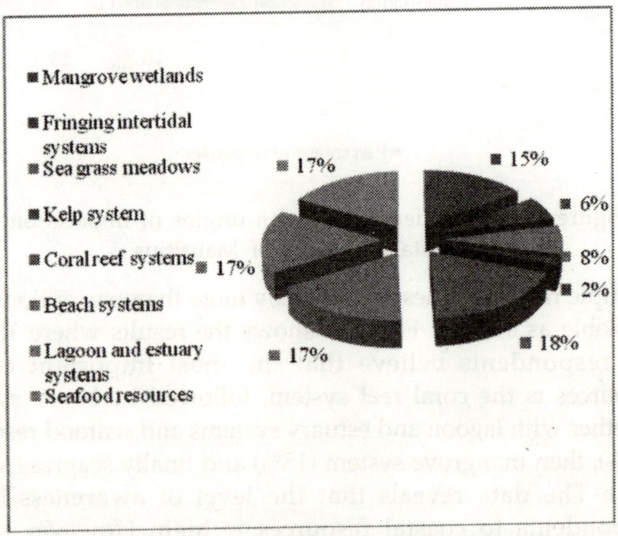

Figure 3 : Perception of respondents about coastal resources of Mauritius.

Respondents were asked about the awareness of the Mauritian coastal resources. It should be noted that this was a

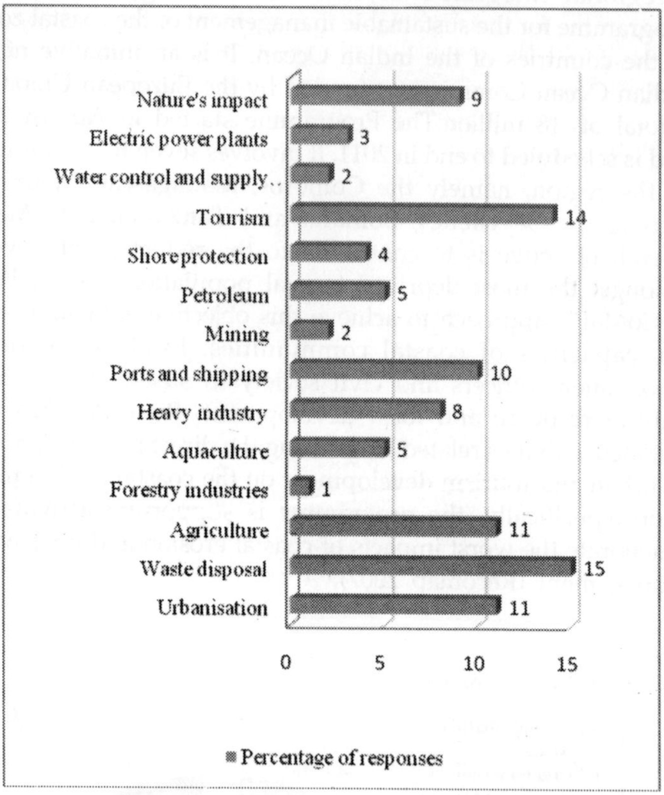

Figure 4 : Respondents' views on origins of impacts on the coastal resources of Mauritius.

multiple response question whereby more than one options were available as answer. Figure 3 shows the results where 18 % of the respondents believe that the most important coastal resources is the coral reef system, followed by beach systems together with lagoon and estuary systems and seafood resources (17%), then mangrove system (15%) and finally seagrass system (8%). The data reveals that the level of awareness of the respondents to coastal resources is high. However people seemed to be unaware of the kelp system (2%). Actually

Mauritius being a tropical island does not possess the kelp system. Since the major tropical habitats namely mangrove ecosystem, seagrass ecosystem and coral reef ecosystem are very essential for the marine biodiversity and that our beaches are undoubtedly a very important asset for the Mauritian economy, according to this question local people seemed to be aware of gravity of destroying these natural systems.

The mostly chosen option, as origins of impact on the coastal resources, was waste disposal (15%), followed by tourism (14%), finally urbanisation and agriculture (11%). Waste disposal is actually a serious problem around the island. Coastal communities often dump large amounts of solid wastes onto the coastal zone which disfigure the coastal landscape and leach pollutants into the coastal areas. Industrial waste, sewage, poorly run landfills and disposal of toxic chemicals are significant contributors to marine pollution and coastal degradation. Solid waste was traditionally disposed of in open dumps. In the 1980's there were 24 dumping sites, but by 2002 they have been completely eliminated (Anon : Fact sheet on coastal zone). At present most solid waste is disposed of in a single sanitary landfill at Mare Chicose after being compacted at transfer stations located around the country. The present solid waste generated per capita is around 0.93 to 1.0 kg/day, thus making a total of around 430 000 tonnes annually. Dumping are overflowing with heterogeneous wastes and open dumps have obvious disadvantages with regards to emission of air pollutants and odours, contamination of the water table and contamination of aquatic resources like rivers and the lagoon due to seepage and run-off respectively. Lagoon is polluted with all types of solid waste such as old vehicle tyres, broken glass bottles, plastic containers and so on (Anon. 2005).

Sewerage systems are confined to parts of Port-Louis and Plaines Wilhems districts only. Presently only 21% of the whole island is sewered. The remaining 79% either disposed of their liquid waste in the absorption pits or via septic tanks. The island of Mauritius being volcanic in nature, its geology is characterized by highly fissured lava with relatively high permeability. Hence leachates easily find its way into our water bodies in case absorption pits are not well located. Where sewerage systems exist, illegal connections between the storm and sewerage systems and direct leakage cause contamination of adjacent watercourses. Sometimes un-sewered industries discharge effluent directly to surface water often with little or no wastewater treatment. The most polluting sectors are the sugar industries and animal food processing industries. Moreover contaminant metals are potentially toxic which accumulate in the sediment and ultimately are bioconcentrated by organisms and may cause health problems to humans via the food chain. Discharge of industrial wastewaters and untreated sewage pose a particularly serious threat to the coastal environment of Mauritius (Anon.: Fact sheet on coastal zone).

Tourism, being the third pillar of the Mauritian economy after the manufacturing sector and agriculture, has been a key factor in the overall development. But unplanned tourism has caused many negative impacts on the coastal zone of Mauritius. Construction of hotels, recreation and other facilities often leads to increased sewage pollution. The development of tourism facilities, water supplies, restaurants and recreation facilities around the coast involved sand mining, beach and sand dune erosion, soil erosion and extensive paving. Development of marinas and breakwaters can cause changes in currents and coastlines. Tourism causes loss of biological diversity. Urbanization of the coastal area also leads to coastal zone degradation especially construction of hotels and bungalows (Anon. 2007). It should be noted that agriculture for instance sugar cane has also contributed to the degradation of the coastal resources in Mauritius. Land clearing and farming causes massive erosion of soil which ends up in coastal waters. Nature's impact is also inevitable. The sea is not altered by man alone; nature is implicated in many serious disturbances such as intense tropical cyclones, heavy rainfall and red tides. Only 2% said sand mining, but sand mining has contributed a lot to the degradation of our resources. Just recently it has been banned. Port infrastructure supports many activities such as oil refinery, oil storage, shipping industry and so on. Oil and chemical spills become a major threat (Ramessur course manual).

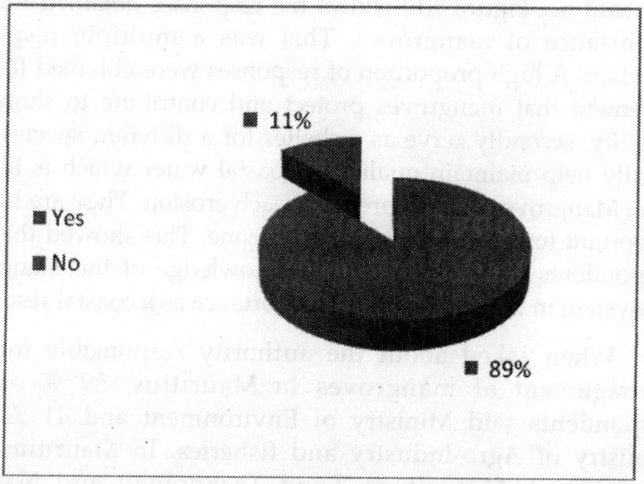

Figure 5(a) : Awareness of respondents on the importance of mangroves

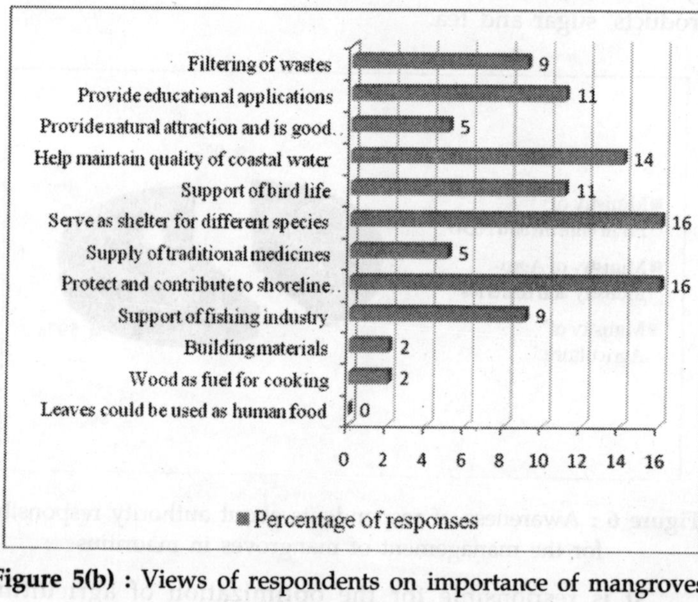

Figure 5(b) : Views of respondents on importance of mangroves.

It can be seen that form Figure 5(a) that 89% of the

respondents are aware of the importance of mangroves. Very few said no. Figure 5(b) shows the responses obtained for the importance of mangroves. This was a multiple response question. A high proportion of responses were obtained for the statement that mangroves protect and contribute to shoreline stability, secondly serve as a shelter for a different species and thirdly help maintain quality of coastal water which is highly true. Mangroves help to prevent beach erosion. They are highly important for the Mauritian coastal zone. This showed that the respondents do have a scientific knowledge of the mangrove ecosystem in Mauritius and its importance as a coastal resource.

When asked about the authority responsible for the management of mangroves in Mauritius, 59 % of the respondents said Ministry of Environment and 41 % said Ministry of Agro-Industry and fisheries. In Mauritius, the Ministry of Agriculture, Food Technology and Natural Resources acts as a facilitator and a catalyst for promoting agricultural production such as milk and meat, horticultural products, sugar and tea.

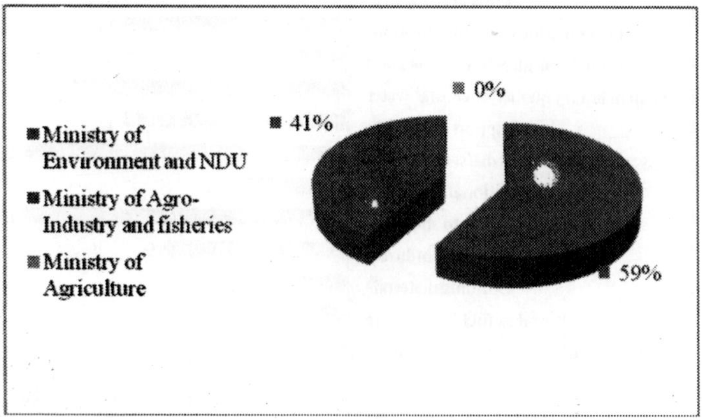

Figure 6 : Awareness of respondents about authority responsible for the management of mangroves in mauritius

It is responsible for the optimization of agricultural production through rational use of resources, ensuring the sustainability of the agricultural sector and the preservation of agricultural lands as well as conservation of natural resources such as forests, fauna and flora. The fisheries division provides an enabling environment for the promotion of sustainable development of the fisheries sector and to ensure continued economic growth and social development within the framework of good governance (www.gov.mu.). The IZCM division in the Ministry of Environment is responsible for management of mangroves. From the results it can be seen that respondents are not well informed or have little awareness regarding the responsibilities of a particular ministry. It should be noted that for a better functioning of a particular society stability is essential. A proper stability refers to good governance with updated laws and regulations.

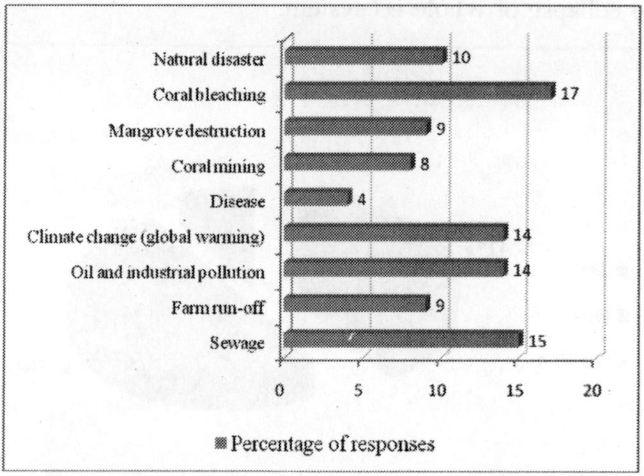

Figure 7 : Respondents views about origins of coral reef destruction in Mauritius

Respondents were asked about the causes of coral reefs destruction around Mauritius. Figure 7 shows the results obtained. This was a multiple response question hence respondents had more than one option. The majority of the respondents quoted coral bleaching as the origins of destruction of coral reefs around Mauritius (17 %), followed by sewage (15 %). And rightly so, corals around Mauritius are dying due to coral bleaching as a result of higher sea surface temperature.

Leaked oil and chemicals (e.g. from detergents, paints) flowing into the sea from factory outlets are a key threat. Chemical fertilizers are yet another pollutant. Very few respondents (4 %) quoted disease as the origin of destruction. It should be noted that coral diseases and syndromes generally occur in response to biotic stresses such as bacteria, fungi and viruses, and abiotic stresses such as increased sea water temperatures, ultraviolet radiation, sedimentation and pollutants. One type of stress may exacerbate the other (Clark, J.R. 1998). Coral mining was once a very frequent activity in Mauritius. As a result the coral reef system was deteriorated. Studies have shown that climate change is responsible for coral bleaching. If there is no coral ecosystem there will be no life in water and hence collapse of whole ecosystem.

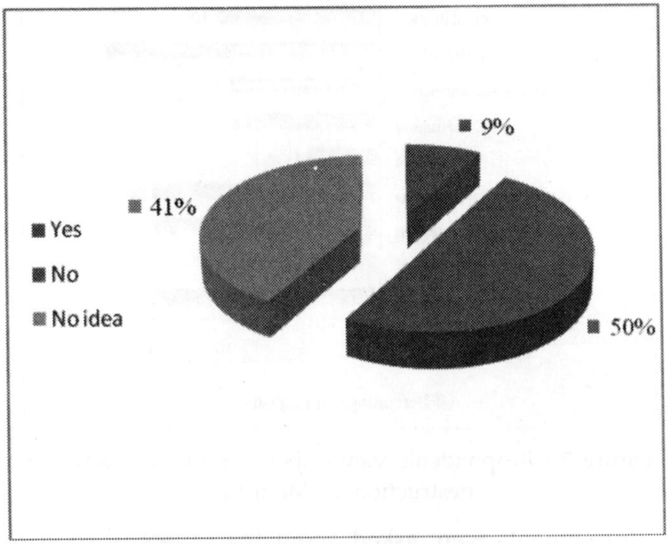

Figure 8 : Respondents views about proper coordination and cooperation among the various stakeholders involved in the coastal development in Mauritius

From Figure 8 it can be seen that 50 % of the respondents said that there is no proper coordination and cooperation among various stakeholders involved in the coastal development in Mauritius and 41 % have no idea. Enforcing laws and regulations would not be enough. It is imperative to educate the local population as wells as to encourage the stakeholders to participate actively to be able to impart the right knowledge, values, attitudes, wise practices and know-how and thus to contribute to the sustainable development of the coastal areas in Mauritius. There are cases which provide good examples of governments involving a wide range of participants in their policy making processes. Public hearings have been in common practice in many countries for a long time, but the new trend is to go beyond simple hearings to inviting the public, private enterprise, and NGOs to actively participate in developing policy options and implementing programmes on the environment. The goal of participatory programmes is to make stakeholders responsible for the decision making process and for their behaviour, which have a significant influence on ways they use their resources.

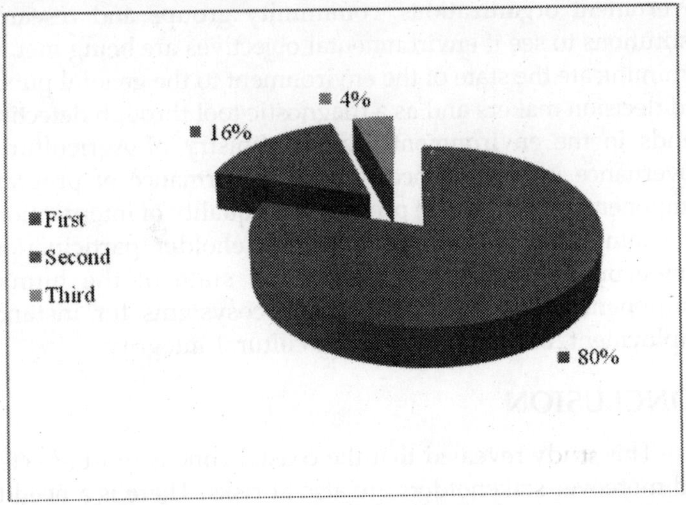

Figure 9 : Priority listed by respondents about environmental indicators to monitor the state of the coastal and marine environment in Mauritius (First -Environmental indicators, Second- Socio economic indicators Third- Governance indicators).

Respondents were asked about the indicators most important to monitor the state of the coastal and marine environment in Mauritius. 80 % of the respondents quoted environmental indicator in the first position. This shows the importance of the environment to the respondents. An indicator can be defined as something that helps us to understand where we are, where we are going and how far we are from the goal. An indicator quantifies and simplifies phenomena and helps us understand complex realities. Environmental indicators are simple measures that tell us what is happening in the environment. Since the environment is very complex, indicators provide a more practical and economical way to track the state of the environment than if we attempted to record every possible variable in the environment for example, coastal water quality. Environmental indicators are statistics that are scientifically credible and representative of an environmental issue and can help us to keep track of trends in the state of the environment and measure progress towards sustainable development. Environmental indicators are used by governments, non-government organizations, community groups and research institutions to see if environmental objectives are being met, to communicate the state of the environment to the general public and decision makers and as a diagnostic tool through detecting trends in the environment (Anon. Ministry of Agriculture). Governance indicators measure the performance of program components as well as the progress and quality of interventions for example, legislations policies, stakeholder participation. Socioeconomic indicators reflect the state of the human component of coastal and marine ecosystems for instance employment, economic value and cultural integrity.

CONCLUSION

This study revealed that the coastal zone is most affected and moreover stakeholders are also at risks. There is a need to empower the generation, especially with disastrous consequences of climate change. Young people have the potential to lead the way in the emerging grassroots climate movement to demand and create a climate friendly future. Climate change will not only affect people, nature but also business of people and hence the long-term sustainable development of the island.

RECOMMENDATIONS

ICZM is essential to identify the strength and weaknesses of the coastal area as well as stakeholder participation. Coastal communities require support and guidance to manage the challenges of climate change and the environmental impacts it will have on coastal communities. Specifically support is required in the area of land use planning (both strategic and statutory), strategic development, insurance, infrastructure management, natural environment (habitat, erosion, beaches and so on) and research and investment into professional services to develop management plans. There is a need to promote capacity-building at various levels, from public awareness programmes to research and development, in order to remain in phase with new developments in coastal management techniques and technology. Local creativity and innovation should also be encouraged. Increased adverse weather events resulting from climate change may also require reassessing existing reduction strategies, for example, the adequacy of drainage system, regular cleaning of rivers, or building sea wall protection in certain vulnerable coastal areas. Local Government as the closest sphere of government to the community works on the front line for delivery of local, state and federal climate change agendas. They have a major leadership role to play in the delivery of programs and as such need to work closely on cooperative and collaborative programs with the stakeholders and state government.

There should be a better management of the public beaches. Legislation should be revised and the laws should be enforced regarding beach degradation. There is a need for clean-up campaign as well as sensitization campaign at regular basis to make people more conscious about the beaches. Public awareness is very important regarding coastal degradation. A better sewage treatment program is essential. Management of the coast is inextricably linked to managing society's use of the coast. The success of coastal management depends on the support and participation of the many stakeholders (individuals and organizations from both public and private sector) who have a direct interest in actions or decisions. A need for coastal management usually arises from development pressures such as conflicting uses of coastal resources, urbanization, access, pollution and environmental degradation, competing industrial development for instance tourism and aquaculture. Problems can also arise where there is misunderstanding, competition and conflict between those who make and implement decisions and those affected by them. In many instances technical questions are only part of the issue, and other elements, such as differing perceptions of the problem, may present a greater barrier to successful resolution.

Any management activity is fundamentally a negotiation process between stakeholders that should be initiated from the start and maintained not only during the project initiative but in the long term through appropriate organizational and institutional arrangements. Stakeholder participation is not a one-time activity in a programme but is used to engage people, organizations, NGOs and government in ICZM. Although building understanding and relationships within and between stakeholders is important, ultimately the power to overcome and resolve coastal problems lies with the organizations and political and social structures that can create the institutional setting for management to exist. Therefore the stakeholder participation process needs to develop trust between organizations and communities, and leadership to influence the start of a transparent and sustainable management system. For water resources more efficient management of demand and supply, improved monitoring and forecasting systems for floods and droughts, desalination of sea water is recommended. For human settlement and infrastructure a hazard mapping, improved forecasting and early warning systems and insurance provision is required. For the public health a development of health surveillance and forecast system, strengthening of data collection and reporting systems is recommended together with vaccination campaigns and health education. For tourism a sustainable development approach such as eco-tourism should be implemented and for better management of coastal zone a sustainable coastal zone resource management is recommended.

Moreover an increase council officers resources and a power for local authorities is recommended to tackle climate change. In fact there need to be officers responsible for tackling climate change issues in each organization. Further examples of good practice and peer networks of officers and members together with external consultancy support are recommended. Climate change strategy or policy needs to be developed.

FUTURE WORK

According to this study there is a need for environmental indicators to monitor the state of the coastal and marine environment in Mauritius. To better understand climate change impacts on the coastal zones there is a need to better understand the coastal resources. Hence such a study can be carried out to assess the relevant indicators for Integrated Coastal zone Management in Mauritius which aims to measure the state of coastal environment. An indicator provides a simplified view of a more complex phenomenon, or provides insights about a trend or event that cannot be directly observed. Thus, indicators both quantify information and simplify information about a trend or event that cannot be directly observed. In ICZM, indicators can be used in two basic, overlapping ways: as a means of communication and as a means for measuring. These uses include: informing decision making, increasing understanding of important issues, assessing conditions and trends, comparing conditions in different geographical areas, projecting trends, measuring performance and results of policies or actions, and showing the connections between environmental, social and economic concerns. Sustainability at the coast can only be maintained if the ecosystem and other natural assets that generate the resources used by man can be managed in a sustainable manner. To manage and maintain these activities indicators are essential to gauge and monitor progress (Belfior, S. 2005). ICZM requires robust indicators of sustainability that gauge the 'health' of the coast in relation to both environmental, social and economic activity, whilst using fewer resources (materials, fuels, or land), and whilst managing the impacts of the resources that are used so that their adverse effects are minimized. Such indicators are essential tools for monitoring the state of the coastal environment, to inform managers and policy makers of the effectiveness of strategies in achieving sustainability. Indicators give a quick scan of the coastal zone, they can certainly help to install a sustainable coastal zone management.

REFERENCES

Anon. (2003) : "Study on coastal erosion in Mauritius, final technical report", *Ministry of Environment and National Development Unit*. Anon., "Fact sheet on coastal zone", Ministry of Environment and NDU.

Anon. (2005) : "Staking out the future", Ministry of Environment and NDU.

Anon. (2007) : "Mauritius Tourism Development Plan", Ministry of Tourism. Anon., "Global Climate Change Response Measures and Adaptation in Agriculture", Ministry of Agriculture.

Belfior, S. (2005) : In "A Handbook for Measuring the Progress and Outcomes of Integrated Coastal and Ocean Management", *UNESCO*.

Burnett, W.C, Aggarwal, P.K, Bokuniewicz, H, Cable, J.E, Charette, M.A., Kontar, E., Krupa, S., Kulkarni, K.M., Loveless, A., Moore, W.S., Oberdorfer, J.A., Oliveira, J., Ozyurt, N., Povinec, P., Privitera, A.M.G., Rajar, R., Ramessur, R.T., Scholten, J., Stieglitz, T., Taniguchi, M., Turner, J.V. (2006): Quantifying Submarine Groundwater Discharge in the Coastal Zone via Multiple Methods, *Science of the Total Environment*, 367 : 498-543.

Clark. J.R. (1998): In "Coastal Seas, The Conservation Challenge" : pp1-77.

Jowaheer, V., Lalbahadoor, V., Ramessur, R., Dosoruth, L. (2010) : A multifactor statistical model for analyzing the physico-chemical variables in the coastal area at St Louis and Tamarin, Mauritius, *Journal of Applied Quantitative methods*, 5 (1).

Ramessur, R.T. (2002) : Anthropogenic driven changes with focus to the coastal zone of Mauritius, South Western Indian Ocean, *Regional Environmental Change*, 3 (13): 99- 106.

Ramessur, R.T., Boodhoo, K., Balgobin, J., Aurelio, A. (2010) : Statistical comparison between consecutive winter and summer concentrations in zinc and lead from sediments in a tropical urban estuary in Mauritius, *Environmental Monitoring and Assessment*, 168 (1) : 345-351.

Ramessur, R.T. (2010) : Trace metal contamination in urbanized estuaries and implications of submarine groundwater discharge for coastal zone management in Mauritius. In: Natural and Man Made Disasters Section-4 Vol 1 : Chapter 16 ed. by Singh, K.K and Singh, A.K, *Times Press*, New Delhi, pp 329-342. Ramessur. R.T., "Integrated Coastal Zone Management, BEVS 3231", Course Manual.

ReCoMap (2007) : 1st Call for Proposals, *Practical Advice for non-state actors* : pp 1-8.

Swayne, D., Denko, M., Fagoonee, I., Marshall, L., Molina, G., Ramessur, R.T., Solem, K.E., Woldeyes,T. (2007) : ICT Issues for Environmental monitoring and management. Chapter 6 : Environment. IFIP WITFOR 2007 (World Information Technology Forum), *IFIP Press Austria*, pp: 97-110.

8

Distribution Pattern of Benthic Organisms in Sahastradhara Stream At Dehradun (Uttarakhand)

D.S. Malik and Umesh Bharti

ABSTRACT

The chapter deals with diversity of benthic population of Sahastradhara stream at Dehradun. The composition and abundance of benthic organisms are commonly used as bio-indicators to determine the impact of water pollution on physico-chemical characteristics in streams & river water and indicates the changing pattern of biotic integrity and other higher trophic communities in aquatic ecosystem. The present study was carried out during 2009-10 in Sahastradhara stream and is one of the famous sulphur stream and major tributary of river Song, situated on the latitude, 29.57°-31.20°E and longitude 77.35°-79.20°N, in Garhwal region of Uttarakhand. The four different sampling sites were selected on different stretches in Sahastradhara stream. The 7 faunal groups and 23 genera were observed from four different sampling sites. Among faunal groups, *Oligochaeta* (3 genera), *Plecoptera* (3 genera), *Trichoptera* (3 genera), *Diptera* (4 genera), *Ephemeropetra* (5 genera) *Odonata* (2 genera) and *Hemiptera* (3 genera) were recorded. The total benthic organisms (**ind. m^{-2}**) occurred as 953, 415, 918 and 1265 on S_1, S_2, S_3 and S_4 sampling sites respectively. Diversity indices were also calculated to measure the diversity level of each faunal group at each level of stream stretches in Sahastradhara stream. The physico-chemical characteristics recorded seasonally as the variables *i.e.* temperature (19.7°C – 18.7°C), velocity (0.59ms –0.49m/s), pH (8.1–7.8), alkalinity (43.7mg/l–37.33mg/l), DO (8.63mg/l–7.68mg/l), calcium (84.64mg/l–84.4mg/l) and magnesium (52.14mg/l–51.24mg/l) in the Sahastradhara stream. The existing water quality characteristics have reflected the seasonal as well as sampling sites variability in distribution pattern, population structure and density of benthic organism in the Sahastradhara stream.

Keywords

Benthic population, biotic integrity, physico-chemical characteristics, distribution pattern, benthic organism.

INTRODUCTION

The benthic flora and fauna of fresh water has been the subject of intensive ecological research because they play important role as indicators of water pollution (Mason, 1991). The macro-invertebrates act as the secondary producers in the aquatic ecosystems. They are an important link in the food web of aquatic ecosystem. Aquatic insects play a role in the processing of organic matter (Cummins 1974, Cummins *et al.*, 1984). Aquatic macro-invertebrates are important food for fish and waterfowls. They also play an important role in transferring energy from the first trophic level to second trophic level in freshwater ecosystem. Benthic fauna are sensitive indicators of environmental changes in streams because they express long-term changes in water and habitat quality rather than instantaneous conditions (Johnson *et al.*, 1993). Therefore, a maiden attempt made towards the distribution pattern of benthic macro-invertebrate in rock surfaces, sandy sediments, crevices in gravel and pebbles, in Sahastradhara, one of the important tourist place, famous as Sulphur spring in Doon Valley at Dehradun, India.

STUDY AREA

Sahastradhara stream originated basically from the hills of Mussoorie. Downstream of Sahastradhara is connected with river Badli near village Kalagaon. River Badli is dividing into two branches Jhandu Nala and Dhadala at 840-1310 msl and lastly confluences along with Song river tributary of river Ganga and well known for sulphur springs that gush out from mountain cavities in as spectacular way, attract maximum tourists for bathing and other adventurous activities. Except tourism, other natural and anthropogenic activities also caused the degradation of water quality and biotic environment of the stream. Four sampling sites were identified depending upon the accessibility and the anthropogenic interventions. The first site (S_1) upstream was identified near a small village (**Karligarh**) dominated by pebbles and cobbles. The second sampling site (S_2) was identified on main tourist spot (Sahastradhara), famous for sulphur spring. Anthropogenic activities like washing and bathing are common at this site because of several fait to this water to removing the skin disorder. The substrates at this site were dominated by carbonate rocks and sandy sediment with some cobbles. The third sampling site (S_3) downstream was selected at **Kalirav** 5-7 Km down from the site S_2, which was less interfered by anthropogenic activities and substrates, is dominated by small boulders at this site. The site fourth (S_4) was selected near **Kalagaon**, a village, devoid of less anthropogenic activity. Composition of stream bottom substrates is small boulders and sand. Here, the growth of macrophyte is maximum than other sites.

METHODOLOGY

Geographically, Garhwal has almost central position in the long Himalayan sweep. The Doon Valley lies between latitude 20° 35′ 30° to 30° 30′ 30° N and longitude 77° 35′ to 78° 19′ E in India. The maximum rainfall ranges between 600-750mm during July-September and minimum during April-May. The maximum temperature reaches upto 40 °C during the months of May and June, whereas minimum ranges 2-3 °C during the months of December–January. The present study was carried out from May 2009–April 2010. The water samples were taken seasonally from four different sites (S_1, S_2, S_3 and S_4) at about 15 to 18 km stretches of the stream. The benthic macro-invertebrates collection was done at day time between 8:00 to 10:30 AM. All benthic macro-invertebrates were collected with the help of Ekman's Dredge sampler (US No 60cm mesh net size) to a depth of about 15cm in a quadrate by hand picking. Quantitative estimation of benthic macro-invertebrate was based on numerical counting, *i.e.*, unit per square meter (**ind. m^{-2}**). The benthic macro-invertebrates were preserved in 4% formalin & identified to their possible lowest taxonomic level. The qualitative analysis of macro-invertebrate samples was made with the help of Edmondson (1992) and Needham & Needham (1962).

STATISTICAL TREATMENT OF THE DATA

For statistical treatment of the data, species diversity indices were calculated using the Shannon-Wiener (1946), Simpson (1949), and Evenness index.

$$\text{Shannon – Wiener Index}(H') = -\sum \frac{ni}{N} \ln \frac{ni}{N}$$

$$\text{Simpson's Index } C = 1 - \sum \frac{(ni)^2}{N}$$

Where H′= Shannon-Wiener index of diversity, ni = total number of individual of a taxon, N = total number of individuals of all taxa, ln = $\log_e$ (logarithm to the base)

Karl Pearson's correlation coefficient (r value) was used to quantify relationships between various hydrological attributes and the influence of hydrological attributes on biological communities.

Table 1 : The Distributional Pattern of Benthic Organisms (Ind./m²) in Sahastradhara Stream at Dehradun

OLIGOCHAETA	S_1	S_2	S_3	S_4
Tubifex sp.	++	+++	++	+
Branchiora sowerbyii	++	++	+	++
Limnodrillus hoffmeisteri	+	+++	+++	++
PLECOPTERA				
Pteronarcys	+	-	++	++
Acroneuria	++	-	+++	++
Isoperia	++	-	++	++
TRICHOPTERA				
Hydrosyche	+	+++	++	++
Leptocella	++	+++	++	+
Ochrotrichia	++	++	++	+
DIPTERA				
Chironomous plumosus	++	+++	+++	++
Tendipestentans	+	++	++	++
Culicoides	++	+++	++	++
Alabesmyia	+	++	++	++
EPHEMEROPETRA				
Adult Mayfly	+	+++	++	++
Stenononema	++	++	++	++
Leptophlebia	+	++	++	++
Ephemeralla indica	+	+++	+++	++
Cinygma	++	++	++	++
ODONATA				
Epicordulia (Dragonfly)	+	+++	++	++
Macromia	++	++	+++	++
HEMIPTERA				
Aquarius remigis (Water striders)	+	+++	+++	++
Sigara mckinstryi (Water boatmen)	+	++	+++	+
Notonecta unifasciata (Backswimmers)	+	++	++	++

+++ : Abundant, ++ : Common, + : Rare, -: Absent.

Table 2 : Macro-Benthic Diversity Indices of Sahastradhara Stream

Organisms	*Shannon-Wiener Diversity Index (H')*					*Simpson's Index*					*Evenness Index*				
	S_1	S_2	S_3	S_4	*S.D.*	S_1	S_2	S_3	S_4	*S.D.*	S_1	S_2	S_3	S_4	*S.D.*
Oligochaeta	3.2	2.5	1.9	2.1	±0.57	**0.49**	0.38	**0.29**	0.32	±0.09	2	**1.5**	1.2	**1.3**	±0.35
Plecoptera	2.8	Nil	2.6	2.4	±0.2	0.43	0.51	0.39	0.45	±0.05	1.8	Nil	1.6	1.5	±0.15
Trichoptera	3.2	6.9	2.4	1.4	±2.39	0.48	**0.84**	0.41	0.76	±0.21	**1.6**	**6.9**	**1.2**	7	±3.21
Diptera	**1.3**	**1.3**	**4.6**	6.3	±2.49	0.72	0.82	**0.86**	0.76	±0.06	6.5	6.3	2.3	3.2	±2.19
Ephemeropetra	1.4	**9.5**	**1.5**	**1.5**	±4.02	0.76	0.82	0.79	**0.78**	±0.02	6.1	4.1	**6.5**	6.1	±1.15
Odonata	**6.9**	2.1	3.1	6.8	±2.49	**0.5**	**0.32**	0.46	0.49	±0.08	**6.9**	2.1	3.1	6.8	±2.49
Hemiptera	2.3	4.3	4.4	**8.7**	±2.69	0.49	0.51	0.85	**0.16**	±0.21	2.3	4.1	4.4	**8.7**	±2.71

Table 3 : Physico-Chemical Characteristics of Sahastradhara Stream (Mean Values of Each Season During 2009-10)

Parameters	*Karligarh Upstream* S_1	*Main Tourist Spot* S_2	*Kalirov Downstream* S_3	*Kalagaon Downstream* S_4	*S.D.*
Temperature °C	18.7	19.1	19.2	**19.7**	0.4113
Velocity m/s	0.59	0.59	0.53	0.49	0.0489
pH	**8.1**	7.9	7.8	7.8	0.1414
Alkalinity mg/l	37.33	38.3	**43.7**	38.9	2.8364
DO mg/l	8.63	8.08	7.81	**7.68**	0.4211
Calcium mg/l	84.41	84.55	84.4	**84.64**	0.1158
Magnesium mg/l	**51.78**	52.14	51.24	51.66	0.3711

Table 4 : Statistical Correlation Computed Between Different Physico-Chemical Parameters of Sahastradhara Stream

Parameters	*WT*	*WV*	*pH*	*Alk*	*DO*	*Ca*	*Mg*
Temperature °C	1						
Velocity m/s	–0.89						
pH	–0.40	0.77					
Alkalinity mg/l	0.15	–0.44	–0.66				
DO mg/l	–0.49	0.84	0.99	–0.58			
Calcium mg/l	0.38	–0.47	–0.45	–0.37	–0.54		
Magnesium mg/l	–0.54	0.59	0.42	–0.83	0.39	0.43	1

Table 5 : Pearson's Correlation Coefficient (r-values) Calculated Between Physico-Chemical Parameters and Diversity of Macro-Invertebrates Occurred in the Sahastradhara Stream Stretches at Dehradun, India

Parameters	*WT*	*WV*	*pH*	*Alk*	*DO*	*Ca*	*Mg*
Oligochaeta	– 0.39	– 0.93	– 0.92	0.88	– 0.76	0.69	0.74
Plecoptera	0.76	0.99	0.77	– 0.99	0.53	– 0.30	– 0.87
Trichoptera	0.98	0.86	0.25	– 0.78	– 0.01	0.37	– 0.74
Diptera	– 0.77	– 0.24	– 0.33	0.49	– 0.29	– 0.45	0.09
Ephemeroptera	0.96	0.92	0.32	– 0.84	0.04	0.25	– 0.82
Odonata	– 0.29	– 0.85	0.02	0.43	0.29	0.14	0.84
Hemiptera	0.55	0.21	0.77	– 0.13	0.77	– 0.11	– 0.13

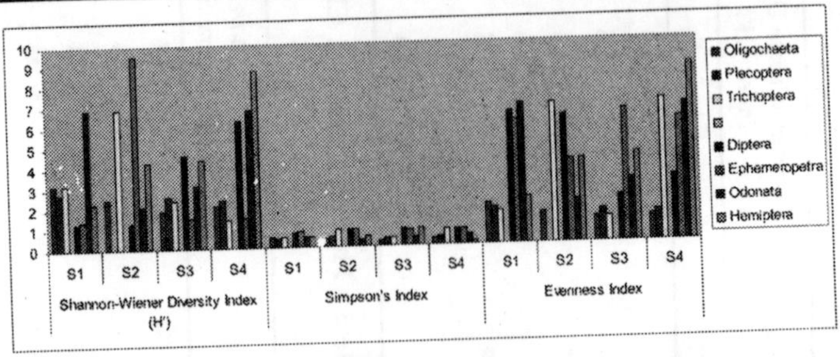

Figure 1 : Shannon-Wiener Diversity Index (H'), Simpson's index and Evenness index at different sampling stations in Sahastradhara stream.

RESULTS

The physico-chemical characteristic of Sahastradhara stream at all four different sites are presented in table 3. The water temperature was observed in between 18.7 to 19.7 °C ± 0.41°C. The water velocity was recorded highest 0.59 m/s and lowest 0.49 ± 0.05 m/s. The pH was found maximum (8.1) at site 1st (Upstream) and minimum (7.8 ± 0.14) at downstream site 4th. The Alkalinity was highest (43.7 mg/l) at downstream site 3rd and minimum (37.33 ± 2.84mg/l) at upstream site 1st. Dissolved oxygen showed relatively higher concentration (8.63 mg/l) at upstream site 1st and lower (7.68 ± 0.42mg/l) at downstream site 4th. Calcium ion concentration recorded higher (84.64 mg/l) at downstream site 4th and minimum (84.4 ± 0.115mg/l) at downstream site 3rd. Similarly Magnesium ion concentration was recorded maximum (51.78 mg/l) at upstream site 1st and lowest (51.24 ± 0.37mg/l) at downstream site 3rd.

During the study period, a total of 23 genera belonging to 7 orders of benthic fauna were recorded. The most abundant order having maximum number of genera recorded was *Ephemeroptera* at each site followed by *Oligocheata, Plecoptera, Trichoptera, Diptera, Odonata* and *Hemiptera*. Table 1st showed the variation in the abundance of benthos for the period of study at each site. From the observations, it is inferred that maximum number of genera were recorded at site 3rd and minimum at site 1st.

The variations in the Shannon-Wiener diversity index and Simpsons index was calculated to revealed the density of aquatic insects. At site 1st Shannon-Wiener index calculated higher (6.9 Ind.m^{-2}) in *Odonata* and minimum in *Diptera* (1.3 Ind.m^{-2}). At site 2nd maximum diversity index was reported to *Ephemeropetera* (9.5 Ind.m^{-2}) and minimum by *Diptera* (1.3 Ind.m^{-2}). The *Plecoptera* was nil at site 2nd due to the canalization of the stream and diversion of the running water in a stagnant body through making small patches of water pound. At site 3rd the *Diptera* were recorded higher (4.6 Ind.m^{-2}) and *Ephemeroptera* showed positively least value of diversity (1.5 Ind.m^{-2}). *Hemiptera* were also showed maximum diversity (8.7 Ind.m^{-2}) than the *Ephemeroptera* (*i.e.* 1.5 Ind.m^{-2}).

Simpsons index was also calculated for the same four sites. Among all four sites, sites 2nd showed the higher diversity (0.84 $Ind.m^{-2}$) of the *Trichoptera* and minimum (0.32 $Ind.m^{-2}$) to the *Odonata*. Evenness index showed richness of species at all sites. At site 1st maximum richness (6.9) by *Odonata* and minimum (1.6) by *Trichoptera*. *Trichoptera* was also higher (6.9) at site 2nd and minimum (1.5) by *Oligochaeta*. At site 3rd *Ephemeroptera* showed the maximum richness (6.5) followed by the least richness of *Trichoptera* (1.2). *Hemiptera* recorded maximum evenness (8.7) and minimum (1.3) by *Oligocheata* at site 4th (Table 2).

Pearsons correlation coefficient was calculated between physico-chemical parameters and diversity of macro-invertebrates occurred in the stream water. The *Plecoptera, Trichoptera, Ephemeroptera* and *Hemiptera* showed the positive relationship with water temperature, 0.76, 0.98, 0.96 and 0.55 respectively. pH positively correlated with *Plecoptera* (0.77), *Trichoptera* (0.25), *Ephemeroptera* (0.32), *Odonata* (0.02) and *Hemiptera* (0.21). *Oligochaeta, Diptera* and *Odonata* showed the positive relation with Alakalinity. Dissolved oxygen showed the positive relationship with *Plecoptera, Ephemeroptera, Odonata* and *Hemiptera*, 0.53, 0.04, 0.29 and 0.77 respectively. The Calcium ion represented the positive correlation with *Oligochaeta* (0.69), *Trichoptera* (0.37), *Ephemeroptera* (0.25) and *Odonata* (0.14). Magnesium ion concentration was reported positively correlated with *Oligochaeta* (0.74), *Diptera* (0.09) and *Odonata* (0.84) (Table 5).

DISCUSSION

Benthic aquatic insects are sensitive indicators of environmental changes in stream because they express long term changes in water and habitat quality rather than instantaneous conditions (Johson *et al.* 1993). Invertebrate communities are also good indicators of water quality conditions (Resh 1995, De Shon 1995). Similarly, studies of temporal variation in the community structure of streams (Fisher *et al.* 1982, Scrimgeour and Winterbourn 1989) have indicated that physical disturbance can be important determinants of community structure in lotic systems (Sousa 1984, Resh *et al.* 1988). On the other hand, physical disturbances in terms of water flow, temperature, ion concentration and substrates are the major factors determining the composition and abundance of benthic invertebrates (Ward and Stanford 1979).

Ward and Stanford (1979) suggested that temperature pattern influences the life cycle phenomenon of insects such as emergence, which leads to an increase in density. Relationship with mean density of benthos and water temperature was found negatively correlated with some organism ($r = -0.39, -0.77$ and -0.29) in the present study at Sahastradhara stream (Table 5). Similar results were also found during the study on the river Tons river (Sharma *et al.* 2009). Maximum abundance as macroinvertebrates was also reported in Song River at Dehradun (Malik *et al.* 2010). The abundance of benthic macro-invertebrates dwelling in Sahastradhara stream was found to be higher at site 2nd and site 3rd and gradually decrease at site 1st and 4th. The abundance of macro-invertebrates in the Sahastradhara stream was found minimum site 1st and 4th, due to high water velocity and low channel width. Many aquatic populations living in harsh environments of unpredictable flow suffer high mortality directly through physical disturbance (Erman and Mohoney 1983 and Cushman 1985).

Temperature played a vital role in the diversity of macro-benthos that was possible during the summer season that may be due to a substantive increase in water temperature. The water temperature was positively and inversely related with the diversity index of occurred organism (*Oligochaeta, Plecoptera, Trichoptera, Diptera, Ephemeroptera, Odonata* and *Hemiptera*) (Table 5). Several studies have also found that insect taxa richness may be influenced the stream ecosystem by annual water temperature variations (Stanford and Ward, 1982). Dissolved oxygen was found to be positively correlated with diversity indices of benthos at all sites of Sahastradhara stream. Similarly, Neleson *et al.* (2000) shown that dissolved oxygen has a strong influence on macroinvertebrate community structure. The stream consisted of bottom substrates mainly, sand and big boulders. The stream was famous for the stone mining therefore, the concentration of Calcium and Magnesium ion was found high at site 4th that may be due to the continue deposition of silt and clay from upstream to downstream. Angradi (1996) supported that physical substrate types like leaves, gravel, wood & macrophytes helps more diversity than structurally simple substrate such as sand bedrock.

The diversity of macro-benthos in the stream at different sites was found to be in the order $S_3 > S_2 > S_4 > S_1$, along the river gradient. The Sahastradhara stream is very important in the mountainous region of Garhwal Himalayas, as it provides a suitable habitat for the aquatic insects. These aquatic insects provide a very good natural food for most of the insectivorous coldwater fishes. The stream ecosystem was encountered with many natural (sedimentation, debris deposition) and anthropogenic pressure (Bathing activities, adjacent intensification of agriculture, increasement in the human settlement, hotel constriction, soil erosion, Pebbles and stones in the catchments area). These factors directly and indirectly impact on the diversity of aquatic insects. At sites S_3 the stream side is dominated by macrophytes, thus providing a suitable niche for the macro-invertebrate. In spite of the differences in substrate composition and other anthropogenic activities, the benthic distribution at all four sites remains fairly good, suggested that the quality of stream is fairly good and supports diverse and well balanced macro-benthic communities in Sahastradhara stream. Metcalfe-Smith (1996) also observed that the higher value of Shannon-Wiener index is the greater the diversity and supposedly the healthier environment.

REFERENCE

Angradi, T.R. (1996) : Inter-habitat variations in benthic community structure, function and organic matter storage in three Appalachian headwater streams. J.N. Am Benthol Soc USA 15 : 42–63.

Cummins, K.W. (1974) : Structure and function of stream ecosystems. *Bioscience* 24 : 631–641.

Cummins, K.W., Merritt, R.W. and Burton, T.M. (1984) : The role of aquatic insects in the processing and cycling of nutrients. In : Resh

VH, Rosenberg DM (eds) The ecology of aquatic insects. Praeger, New York, pp : 134–163.

Cushman, R.M., (1985) : Review of ecological effects of rapidly varying flows downstream from hydroelectric facilities. N Am J Fish Manage 5 : 330–339.

De Shon, JE., (1995) : Development and application of the invertebrate community index (ICI). In : Davis W.S., Simon T.P. (eds) Biological assessment and criteria. Lewis Publishers, England, pp : 217–243.

Edmondson, W.T. (1992): Fresh water biology, second edition, pp : 1248.

Erman, D.C., and Mahoney, S.D. (1983) : Recovery after logging in streams with and without buffer stripes. California Water Research Center, University of California, Northern California, pp : 50.

Fisher, S.G., Gray, L.J., Grimm, N.B., and Busch, D.E., (1982) : Temporal succession in a desert stream ecosystem following flash flooding. *Ecol Monogr*, 52 : 93–110.

Johnson, R.K., Wiederholm, T., and Rosenberg, D.M. (1993) : Freshwater biomonitoring using individual organisms, populations, and species assemblages of benthic macro invertebrates. In: Rosenberg DM, Resh VH (eds) Freshwater biomonitoring and benthic macro-invertebrate. Chapman and Hall, New York, pp : 40–158.

Malik, D.S., Bharti, U. and Kumar P. (2009) : Macro-benthic diversity in relation to biotic indices in Song river at Dehradun, India. *Environment Conservation J.*, 11(1&2) 99–104.

Mason, C.F., 1991 : *Biology of freshwater pollution*. Longman Scientific & Technology. London : 233-239.

Metcalfe-Smith JL (1996): Biological water-quality assessment of rivers: use of macr-invertebrate communities. In: Petts G, Calow P (eds) River restoration. Blackwell Science, Oxford, pp : 17–43.

Nelson, MS., Roline, RA., Thullen, JS., Sartoris, JJ., and Boutwell, JE. (2000) : Invertebrate assemblage and trace element bioaccumulation associated with constructed wetlands. *Wetlands* 20 : 406–415.

Resh,VH. (1995) : Freshwater benthic macro-invertebrates and rapid assessment procedures for water quality monitoring in developing and newly industrialized countries. In: Davis WS, Simon TP (eds) Biological assessment and criteria. Lewis, Chelsea, pp : 167–177.

Scrimgeour, GJ., Winterbourn, MJ., (1989) : Effects of floods on epilithon and benthic macro-invertebrate populations in an unstable New Zealand river. *Hydrobiologia* 171 : 33–44.

Shannon, C.E. and Wiener, W. (1949) : The mathematical theory of communication. The University of Illionois press, Urbana IL.

pp : 117.

Sharma, R.C., Arambam R. and Sharma R. (2009) : Surveying macro-invertebrate diversity in the Tons river Doon Valley, India. *Environmentalist*, 29 : 241–254.

Simpson, E.H. (1949) : Measurement of diversity. *Nature*, 163 : 688.

Sousa WP (1984) : The role of disturbance in natural communities. Annu Rev Ecol Syst, 15 : 353–391.

Stanford, J.A., and Ward, J. V. (1982) : Insect species diversity as a function of environmental variability and disturbances in stream system. In: Barnes JR, Minshal W (eds) Stream ecology: application and testing of general ecological theory. Plenum, New York, pp : 265–278.

Ward, JV. and Stanford, JA. (1979) : Ecological factors controlling stream zoo-benthos with emphasis on thermal modification of regulated streams. In: Ward JV, Stanford JA (eds) Ecology of regulated streams. Plenum, New York, pp : 35–53.

9

Effects of Temperature and UV-B Radiation on Antarctic Micro Flora and Their Adaptive Responses

Anand K. Dubey and Jaswant Singh

ABSTRACT

Cryptogams are the most advanced plants found on continental Antarctica. To understand, how climate change will affect Antarctic plant communities we should know how changes in global temperature will affect plant metabolic processes. If the other species that co-occurs with cryptogams are tolerant, then it is likely to be disadvantaged under future climate and may loose its dominance in these communities. These effects were evaluated with open top chambers (OTCs) and UV filter frame studies and found that plant physiology have changed. The UV-B radiation depressed the photosynthetic rate in cryptogams which were adapted with formation of secondary metabolites. The photosynthesis in Antarctic cryptogams enhanced due to increase in temperature while its inverse in case of UV-B radiation exposure studies.

Keywords

Adaptive responses, Antarctica, Ecology, Temperature, UV-B Radiations.

INTRODUCTION

The rising trend of greenhouse gas concentrations in the

atmosphere, and the subsequent temperature increase on the Earth's surface is an accepted phenomenon (IPCC, 2007). Antarctica is the coldest, driest, windiest and highest continent on earth. Due to harsh climate Antarctic vegetation mainly consists of cryptogams (mosses and lichens). The Antarctic ecosystem is one of the most simple ecosystems on the planet. In Antarctic continent most water is permanently locked up as ice and snow while large areas are accurately described as frigid desert receiving low or no direct precipitation and experiencing chronically low relative humidity. India's first research station 'Dakshin Gangotri" commissioned in the year 1984 and second research station "Maitri" commissioned in the year 1988-89 over the exposed rocks of Shirmacher Oasis of Queen Maud Land area in East Antarctica. The Shirmacher Oasis (70° 45′ 58"S) is one of the small, snow/ice covered during the winter (June-August) and spring (September-November) seasons, typical desert near the shore of East Antarctica, with a width of 1.6 Km and length of about 20 Km with east west orientation. Its composed of with mainly high grade of metamorphic rocks, the fresh water lakes, ponds, pools cover a total area of 35 square Km, the altitudes varies from 0 to 228m with an undulated surface. Recently India started to build its third station at Larsemann Hills (69° 20′S to 69° 30′S Lat; 75°55′E to 76°30′E) in Prytz Bay, is an ice free oasis on the Ingrid Christensen coast of East Antarctica. Organisms living in Antarctica must therefore be able to survive long periods of freezing and consequent drought, year around. In summer water may become available from snow melt and around melt lakes, however this water supply is transient and repeated freeze thaw events still occur. In the context of climate change three elements are fundamental to the biology of Antarctic terrestrial organisms, temperature, water and solar irradiance. Mean air temperatures even during the warmest summer months are low (below 0°C in the continental, 0-2°C in the maritime and 5-10°C in the sub Antarctic (Walton, 1984; Convey, 1996). The greater rates of warming have been reported in some Antarctic freshwater ecosystems (Quayle, *et al.* 2002; 2003). It has been recognized since the late 1980s, evaluating the impacts of climate variables on the reproductive success of autochthonous and alien phanerogams was the key targets for plant ecophysiologists in the sub arctic and Antarctic to determine climate change effects.

Antarctic plants are exposed to a highly seasonal, rapidly variable and often extreme climate with frequent subzero temperatures and an intermittent water supply. These severe environmental conditions limit plant life of small, non vascular plants such as mosses and lichens. Global climate change has the potential to increase temperature and increased availability of water as snow melt to Antarctic organisms. Several communities are nutrient limited and this is likely to limit their ability to respond to climate change. Climate change is thus likely to influence biodiversity in Antarctica with severe consequences for some endemic species (Bokhorst, *et al.* 2007). Recently scientists from all over the world have focused their studies on climate change and its impacts on Antarctica. Antarctic flora is experiencing some of the major changes especially in relation to increasing UV-B radiation due to ozone depletion and temperature.

In recent years, considerable effort has been invested in determining the responses of polar vegetation, especially, to UV-B radiations. Some previous warming studies have simulated increase in above ground vegetation biomass (Parsons, *et al.* 1995; Rustad, *et al.* 2001; Aerts, *et al.* 2006). This observation is consistent with the view that riccionidin acts to block visible radiation of 500nm from reaching the chloroplasts of C. *varians,* reducing the possibility of photoinhibition (Post and Vesk, 1992). Phoenix, *et al.* (2002), have used polycarbonate filter that absorb both UV-A and UV-B radiation, in this experimental model the effect of ultraviolet radiation, in sub arctic shrubs, reductions in UV-B radiation resulted in increased levels of UV-B absorbing compounds which was attributed to either increased secondary metabolism in absence of UV-B radiation or UV photo-oxidation of these pigments. Thus, as with the anthraquinone, parietin in the thalli of *Xanthoria parietina* (Solhaug, *et al.* 2003; Solhaug and Gauslaa, 2004), although the synthesis of riccionidin is apparently induced by exposure to UV-B radiation, the main function of the pigment in the thalli of *C. varians* appears to be in attenuating short wavelength visible radiation.

Mosses and lichens are simple, small plants that can dehydrate and therefore survive the winter in a dormant state. The bryophytes in polar regions responding to UV-B radiation arising from ozone depletion could thus be significant, given that their carbon fixation is limited by low temperature, desiccation and reduced light availability owing to snow cover (Snell, *et al.* 2007). In plants the photosynthetic pigments like chlorophyll can be negatively affected. The recent finding that fewer DNA photoproducts accumulate in desiccated, compared with hydrated, mosses also suggests that screening and passive defense mechanisms are well developed in Antarctic bryophytes (Turnbull, *et al.* 2009). However, this probably reflects a species-specific response, (Lappalainen, *et al.* 2008). This chapter focuses on the impact of increasing UV-B radiation as a result of ozone hole and the works on many aspects of climate change on Antarctic plants.

ANTARCTICA OZONE DEPLETION AND UV-B RADIATIONS

Antarctic ozone depletion was demonstrated at Hally Bay region due to anthropogenic effects (Farman, *et al.* 1985). The depletion of Earth's protective stratospheric ozone layer is caused annually in the austral spring over Antarctica by a light catalyzed reaction between ozone and pollutants concentrated at high altitudes and latitudes by atmospheric circulation patterns resulted increased UV irradiance at Earth's surface (Frederick and Snell, 1988; Karentz, 1991; Roy, *et al.* 1994). The effects of ultraviolet radiation on the photosynthetic and UV screening pigments in needles of Scots pine (*Pinus sylvestris*) were studied in UV exclusion field chamber showed, that the increase in chlorophyll and the concentration of neoxanthin, ±-carotene increased in exposure to UV, while that of carotene, lutein, violaxanthin, antheraxanthin, and zeaxanthin decreased, which explained the decrease in the total carotenoid concentration (Martz, *et al.* 2007). Chlorophyll content was significantly lower in plants grown under near ambient UV, while the relative proportions of photoprotective carotenoids, especially *b*-carotene and zeaxanthin, increased (Robinson, *et al.* 2005). In sub arctic shrubs, reductions in UV-B radiation resulted in increased levels of UV-B absorbing compounds (Phoenix, *et al.* 2002), which was attributed to either increased secondary metabolism in the absence of UV-B radiation or UV photo-oxidation of these pigments.

RESPONSE OF ENHANCED TEMPERATURE ON ANTARCTIC PLANTS

Concern about the regional impacts of climate change has motivated an increased focus on the relationship between species and ecosystem structure and function and temperature over the past decade. Antarctic terrestrial vegetation is subject to one of the most extreme climates on Earth, as parts of Antarctica are one of the fastest warming regions on the planet. Bokhorst, *et al.* (2007), investigated the effect of experimental warming on the diversity and abundance of coastal plant communities in the Maritime Antarctic region (cryptogams only) and the Falkland Islands (vascular plants only) during three growing seasons. Communities from the Falkland Islands (51°S, mean annual temperature 7.9°C), with those of Signy Island (60°S, –2.1°C) and Anchorage Island (67°S, –2.6°C), and experimental temperature manipulations at each of the three islands were compared by using Open Top Chambers and concluded that small temperature increases may rapidly lead to decreased soil moisture, resulting in more stressful conditions for plants.

In comparative cryptogams ecology would be particularly timely, since bryophytes and lichens are thought to be facing great changes in their abundance biomass and composition as a consequence of global environmental changes (Bates and Farmer, 1992; Callaghan, *et al.* 2004). Lichens and probably non *Sphagnum* bryophytes may decline in response to vascular plant expansion as induced by climate warming in some of the cold biomes (Cornelissen, *et al.* 2001; Van Wijk, *et al.* 2004). Micro flora (Arctic) typically have low-growing dominated by mosses and lichens, because dry ecosystems tend to be more strongly nutrient-limited than wet ecosystems (Shaver and Chapin, 1991), the photosynthetic response to warming may be weakest in dry sites because leaf area and photosynthesis can only increase if nutrients are available to support them. Experimental warming caused a 26-36% decrease in species richness, a response that was generally dampened by experimental grazing. Higher species losses occurred at the drier sites where nitrogen was less available. Moreover, observed an indirect effect of climate change on species richness as mediated by plant–plant interactions. Heat stress and warming-induced litter accumulation are potential explanations for the species responses to experimental warming (Klein, *et al.* 2004).

Antarctic and sub Antarctic photosynthetic autotrophs maintain photosynthesis at low temperatures, with being active well below 0°C. Photosynthetic activity has been reported in continental Antarctic lichens below –10°C (Schroeter, *et al.* 1995). Some species additionally show plasticity in their optimal temperature, when grown at temperature above those experienced in the field, the optimal temperature for *Sanionia uncinata* remained at 15°C while that for *Polytrichum alpestre* increased from 5-10°C to 15°C (Collins, 1977). Prevailing low temperatures throughout the Antarctic biome are generally considered to limit net photosynthesis for most of the growing season (Xiong, *et al.* 1999). Increased in net photosynthesis under elevated temperature have been demonstrated in three continental moss species (Smith, 1999) and the two maritime vascular species (Xiong, *et al.* 2000). However, (Kappen, *et al.* 1989), reported decrease in net photosynthesis under increasing temperature might also be due to consequential increase in photo inhibition has been shown in some continental Antarctic mosses. In the two maritime Antarctic vascular net photosynthetic rates are negligible at canopy air temperature greater than 20°C with temperature rather than high irradiance responsible for this photosynthetic depression.

METABOLIC RESPONSES IN ANTARCTIC FLORA

The study of the effects of sustained heating on the ecosystems of the world is necessary so that might predict and respond to coming changes on both large and small spatial scales (Aronson and McNulty, 2009). During summer photosynthetic organs achieves temperatures above zero and optimum temperatures for photosynthesis are corresponding higher, ranging from 10-12°C in *Deschampsia antarctica* from Signy Island and Antarctic Peninsula (Edwards and Smith, 1988, Xiong, *et al.* 1999). However, it has also been demonstrated for cryptogamic species that increasing vegetative growth overweighs decreases in photosynthetic rates under 20°C at daytime temperatures (Xiong, *et al.* 2000).

Selkirk and Skotnicki, (2007) have measured the growth rate of three moss species *Bryum pseudotriquetrum, Schistidium antarctici* and *Ceratodon purpureus,* and the results provided the first long-term growth measurements for plants in Antarctica, confirming that moss shoots grow extremely slowly in Antarctica, elongating between 1 and 5 mm per year. Leading to these types of studies Robson *et al.* (2003) used polyester film for the attenuation of ultraviolet radiation, which transmit a large percentage of the UV-A radiation and found that photosynthetic pigment altered with UV-B radiations. Thus, despite the estimate made here indicating that *C. varians* might only need to expend a few percent of its photosynthetic output on synthesizing flavonoids in response to UV-B exposure, a small reduction in output could have cumulative impacts on the performance of the liverwort. Such estimates might be made for species of *Bryum* and *Lycopodium,* which are known to synthesize flavonoids in response to elevated UV-B radiation arising from stratospheric ozone depletion (Dunn and Robinson, 2006; Lomax, *et al.* 2008; Ryan, *et al.* 2009). The phenolic compounds were also protecting the plants under the exposure of UV-B, it may contribute to the decrease in active oxygen species by acting as anti oxidants (Markham, *et al.* 1998).

CONCLUSION

The knowledge on the basis of research conducted on climate change at Antarctic ecosystems indicates minor increase in temperature. This can significantly affect the organisms surviving under already existing harsh climatic conditions.

Apparently small changes are taking place *e.g.*, the limited photosynthetic response to a minor increase in mean temperature may assume much greater significance in the context of pre-existing limitations on physiological activity in Antarctic habitats. In contrast, current evidence suggests that Antarctic plants may be considerably more vulnerable to climate change effects.

ACKNOWLEDGEMENTS

We gratefully acknowledge the financial support of the National Centre for Antarctic and Ocean Research (NCAOR, Goa), Ministry of Earth Sciences, New Delhi, India.

REFERENCES

Aerts, R. Cornellisen, J.H.C.; Dorrepaal, E. (2006): Plant performance in a warmer world: General responses of plants from cold, northern biomes and the importance of winter and spring events. *Plant Ecology,* 182 : 65-77.

Arft, A.M.; Walker, M.D.; Gurevitch, J. Alatalo, J.M.; Bret-Harte, M.S.; Dale, M.; Diemer, M. Gugerli, F.; Henry, G.H.R.; Jones, M.H.; *et al.* (1999) : Responses of tundra plants to experimental warming: Meta analysis of the international tundra experiment. *Ecological Monographs,* 69 : 491-511.

Aronson, E.L. McNulty, S.G. (2009) : Appropriate experimental ecosystem warming methods by ecosystem, objective, and practicality. *Agricultural and Forest Meteorology,* 149 : 1791–1799.

Australian Antarctic Division (2004) : A new plant on Heard Island. www.aad.gov.au/default.asp?casid=13711.19/02/2004.

Australian Antarctic Division (2005): www.aad.gov.au/default.asp?casid=19987.19/05/2005.

Bates, J.W. and Farmer, A.M. (1992) : Bryophytes and lichens in a changing environment. Oxford: Clarenden Press.

Bokhorst, S. Huiskes, A. Convey, P. and Aerts, R. (2007): The effect of environmental change on vascular plant and cryptogam communities from the Falkland Uslands and the Maritime Antarctic. *BMC Ecology,* 7 : 1-13.

Callaghan, T.V.; Bjorn, L.O.; Chernov, Y. Chapin, T. Christensen, T.R.; Huntley, B. *et al.* (2004): Responses to projected changes in climate and UV-B at the species level. *Ambio,* 33 : 418-435.

Convey, P. (1996) : Reproduction of Antarctic flowering plants. *Antarctic Science,* 8 : 127-134.

Cornelissen, J.H.C.; Callghan, T.V.; Alatalo, J.M.; Michelsen, A. Graglia, E. Hartley, A.E. *et al.* (2001) : Global change and Arctic ecosystems: is lichen decline a function of increases in vascular plant biomass. *Journal of Ecology,* 89: 984-994.

Collins, N.J. (1977) : The growth of mosses in two contrasting communities in the maritime Antarctic: measurement and prediction of net annual production, in G.A. Llano (ed.), *Adaptations within Antarctic ecosystems,* Gulf Publishing, Houston, Texas, pp : 921-933.

Cook, A.J.; Fox, A.J.; Vaughan, D.G.; Ferrigno, J.G. (2005) : Retreating glaciers fronts on the Antarctic Peninsula over the past half century. *Science,* 308 : 541-544.

Dorrepaal, E. Aerts, R. Cornelissen, J.H.C.; Callaghan, T.V.; van Logtestijn, R.S.P. (2004) : Summer warming and increased winter snow cover affect *Sphagnum fuscum* growth, structure and production in a sub-arctic bog. *Global Change Biology,* 10 : 93-104.

Dunn, J.L.; Robinson, S.A. (2006) : Ultraviolet B screening potential is higher in two cosmopolitan moss species than in a co-occurring Antarctic endemic moss : implications of continuing ozone depletion. *Global Change Biology,* 12 : 2282–2296.

Edwards, J.A. and Smith, R.I.L. (1988) : Photosynthesis and respiration of *Colobanthus quitensis* and *Deschampsia antarctica* from the maritime Antarctic. *British Antarctic Survey Bulletin,* 81 : 43-63.

Farman, J.C.; Gardiner, B.G. and Shanklin J.D. (1985) : Large losses of total ozone in Antarctica reveal seasonal ClO_x/NO_x interaction. *Nature,* 315 : 207-210.

Frederick, J.E.; Snell, H.E. (1988) : Ultraviolet radiation levels during the Antarctic spring. *Science,* 241 : 438-441.

Intergovernmental Panel on Climate Change (IPCC). (2007) : Climate Change 2007 : The Physical Science Basis. Summary for Policymakers.

Kappen, L.; Smith, R.I.L. and Meyer, M. (1989) : Carbon dioxide exchange of two ecodemes of *Schistidium antarctici* in continental Antarctica, *Polar Biology,* 9 : 415-422.

Karentz, D. (1991) : Ecological consideration of Antarctic ozone depletion. *Antarctic Science,* 3 : 3-11.

Klein, J.A.; Hartel, J. and Zhao, X.Q. (2004) : Experimental warming causes large and rapid species loss, dampened by simulated grazing, on the Tibetan Plateau. *Ecology Letters,* 7 : 1170 1179.

Lappalainen, N.M.; Huttunen, S. and Suokanerva, H. (2008) : Acclimation of the pleurocarpous moss *Pleurozium schreberi (Britt.) mitt.* to enhanced ultraviolet radiation in situ. *Global Change Biology,* 14 : 321-333.

Lomax, B.H.; Fraser, W.T.; Sephton, M.A. *et al.* (2008): Plant spore walls as a record of long term changes in ultraviolet-B radiation. *Nature Geoscience,* 1 : 592–596.

Markham, K.R.; Tanner, G.J.; Caasi-Lit, M. Whitecross, M.F.; Nayudu, M. Mitchell, K.A. (1998) : Protective role for 3, 4-dihydroxyflavones induced by UV-B in tolerant rice 15 cultivar. *Phytochemistry,* 49 : 1913-1919.

Mirecki, R.M. and Teramura, A.H. (1994) : Effects of Ultraviolet-B irradiance on Soybean. *Plant Physiology,* 74 : 475-480.

Martz, F. Sutinen, M.L.; Derome, K. Wingsle, G. Tiitto, R.J. and Turunen, M. (2007) : Effects of ultraviolet (UV) exclusion on the seasonal concentration of photosynthetic and UV-screening pigments in Scots pine needles. *Global Change Biology,* 13 : 252-265.

Parsons, A.N.; Press, M.C.; Wookey, P.A.; Welker, J.M.; Robinson, C.H.; Callaghan, T.V. and Lee, J.A. (1995): Growth Responses of *Calamagrostis lapponica* to Simulated Environmental Change in the Sub-Arctic. *Oikos,* 72 : 61-66.

Phoenix, G.K.; Gwynn-Jones, D. Lee, J.A.; Callaghan, T.V. (2002): Ecological importance of ambient solar ultraviolet radiation to a sub-arctic heath community. *Plant Ecology,* 165 : 263–273.

Post, A. Vesk, M. (1992) : Photosynthesis, pigments and chloroplast ultrastructure of an Antarctic liverwort from sun-exposed and shaded sites. *Canadian Journal of Botany,* 70 : 2259–2264.

Quayle, W.C.; Peck, L.S.; Peat, H. Ellis-Evans, J.C. and Harrigan, P.R. (2002): Extreme responses to climate change in lakes, *Science,* 295 : 645.

Quayle, W.C., Convey, P. Peck, L.S., Ellis-Evans, C.J., Butler, H.G. and Peat, H.J. (2003): Ecological responses of maritime Antarctic lakes to regional climate change, in E. Domack, A. Burnett, A. Leventer, P. Convey, M. Kirby and R. Bindschadler (eds.), *Ecological responses of maritime Antarctic lakes to regional climate change. Antarctic Research Series* vol. 79, American Geophysical Union, U.S.A., pp : 159-170.

Robinson, S.A.; Turnbull, J.D.; Lovelock, C.E. (2005) : Impact of changes in natural ultraviolet radiation on pigment composition, physiological and morphological characteristics of the Antarctic moss, *Grimmia antarctici. Global Change Biology,* 11 : 476-489.

Robson, T.M.; Pancotto, V.A.; Flint, S.D.; Ballare, C.L.; Sala, O.E.; Scopel, A.L. and Caldwell, M.M. (2003) : Six years of solar UV-B manipulation affect growth of Sphagnum and vascular plants in Tierra del Fuego peatland. *New Phytologist*, 160 : 379–389.

Roy, C.R.; Gies, H.P.; Tomlinson, D.W. (1994): Effects of ozone depletion on ultraviolet radiation environment at the Australian stations in Antarctica. Ultraviolet radiation in Antarctica : measurements and biological effects. *Antarctic Research Series*, 62 : 1-15.

Ruhland, C.T. and Day, T.A. (2001) : Size and longevity of seedbanks in Antarctica and the influence of ultraviolet-B radiation on survivorship, growth and pigment concentrations of *Colobanthus quitensis* seedlings. *Environmental and Experimental Botany*, 45 : 143-154.

Rustad, L.E.; Campbell, J. Norby, R.J.; Marion, G.M.; Mitchell, M.J.; Hartley, A.E Cornelissen, J. H.C.; Gurevitch, J. *et al.* (2001) : A Cross-Biome Synthesis of Ecosystem Response to Global Warming. *GCTE-NEWS*

Ryan, K.G.; Burne, A. Seppelt, R.D. (2009) : Historical ozone concentrations and flavonoid levels in herbarium specimens of the Antarctic moss *Bryum argenteum*. *Global Change Biology*, 15 : doi : 10.1111/j.1365-2486.2009.01885.x.

Schroeter, B. and Scheidegger, C. (1995) : Water relations in lichens at subzero temperatures: structural changes and carbon dioxide exchange in the lichen *Umbilicaria aprina* from continental Antarctica. *New Phytologist*, 131 : 273-285.

Shaver, G.R. and Chapin, F.S. (1991) : Production: biomass relationships and element cycling in contrasting arctic vegetation types. *Ecological Monographs*, 61 : 1–31.

Solhaug, K.A.; Gauslaa, Y. Nybakken, L. Bilger, W. (2003) : UV induction of sun-screening pigments in lichens. *New Phytologist*, 158: 91–100.

Solhaug, KA.; Gauslaa, Y. (2004) : Photosynthates stimulate the UV-B induced fungal anthraquinone synthesis in the foliose lichen *Xanthoria parietina*. *Plant, Cell and Environment*, 27 : 167–176.

Sonesson, M. Carlsson, B.A.; Callaghan, T.V.; Halling, S. Bjorn, L.O.; Bertgren, M. Johanson, U. (2002) : Growth of two peat forming mosses in sub arctic mires: species interactions and effects of simulated climate change. *Oikos*, 99 : 151-160.

Selkirk, P.M.; Skotnicki, M.L. (2007) : Measurement of moss growth in continental Antarctica. *Polar Biology*, 30 : 407–413.

Snell, K.R.S.; Convey, P. Newsham, K.K. (2007) : Metabolic recovery of Antarctic liverworts *Cephaloziella varians* during spring snow melt. *Polar Biology*, 30 : 1115-1122.

Smith, R.I.L. (1999) : Biological and environmental characteristics of three cosmopolitan mosses dominant in continental Antarctica. *Journal of Vegetation Science,* 10 : 231- 242.

Turnbull, J.D.; Leslie, S. Robinson, S.A. (2009) : Desiccation protects two Antarctic mosses from ultraviolet-B induced DNA damage. *Functional Plant Biology,* 36 : 214–221.

VanWijk, M.T.; Clemmensen, K.E.; Shaver, G.R,. Williams, M. Callaghan, T.V.; Chapin, F.S. *et al.* (2004): Long term ecosystem level experiments in Toolik lake, Alaska, and at Abisko, Northern Sweden: generalizations and differences in ecosystem and plant type resposes to global change. *Global Change Biology,* 10 : 105-123.

D.W.H. Walton (1984): *Terrestrial environment.* In : Laws R.M. (ed), Antarctic ecology, Vol 1, Academic Press, London, pp : 1-60.

Xiong, F.S.; Mueller, E.C. and Day, T.A. (2000): Photosynthetic and respiratory acclimation and growth response of Antarctic vascular plants to contrasting temperature regimes, *American Journal of Botany,* 87 : 700- 710.

Xiong, F. Ruhland, C. and Day, T.A. (1999): Photosynthetic temperature response of the Antarctic vascular plants *Colobanthus quitensis* and *Deschampsia antarctica, Physiologia Plantarum* 106, 276-286.

10

Algal Weed : A Pollution Indicator

Rajan K. Gupta, V. Prasad, D. Vyas,
G.K. Dhingra and Mukesh Kumar

ABSTRACT

Algal population causing adverse effects are referred to as nuisance, are called algal weeds. Both macrophytes and microphytes may be sea weed as well as freshwater algal blooms. Terrestrial algal mats or aquatic algal blooms, inefficient to atmospheric N_2-fixation may hamper the agriculture and consequently economy of farmers. Some weeds or harmful algal blooms (HABs) may cause adverse effect on animals and human health. But, some algal weeds may be useful too for various pharmaceutical, sanitational, food and feed, bioremediation, useful chemicals, minerals, antibiotics and energy production purposes.

Algal weed management by six active ingredients of the algicides are common practice. Alteration of environmental conditions such as nutrient reduction or darkness may be the most effective management tool for macroalgae. At present the management by biological agents, water fowl, snails, cray fish, herbivour fish and grass carp, is the most appreciated ecofriendly management tool. Pharmacological screening aimed at discovering novel pharmaceuticals has largely been carried out with higher plants and non-cyanobacterial microorganisms. Algae, in general, and cyanobacteria, in particular, have not yet received much attention. Till now, our knowledge of cyanobacteria-derived pharmaceuticals is based on few species. The chance of discovering pharmacologically active compounds from these underexplored organisms seems to be high. Because of their rapid growth rate, simple growth requirements, amenability to controlled laboratory culture, ubiquity and diversity, they appear to be suitable candidates for the purpose.

Keywords

Algae, weed, HABs, management, environmental alteration, biocontrol.

INTRODUCTION

Wetland Ecosystem

Global population increase has led a serious food problem and consequently scientists are attracted towards the technologies for higher and higher yield with sustainable agricultural system. Weed management is one of the important technology of higher yield. The organisms producing unavoidable problem and sharing the benefit of others by reducing productivity and causing adverse effects are called weeds (Sen, 1998). They usually emerge earlier, absorb available water and nutrients and suppress the crop growth. Grain yield is drastically reduced if crop is not weeded out during early growth stages (Pande *et al.* 1994). Farmer should keep the field weed free in the early stage upto 42 days after the transplanting rice (Thapa and Jha, 2002). Weed management by modern chemicals must be complemented by biological practices developed by biotechnological innovations and environmentally safe products for sustainable agriculture system (Bhowmik, 1998). The two third area of earth is covered by water and thus posses numerous type of algae in marine and freshwater wetland as aquatic ecosystem as well as agricultural lands. Some of them are very usefull for welfare of human beings but some algal blooms have adverse impact to crop, human and environmental health. Algal population that cause adverse effect referred to as nuisance are called algal weed (Kumar, 1999).

Table 1 : Algal Weed Species of Marine and Freshwater Forms Algal Weed Type

Marine Species	*Freshwater Species*
Laminaria sp.	*Chlorella* sp.
Undaria pinnatifida	*Scenedesmus* sp.
Prophyra crispate	*Spirulina* sp.
Ulva lactuca	*Microcystis* sp.
Enteromorpha proliferac	*Oscillatoria* sp.
Caulera racemosa	*Phormedium* sp.
Gelidiella acerosa	*Lyngbya* sp.
Gracilaria sp.	*Cladophora* sp.
Monostroma oxyspermum	*Stigeoclonium* sp.
Ascophyllum nodasum	*Oedogonium* sp.
	Ulothrix sp.
	Pithophora sp.
	Chara sp.

NUISANCE TYPES AND ITS IMPACTS

The three basic nuisance classes are attributable to certain bloom forming phytoplankton taxa such as (*i*) some water quality deterioration (*ii*) chronic or intermittent health hazards including toxicity (*iii*) losses of aesthetic and recreational value of affected water.

The first nuisance may include physio-chemical alteration of water such as deoxygenation, toxicity, transparency, chemical condition and sift from plankton based food webs to benthic or detrial ones (Pearl, 1988). The second nuisance encompasses growth and stimulation of pathogenic organisms and third one includes a wide variety of undesirable aesthetic impacts such as foul adours, bad tastes discoloration and poor fishing.

Table 2 : Nuisance Bloom Forming Cyanobacterial and Dianoflagellates Taxa (Source: Kumar, 1999).

Genus Properties	*Habitat*	*Bloom*	*Toxic*
Cyanobacteria			
Anabaena flosaquae	River	Surface scum & Dispersed colonies	Anatoxin A
Lyngbya sp. Gastrointestinal	River	Suspended filaments	Disorders
Microcystis aeruginosa	River/Pond	Surface scum & Dispersed colonies	Peptides
Nodularia sp.	Oligohaline	Surface scum & Dispersed colonies	Not known
Oscillatoria sp. Gastrointestinal	Rivers/ Mesotrophic lakes/ Oligohaline estuaries	Surface scum and Dispersed	Anatoxin A colonies Disorders
Dianoflagellates			
Amphidinium sp.	Stable marine	—	Choline
Exuviella sp.	Coastal regions	—	Hepato-neurotoxine
Gonyaulax sp.	Coastal, tropical & subtropical water	--	Saxitoxin
Gymnodinium sp.	Coastal water	—	Neurotoxin/ paralytic hemolytic toxin
Noctiluca sp.	Coastal Korea & Japan	—	Amonical toxin
Peridinium sp.	Freshwater	—	Not known.

Categories of Algal Nuisance

(*i*) Water quality deterioration

- deoxygenation
- chemical condition
- toxicity
- food web shift
- transparency

(*ii*) Chronic or intermittent health hazards including toxicity such as growth stimulation of pathogenic organism.

(*iii*) Losses of aesthetic and recreational value.

HABITAT FORMS OF ALGAL WEEDS

Table 3 : Habitat Forms of Algal Weeds (Souce : Kumar, 1999; Verma *et al*. 2003; Gupta *et al*. 2006)

Habitat forms	Algal weeds
1. Hydrophytes	
Freshwater : abundant in freshwater	*Microcystis, Gloeotrichia, Anabaena,* Spirulina, pithophora, Ulothrix, Hydrodictyon
Benthophytes : (Hyperbenthic) at the bottom	*Chara, Codium, Oedogonium, Coleochaete, Halimedia*
Endobenthic	*Hyella*
Epactiphytes along the shore	*Scytonema, Rivularia*
Thermophytes at high temperature	*Synechococcus, Oscillatoria, Aphanocapsa Aphanothece*
Planktophytes : Float on the bottom	*Nostoc, Rivularia, Chlorella, Cylindrospermum, Gloeotrichia*
Epiphytes : Upon other livings	*Microspora, Oedogonium*
Halophytes : Saline water	*Spirulina, Oscillatoria, Microcystis*
Marine (abundant in sea water)	*Porphyra, Ulva, laminaria, Enteromorpha, Calothrix, Gloeocapsa, Havella*
2. Terrestrial	
Endophytes : upon or inside the Surface of the earth.	
Cryptophytes : inside the Soil	*Microcoleus, Oscillatoria Cyanobactria*
Saprophytes : upon the soil surface	*Nostoc, Aulosira,Vaucheria.*
3. Aerophytes	
Epiphyllophytes : epiphytic upon leaves	*Phyllosiphon*
Epiphyllophytes : bark algae *Haplosiphon*	*Phormidium, Scytonema Schizothrix*
Epizoophytes : on the body of land animals	*Chaetophora*
Lithophytes : on rocks/walls	*Scytonema, Ulothrix, Stigeoclonium*

Cont...

4. Gyophytes : on snow/ ice ***Glecocapsa, Pleurococcus Chlamydomonas.***

5. Symbionts (endophytes)

Habitat Forms	Algal Weeds
Lichen –	*Nostoc, Gleocapsa, Scytonema, Microcystis, Rivularia*
Azolla	Anabaena azallae
Cycas	*A. cicadae*
Sphagnum	*Nostoc*
Anthoceros	*Nostoc*
Notothylas	*Nostoc*

6. Endozoophytes: Inside the body of animals ***Oscillatoria, Nostoc***

7. Parasites: Parasite upon other plants ***Cephaleuros, Ceratocolax Jjanczewskia.***

Beneficial Activities of Algal Weeds

Table 4 : Beneficial Activities of Fresh Water Algal Weeds

Freshwater 1 : Soil health enrichment

Benefits	*Algal Weeds*	*References*
Additional of minerals fertilizer Green manure	*Tolypothrix, Aulosira, Anabaena Anabaenopsis Nastoc Microcystis*	Prabhakaran, 1992 Kumar & Singh,1995
Pharmaceutical	*Tolyphothrix, Scytonema Spirulina*	Patterson *et al.* 1994.
Phytoremediation	*Oscillatoria, Microcystis Chara, Plectonema, Spirulina and Chlorella*	Singh *et al.* 2006
Energy production (H_2 CH_4)	*Synecococcus, Microcystis, Oscillatoria*	Kumar and Singh, 1995
Pollution indicator	*Chlorella,Cladophora Chlamydomonas Anabaena Stigeoclonium*	Rai *et.al.* 1994 Cains Caines 1984 Nirmala *et al.* 2003
Cosmetic	*Spirulina*	Singh *et al.* 2006

Secondry Melabolites

Spirulina (Phycobiliprotein)	Jha *et.al.* 2003
Duraloiella SosD (glycerol)	Borowitzka, 1998
Chlorella(B-carotene)	Shalini *et.al.* 2006

Cont...

Diatoms (Dynamite)		Kumar 1999
Agar (*Gelidium Gracilaria*)		Glombitez and Kuch, 1989
Growth Hormones		Borowitza, 1988
Wasteland reclamation		Blue Green algae Kumar, 1999
Sanitational *Chlorella, Chlamydomonas*		Kumar,1992
Euglena, Anacystis		Uma Subra
Halabacterium, Aphanocapsa		Manian,1990
Phormidium		
Marine		
1. Food	*Prophyra, Ulva, Laminaria Enteromorpha etc*	Kumar 1999
2. Feed	*Ascophyllum Fucus Laminaria Alaria*	Kumar 1999
3. Pharmaceutical	*Laminaria, Porphyra, Enteromorpha, Eisenia*	Kumar 1999
4. Energy Production	*Saragassum*	Serma *et al.*, 1984
5. Organic compound	*Laminaria*	Kumar, 1999

Algae as Ecosystem Engineer

Algae, including many organisms, play role in creation, modification and maintenance of habitats, they direct or indirectly modulate the availability of resources to other species by causing physical change in biotic or abiotic materials. These organisms are called ecosystem engineers (Kumar, 1999). Marine algal blooms absorb light in the upper layer of the ocean and enhance warming on the surface water and develop thermocline. Freshwater algae in lakes, ponds also intercept light in upper water colume and leads to shallower mixing depth (Johnes *et al.* 1994).

Cyanobacterial communities also act as engineers by exuding mucilaginous organic compounds in deserts and arid lands. The slimy exudates change infilteration, percolation, retention and evaporation of water and reduce soil erosion (Mazumdar *et al*; 1990). Crustose coralline algae, *Lithophylum* found in coral reefs overgrow and cement together detritus on outer algal ridge of barrier reef. This tends to break force of water and protect corals against strong wave action (Anderson 1992).

Biological crusts on soil in arid and semi arid shrub lands, grassland, and woodland contribute to soil stability, erosion and atmospheric nitrogen fixation nutrient contribution to plants, soil-plant-water relations infiltration seeding germination and plant growth (Tirkey and Adhikary, 2006). Various cyanobacteria exposed to intense solar radiation produce scytonemin and mycosporin amino acid like substances (MAAS) to protect organism from harmful UV radiation (Xiong *et al.* 1997).

Harmful Algal Blooms (HABs) and Their Impacts

Table 5 : Harmful Activities of Algae

1. Pollution

Impacts	*Organism*	*References*
Eutrophication in chilka lake Orissa	*Microcystis, Anabaena* *Rivularia, Oscillatoria* *Cladophora, Diatoms* *Lyngbya, Trichodesmium*	Kumar, 1999 Samatarry and Padhi, 1998

2. Recreation

Impacts	*Organism*	*References*
Loss of swimming	*Spirogyra Cladophora* *Microcystis Pithophora*	Masroor and Khan 2006
Abnormal taste and odours	*Synura Anabaena* *Synedra Dinobryon*	
Colorations of water	*Chlorella Oscillatoria* *Euglena Chlamydomonas*	
Skin irritation	*Schizothrix Lyngbya* *Oscillatoria* –	Carmichael and Falconer 1993

3. Parasitism

Impacts	*Organism*	*References*
Chlorella, Protococcus (Hydra)-		Kumar, 1999
Coccomyxa, Chlamydomonas (Paramecium)		
Sponges		
Snails		
Mussels		
Ovdinium (dinoflagellate) (Fish and Marine tunicates)		Kumar, 1999
Cephaleuros (Red rust of tea)		Kumar, 1999

Cont...

4. Water Supply

Impacts	*Organism*	*References*
Corrosion of concrete– and metallic walls of pipes	*Anacystis Chaetophora*	Kumar, 1999
Change in pH, CO_2, bicarbonate and O_2 content of water	*Chlorella* and various algae	Kumar, 1999

Table 6 : Toxicity of Freshwater and Marine Algae and Cyanobacteria

Freshwater Algae

Organism	*Toxin*	*Mode of Toxin*	*Reference*
Anbaena flos-aque	Antatoxin	Neurotoxin	Kumar 1999
Aphanizomenon flos-aque	Saxitoxin	Neurotoxin	Kumar 1999
Chlorella sp., *Lyngbya gracilis*	Dermatotoxin	Dermatitis	Piotois *et al.* 2000
Cylindrospermum, Oscillatoria, Nostoc	Hepatotoxin		Khare and Kumar, 2006
Microcystis aeruginosa	Microcystin	Hepatotoxin	Kumar, 1999
Nodularia spumigena	Cyanotoxin	Neurotoxin	Sipla *et al.* 2001
Oscillatoria		Neurotoxin	Khare and Kumar, 2006

Marine Algae

Organism	*Toxin*	*Mode of Toxin*	*Reference*
Gonyaulax calevella	Saxitoxin	Neurotoxin	Kumar 1999
G. tamarensis	Gonyantoxin	Neurotoxin	Kumar 1999
Gymnodinium sp.	—	Neurotoxin	Kumar, 1999
Lyngbya gracilis	Debromo aphysiotoxin	Dermatitis	Kumar, 1999
Prymnesium parvum	Prymnesin	Neurotoxin	Kumar, 1999

Table 7 : HAB's Causing Disease to Animals

Organisms	*Host*	*Reference*
Microcystis, Anbaena	Sheep	Main *et al.*, 1977
Aphanizomenon	Swine	Beasley *et al.*, 1983
Nodularia	Dog	Codd *et al.*, 1992
Oscillatoria	Bat	Nehring, 1993
	Fish	Tencalla *et al.*, 1994
Chrysochromulina	Fish	Aune *et al.*, 1992
Cyanobacteria	Zooplankton	Demott *et al.*, 1991

Table 8 : HABs Causing Diseases to Animals and Human Beings

Organisms	*Host-(Animals)*	*Reference*
B.G.A.	Cattles	
Anabaena, Microcystis	Sheep	Main et al. 1977
Aphanizomenon	Swine	Beasley et al. 1983
Nodularia	Dog	Codd et al. 1992, Nehring 1993
Oscillatoria	Bat Fish	Pybus and Hobson 1985, Tencalla et al. 1994
Chrysochromulina	Fish	Aune et al. 1992
Cyanobacteria	Zooplankton	Demott et al.1991
	Host- (Human-beings)	
Anabaena, Microcystis, Ocillatoria, Lyngbya	Skin, Irritation, Vomiting, Nausea	Carmichael and Falconer, 1993 Kumar, 2006 & Kumar 1999
Chlorella, Nodullaria	Diarrhea	Asa, *et al*, 1982
Aphanizomenon	Fever Influenza, Sore throat	Glimpes pp, 2008
Schizothrix	Muscle & Joint pain	Carmichael, 1994
Cylindrospermum	Liver damage	Carmichael, 1994
Hapalosiphon	Asthma	Carmichael, 1994
Nostoc	Eyes irritation & redness	Carmichael, 1994
Planktothrix	Pneumonia, Paralysis, Earache & itchiness	Carmichael, 1994
Gymnodinium brevis	Respiratory disorder	Codd *et al*, 1992

Management of Algal Weeds

1. Immediate Management : Should band the water for drinking until identification of Algae by a trained person.

2. Short Term Management

(a) Filtration of Water through activated Carbon.

(b) Cyanobacterial toxins should be removed by coagulation, adsorption, biodegradation, and microfiltration.

(c) Chlorination with Chlorine and Chloramines, (Nicholson *et al.* 1994).

(d) Ozonation, UV irradiation, (Alam *et al.* 2001).

3. Photocatalytic oxidation of Algal and Cyanobacterial toxins.

Biological Management (Biological Control Measures)

4. Nutrients and Artificial Sources : Integrated catchments Management to reduce the nutrients for natural and artificial sources.

5. Biological Control : Promoting natural algal grazers through restoration of riparian, vegetation *i.e.* plant, at the edges of water bodies, selective Fish stocking. (Masrror and Khan, 2006, Glimpses pp : 226).

(*i*) Chytrids (fungal Pathogens) to infert Akinets and Heterocysts of algae and Cyanobacteria.

(*ii*) Myxobacteriales to affect repid lysis with range of unicellular and filaments algae and blue green algae.

Viruses Application to Control

(*iii*) Phage AR-1, to attack *Anabaenopsis*.

(*iv*) Phage SM-1 and Phage AS-1, to control *Synechococcus and Microcystis*.

(*v*) Phage c-1, for Lysis of *Cylindrospermum.*

(*vi*) Virus CPP-1, to control *Lyngbya, Phormidium, Plectonema, Chlorella Sirodotia.*

(*vii*) **Chemical Control :** Kumar-1999, Copper sulphate, Potassium permaganate, Sodium arsenate, Sodium penta chlorophenate, $CuSO_4$-dinofllagelates.

(*viii*) **Preventive Agents (Algicide) :** Kumar-1994, Chlorophyll dimethylurea (CPDMU), Dithionine and 2,5, dichloro-3,4-dinitro thiophene (DCDNT), useful for *Anabaena, Microcystis,* and other Cynophytes, Quinone 2, 3-dichloro-1-4-naphtho-quinone (QDCNQ-1ppm)-cyanophytes.

(*ix*) Bilogical Organisms :

Lemna	*Green Algae and Cynobacteria*
Ochromonas (Chrysophyta)	*Microcystis aeruginosa*
Gymnodinium helveticum (Dinoflagellate)	Cyanobacteria and other algae (Frisch, 1979)
Some strains of Bacteria	*Anabaena flos-aquae* (Carmichael and Graham, 1977)

- Water fowl, (Snails, Crayfish, Herbivore Fish, Grass corp.)———— macroalgae, Cassini and Caton, 1986.
- Darkiness, (Alteration of Environmental condition) ——— Inhibit light activated algicides———— Kumar, 1999, Lembie *et al*, 1988. *Cladophora, Spirogyra, Oedogonium,* (Exept *Pithophora* and *Lyngbya*).

ACKNOWLEDGEMENTS

The author, wish to thank Department of Botany, Banaras Hindu University, Varanasi and to UGC, New Delhi and UCOST, Dehradun for financial help.

REFERENCES

Alam MZB, Otaki M, Furumai H, and Okagaki S, (2001) : Direct and indirect inactivation of *Microcystis aeruginosa* by UV radiation Water Res 35 : 1008-1014.

Asai S, Krazanowaski JJ, Anderson WH, Martin DF, Polson JB, Lockey RF, (1982) : Effect of the toxin of red tide, Phychodiscus brevis, on canine tracheal smooth muscle : a possible new Asthma-triggering mechanism, Journal of Allergy and Clinical Immunology 69 : 418-428.

Aune T, Skulbery OM, and Underdal B, (1992) : A Toxic Phytoflagellate bloom of Chrysochromulina cf. Leadbeateri is coastal water in the North of Norway, Aurbio 21 : 471-474.

Beasley VR, Coppock RW, Simony Ely R, Buck WB, and Corley RA, (1983) : Apparent blue-green algae poisoning in swine subsequent to ingestion of a bloom dominated by *Anbaena spiroides*. Journal of American Vet. Med. Assoc. 182 : 413-415.

Bhowmick PC, (1998) : Future of weed science research in relation to food production and environment, In abstract Int. Confer on Environment and Agriculture, Kathmandu, Nepal pp 9-10.

Borowitzka MA, (1998) : Vitamins and fine chemicals from micro-algae In Borowitzka MA, and Borowitzka L.J., (eds) Microalgal Biotechnology, Cambridge University \Press, New York, 153-196.

Cains JR and Cains RK, (1984) : Effect of five insecticides on zygospore germination and growth of green algae, *Chlamydomonas moewusii*, Bull Environ. Contamin. Toxicol. 33 : 571-574.

Carmichael WW, (1994) : The Toxins of Cyanobacteria. Scientific American 270 : 78-86.

Carmichael WW, and Falconer IR, (1993) : Diseases related to Freshwater algal toxins and control measures in Algal Toxins in sea food and drinking water. IR Falconer (ed.) Academic Press, London.

Carmichael WW, and Gorham PR, (1977) : Factors influencing the toxicity and susceptibility of *Anabaena flosaquae* blooms, of Phycol. 13 : 97-100.

Cassini JR and Caton WE, (1986) : Efficient production of triploid grass carp utilizing hydrostatic pressure, Aquaculture 55 : 43-50.

Chung P, Pond WC, Kingburg JM, Wate EF, and Krook L. (1978): Production and Nutritive values of *Arthrospira pletensis*, a spiral blue green algae grown on swine water, *Journal of Animal Science*, 47 : 310.

Codd GA, Edwards C, Beattie KA, Barr WM, and Gunn GJ (1992) : Fatal attraction to Cyanobacteria, *Nature*, 359 : 110-111.

Demott WR, and Moxter F (1991) : Foraging on Cyanobacteria by copepods responses to chemical defenses and resource abundance, *Ecology*, 72 : 1820-1834.

Glombitza KW, and Koch M (1989) : Secondary metabolites of Pharmaceutical potential in Cresswell RC, Rees TAV and Shah N (eds) Algal and Cyanobacterial Biotechnology. Longman Scientific and Technical, New York, 161-219.

Gupta RK, Kumar M, and Paliwal GS (2006) : Glimpse of Cyanobacteria, Daya Publishingh House New Delhi.

Habib I and Chaturvedi UK (2006) : Cyanobacteria the future food for man, In Gupta RK, Kumar M, and Paliwal GS (eds), Glimpses of Cyanobacteria, Daya Publishing House New Delhi.

Henning M and Kohl JG (1981) : Toxic blue green algae water blooms founds in some lakes in the German Democratic Republic, Int. Review Hydrobiol. 66 : 553-561.

Irish AE (1979) : Gymnodinium helveticum Penard (F. Achrou Sknya) a case of Phagotrophy, Brit. Phycol. J 14 : 11-14.

Jha MN, Sharma SG, Prasad AN and Tripathi KK (2003) : Cyanobacterial Biotechnology, In Verma BN, Kargupta AN and Yadav RN (eds), Phycology Nature and Nurture, Kalyani Pub. New Delhi, India, pp 225-237.

Khare A and Kumar M (2006) : Cyanobacteria in relation to Human and Animal health, , In Gupta RK, Kumar M, and Paliwal GS (eds), Glimpses of Cyanobacteria, Daya Publishingh House New Delhi, pp 182-200.

Kremer BP (1981) : Toxische Plankton algae Natur wissen-schaften 68: 101-109.

Kumar HD (1999) : Introductory Phycology IInd edition, East West Press Pvt. Ltd, New Delhi.

Kumar HD and Singh HN (1995) : A Text Book of Algae IVth edition East West Press Pvt. Ltd, New Delhi.

Lembi CA and Waaland JR (1988) : Algae and Human Affairs Cambridge University Press, Cambridge.

Main, DC., Berry, PH., Peet RL. and Robertson, JP. (1977) : Sheep mortality associated with the blue green algae, *Nodularia spumigena*. Australian Vet. J. 53 : 578-581.

Malin G, Turner, SM. and Liss, PS. (1992) : Sulfur: the plankton/ climatic connections J. Phycol. 28 : 590-597.

Marchant, HJ., Davidson, AT. and Kelly, GJ. (1991) : UV-B protecting compounds in the marine algae *Phneocystis pouchetii*, Mar. Biol. 109: 391-395.

Masroor, M. and Khan, A. (2006) : Various roles of blue greens in human life, In Gupta RK, Kumar M, and Paliwal GS (eds), Glimpses of Cyanobacteria, Daya Publishingh House New Delhi, pp. 213-236.

Materassi, R. and Vincen, Zini. M. (1989) : A general survey of feed and food utilization of autotrophic microorganisms. Proc. IIIrd Int. Conf. on Leaf Pro. 89, Pisa-Perugia-vitorbo, Italy, 366-373.

Nehring S (1993) : Mortality of drugs associated with a mass development of Bodularia spumigena in a brackish lake at the German North Sea coast, *Journal of Plankton Res.* 15 : 867-887.

Nicholson BC, Rositano J and Burch MD (1994) : Destruction cyanobacterial peptide hepotoxins by chlorine and chloramines, *Water Res.* 28 : 1247-1303.

Nirmala R, Simons J and Bharathan S (2003) : Taxonomy, Reproduction, and Ecology of *Stigeoclonium*-A review in Verma BN, Kargupta AN and Yadav RN (eds), Phycology Nature and Narture, Kalyani Pub. New Delhi, pp 123-141.

Norin LL and Waern M (1973) : The zone of low algal standing crops near Stockholm, Oikes (suppl.) 15 : 179-184.

Pande HK, Tran DV, and That TT (1994) : Improved upland Rice forming system, FAO Rome.

Patterson GML, Larsen LK and Moore RE (1994) : Bioactive natural products from blue-green algae. Journal of Applied Phycology 6 : 151-157.

Pearl HW (1988) : Nuisance phytoplankton blooms in coastal, estuarine and inland waters Limnol Oceanogr 33 : 823-847.

Piotois S, Jackson MH and Wood BJB (2000) : Problems associated with the presence of cyanobacteria in recreational and drinking waters, International Journal of Environ Health Res, 10 : 203-218.

Prabharan, D. (1992) : Hydrogen production by marine cyanobacteria, Ph.D. Thesis Bharathidarshan University Tiruchirapalli, India.

Pybus MJ and Hobson DP (1985) : Mass mortality of bats due to probable blue-green algal Toxicity, *Journal Wildlife Diseases*, 22 : 449-450.

Rai LC, Gaur JP and Soeder C J (1994) : Algae and water pollution, Schweizerbartsche Stuttgart.

Samantarays and Padhi S (1998) : Ecological aspects of blue-green algal blooms of Chilka lake, Int. Conf. on Environ. & Agric. Kathmandu, Nepal, pp : 32.

Sen DN (1998) : Role of weeds in Rain-fed cropping in Indian desert, Intern. Conf. on Environ. & Agric. Kathmandu, Nepal, pp : 10.

Seshadri CV and Seshagiri S (1986) : *Spirulina* the wonder gift of nature, A compendium MCPC Pub. Madras, India.

Shalini, Singh B.V. and Gupta RK (2006) : Biotechnological potential of cyanobacteria, In Gupta RK, Kumar M, and Paliwal GS (eds), Glimpses of Cyanobacteria, Daya Publishingh House New Delhi, pp. 305-320.

Singh S, Ganguly A and Tiwari A (2006) : Spirulina : A Blue Green Revolution, In Gupta RK, Kumar M, and Paliwal GS (eds), Glimpses of Cyanobacteria, Daya Publishing House New Delhi, pp. 321-349.

Sipla VO, Kankaanpaa HT and Flinkman J, lahti K and Meriluoto J A (2001) : Time Dependent accumulation of cyanobacterial hepatotoxins in flounders (*Platichthys flesus*) and mussels (*Mytilus edulis*) from the Northern Baltic Sea, Environ. Toxicol, 16 : 330-336.

Ten Calla F G, Detrich DR, and Schlotter C 1994, Toxicity of *Microcystis aeruginusa* peptide toxin to yearling rainbow tront (*Oncorhynchus onykiss*) Aquatic Toxicology 30 : 215-224.

Terry LA and Edyvean RG J (1984) : Influences of microalgae on corrosion of structural steel in Lewis JR and Mercor AD (eds), Corrosion and Marine growth on offshore structures, Ellis Horwood Chichester, 38-44.

Thapa CB and Jha PK (2002) : Paddy Crop weed competition in Pokhara, Nepal. Geobios 29: 47-51.

Thomus W.H. Hollibangh J.T. and Seibert DLR (1980) : Effects of Heavy metals on morphology of some marine phytoplanktons, *Phycologia* 19 : 202-209.

Uma L and Subramanian G (1990) : Effective Use of Cyanobacteria in Effluent Treatment, In Proc. Biol/Nitrogen Fixation IARI, New Delhi, pp. 437.

Venkataraman LV (1993) : Spirulina in India Pure. Nat. Sem. Cyanobacterial Research Indian Science, NFMC BARD Tiruchirapalli pp 92-116.

Verma BN, Kargupta AN and Yadav RN (2003) : Phycology Nature and Nurture, Kalyani Publishers, New Delhi, India.

ADDITIONAL REFERENCES

Algal Weeds as Pollution Indicator

- Johes CG, Lawton JL, and Shachak M (1994) : Organisms as Ecosystem Engineers Eikos 69 : 373-386.
- Mazumdar A, Taylor WD, Mc Queen, DJ and Lean DRS (1990): Effects of Fish and Plankton on Lake temperature and mixing depth, Science: 247 : 312-315.
- Anderson RA (1992) : Diversity of eukaryotic Algae, Biodiversity Conserv, 1: 267-292.
- Tirkey J and Adhikany SP (2006) : Cyanobacteria in the soil crusts and their significance, In Gupta RK, Kumar M and Paliwal GS, (eds), Glimpses of Cyanobacteria, Daya Publishing house, New Delhi, pp 1-25.
- Xiong F, Komenda J, Kopecky J and Nedbal L (1997) : Strategies of ultraviolet-B Protection microscopic algae, Physiol Plant 100 : 378-388.

11

Assessment of Water Quality Index of River Ahar, Udaipur (Rajasthan), India

B.R. Bamniya, Vidhya Kapasya, C.S. Kapoor, G.S. Mahecha and R.S. Meena

ABSTRACT

An alarming rate is of deterioration of water quality of fresh water resources like lakes, ponds and rivers etc is now a global problem. Historical River Ahar of Udaipur is of particular importance in the study of surface water pollution because effluents from municipal sewage, agricultural, industrial effluent and urban run-off are discharged into it which bringing about considerable change in the water quality. In present study Water Quality Index (WQI) is used to evaluate the quality of River Ahar. The WQI for the samples was 223.90. The high value of WQI had been found mainly from the higher of iron, chromium and COD and turbidity. The results of the analysis when compared with World Health Organization (WHO) permissible limits indicated that the river is highly polluted and so the water is not safe for domestic and agriculture uses and would need further treatment.

Keywords

Water Pollution, Water Quality Index, Sewage, Heavy Metals.

INTRODUCTION

Water is one of the precious renewable vital natural resource for survival of all kinds of life on planet earth. It is an elixir of life. More than 97% of the earth's water is saline (La Riviere, 1989) of that which is fresh, most occurs as ice (1.97 %) or ground water (0.61%). Only 0.014% of the earth water occurs in the biosphere. This of available for life is relatively small and distributed patchily over earth's surface (Carpenter *etal.* 1996). Most of anthropogenic activity depends on the availability of water. Its uses may include drinking and other domestic uses, industrial cooling, power generation, agriculture, transportation and waste disposal. Borbos (1986) has estimated that 10 million deaths resulted from 25 million new cases of water related disease reported annually and that half of the world population suffered from infection caused by polluted water. Most of the agricultural activity by excessive application of fertilizers and unsanitary conditions threatened the human health (Okeke and Igboanua, 2003). Water is an essential requirement of human and industrial developments and it is one the most delicate part of the environment (Das and Acharya, 2003). The domestic waste discharge into the pond resulted Eutrophication of pond (Pandey *etal.* 2003) WQI is calculated from the point of view of the suitability of surface water for human consumption (Atulegwu and Njoku, 2004). The fresh water has become a scare commodity is due to pollution (Singh R.P. & Mathur P 2005). River Ayad in Udaipur is of particular importance in the study of surface water pollution because effluents from municipal sewage, pesticide industries and agricultural and urban run-off are discharged into the river bringing about considerable change in the water quality. In the last few decades, there has, been a tremendous increase in the demand for freshwater due to rapid growth of population and the accelerated pace of industrialization These anthropogenic activities on the river Ayad pose a serious threat not only to organisms in the river but also the downstream water users. In addition, once the surface water is contaminated, its quality cannot be restored by stopping the pollutants from the source (Ramakrishnaiah *etal.*, 2009). Water quality index is one of the most effective tools to communicate information on the quality of water to the concerned citizens and policy makers. It, thus, becomes an important parameter for the assessment and management of surface water. WQI is defined as a rating reflecting the composite influence of different water quality parameters. The aim of the present study is to provide information on the physico-chemical characteristics of River Ahar and to discuss it's suitability for human consumption based on computed water quality index values. Fatedar A. *etal.* (2009) also studied water chemistry of Ranbir canal flowing from Akhnoor to Jammu.

MATERIAL AND METHOD

Study Area

The city of lakes Udaipur is located in a hill girdled basin of Aravalli hills in southern Rajasthan. This basin is drained by river Ahar and Sisarma. The natural drainage of these rivers is greatly modified by a series of water bodies forming a chain of lakes.

Udaipur is located at latitude of 23°46′-25°5′ North and longitude of 73°9-74°35 East an average altitude of about 579.4 meter above mean sea level. These lakes are Pichhola, Swaroop Sagar and Udaisagar which formed an integral part of the social, cultural and economic activities since last four centuries. Udaisagar, approximately 18 Km. away from Udaipur, is one of the important sources of industrial and irrigation water supply. Two lakes Pichhola and Fateh Sagar are still water supplying sources of Udaipur city. The study area, river Ahar is a historical river where before 4000 A.D. centuries Ahar civilization was developed on the bank of river. This river was lifeline of the city of Udaipur in past and it connect two lakes Fateh Sagar and Udaisagar.

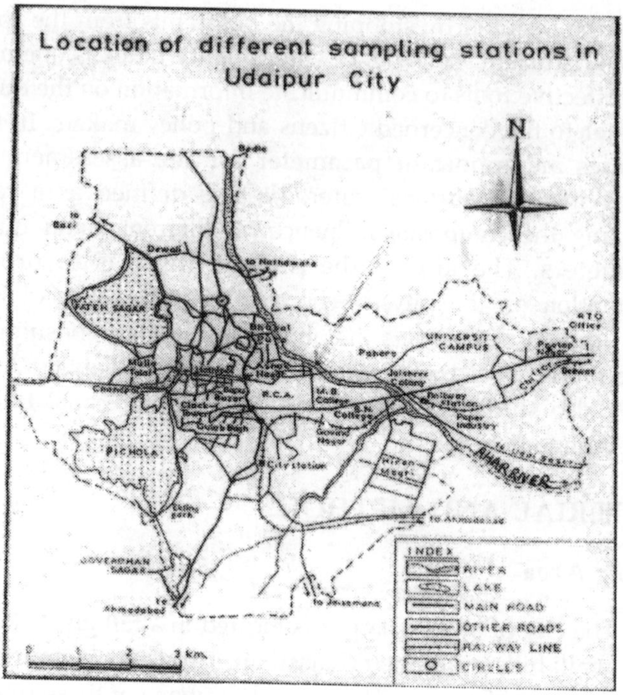

Figure 1 : Map of the study area showing different study sites

Laboratory Analysis

For the present study water samples were collected regularily for two years from January, 2009 to December, 2010. Seasonally water samples have been collected in 2.5 liter capacity cans after rising is properly from each location. In general, the shorter the time that elapses between collection of a sample and its analysis the more reliable will be the analytical results. Various Physico-chemical and metallic parameter like pH, dissolved oxygen, total hardness, alkalinity, Chloride, C.O.D, total dissolved solids, phosphate, nitrate, were analyzed, APHA (1995), Manivaskam (1986) and NEERI, (1988). The heavy metals concentration was determined by digesting the water samples with concentrate HNO_3 and the analyzed by atomic absorption spectroscopy.

Water Quality Index

Seasonal values for each parameter were calculated from monthly sample data. Out of 15 parameters studied 13 parameters were taken for calculating water quality index. A Water Quality Index (WQI) is defined as a rating reflecting the composite influence of different water quality parameters on the overall quality of water.

That is $W_i = K/ S_i$ ... (1)

Where W_i = Unit weight for the parameters.

S_i = (I = 1, 2, 3 ....13) refers to water quality parameters and

K = Constant of proportionality.

Calculation of WQI involves two fundamental steps :

1. Calculation of the quality rating for each of the water quality parameters used in the index and
2. Aggregation of these sub indices into the overall index. The quality rating (or sub indices) q_i corresponding to the parameter. q_i is a number of reflecting the relative value of this parameter in the polluted water with respect to its standard or permissible value. Quality rating may be obtained from

$$q_i = 100 [(V_i - V_{io}) / (S_i - V_{io})] \quad ...(2)$$

Where V_i = Measured value of ith parameter in the polluted water under consideration.

V_{io} = The ideal value of this parameter in pure water.

S_i = Standard or permissible value for the ith parameter.

Since in general, the ideal value $V_{io} = 0$ for the drinking water for most parameter, equation (2) assumes the sample from for these parameters.

$$q_i = 100 (V_i / S_i) \quad ... (3)$$

Equations (2) and (3) ensure that, $q_i = 0$, if the ith parameter is totally absent in the polluted water and $q_i = 100$, if the amount of this parameter is just equal to its permissible Si for the drinking water. But there are the following two exceptions.

(*a*) **For D.O. :** The ideal value may be taken as 14.6 mg/l (the solubility of Oxygen in pure water at 0°C). Since the standard for drinking water is 5.0 mg/l, equation (2) reduces to

$$q_{D.O} = 100\ (V_{D.O} - 14.6)\ /\ (14.6 - 5.0) \qquad \ldots(4)$$

Where $V_{D.O.}$ is the observed value of Dissolved Oxygen

(*b*) **For pH :** The ideal value is 7.0 (for neutral water) and the permissible value is 8.5, so equation 2 for this case may be written as :

$$q_{pH} = 100\ (V_{pH} - 7.0)\ /\ (8.5 - 7.0) \qquad \ldots(5)$$

Where V_{pH} is the observed value of pH.

The overall arithmetic WQI is calculated by aggregating the quality rating q_i linearly.

$$WQI = \sum_{i=1}^{N} \left(q_i W_i \,/\, \sum_{i=1}^{N} W_i \right)$$

In the present study, water quality parameters namely Turbidity, pH, Conductance, D.O, Total Hardness, Alkalinity, Chloride, COD, Total Solids, Nitrate, Iron, Zinc, Lead, Nickel and Cobalt were considered for computing WQI and unit weight (w_i) of each parameter is obtained depending upon its weightage. The standard permissible values of various parameters for drinking purpose are considered as recommended by World Health Organization (WHO) and Bureau of Indian Standards (BIS). Chemical analysis of water

Table 1 : Water Quality Index and Status of the Water Body

WQI	*Status*
0 – 25	Excellent
26 – 60	Good
51 – 75	Poor
76 – 100	Very Poor
101 and above	Unfit for Drinking Purposes.

gives a concept about its physical and chemical composition by some numerical values but for estimating exact quality of water, its better to depend on water quality index which gives the idea of quality of drinking water. The rating of WQI is shown.

Results

The results are obtained for the analysis of water sample of Ahar river, represented in Table 1. The quality of physico-chemical parameters serve as a good index providing it a complete and reliable picture of the condition prevailing in water body. The pH range of the sample is concerned, it is found to be varied between 7.35 to 8.38. The pH of the aquatic system is an important indicator of the water quality and extent pollution. The hardness of natural water depends mainly on the presence of dissolved ions of calcium and magnesium salts (Ikomi and Emuh, 2000).The mean total hardness value in the study area is 218.76 mg L^{-1}, which is above the standard permissible limit.

Table 2 : Physico-Chemical Assessment of River Ahar Water (2009-2010)

Parameters	*Minimum*	*Maximum*	*Mean*	*Standard Deviation*
Chloride	56.8	305.3	165.0333	78.73055
Alkalinity	140	640	285.1667	147.067
Hardness	60	660	366.25	218.7685
pH	7.35	8.38	7.891667	0.445887
T.D.S	680	1320	996	170.2079
Phosphate	0.023	2.0808	0.281692	0.571088
Nitrate	0.0939	1.337	0.398492	0.315837
D.O	3.5	7.2	5.96	1.069693
C.O.D	36	180	86	51.51522
Pb	0.09	0.1	0.098571	0.00378
Fe	0.7	29.5	12.91111	11.93853
Zn	0.1	0.5	0.344444	0.166966
Co	0.01	0.1	0.074286	0.043916
Cr	0.4	1.6	1.088889	0.413656
Ni	0.1	0.3	0.142857	0.07868

Table 3 : Computed Water Quality Index of River Ahar

Parameters	*Standard Permissible Limit*	W_i	q_i	W_iq_i
pH	7.0-8.5	0.117	59.4	6.9498
Chloride	250	0.0040	66.012	0.2640
Hardness	300	0.0033	122.083	0.4028
Alkalinity	120	0.0083	237.638	1.9723
C.O.D	10	0.1000	860	86
D.O	6	0.1666	90	14.994
Nitrate	10	0.1000	0.7969	0.0796
Phosphate	5.50	0.1818	256.08	46.55
T.D.S	500	0.0066	199.2	1.2685
Pb	0.1	10	98.571	985.71
Fe	1	1	4303.70	4303.70
Zn	5	0.2000	6.8888	1.3777
Cr	0.05	20	84.44	1688.8
Σ	31.8876	7138.069		
WQI	223.8509			

Total hardness can be used as an indicator for classify domestic pollution (Unni 1982). Total dissolved solids in river sample ranged between 680 mg/l to 1320 mg/l. Water containing high solids may cause laxative or constipation effects (Egereonu, 2004). The highest concentration of TDS obtained 1320 mg/l due to the intensive irrigation in that area. Chloride content range from 56.8 to 305.3 mg/l. Chloride is widely distributed element in all types of rocks in one or other form (Braide *etal.* 2004). Its affinity towards sodium is high. High concentration of chloride makes water non-potable and, therefore unfit for drinking or livestock watering (Egereonu, 2004). High concentration can occur near sewage, irrigation drains and waste outlets. Alkalinity is measure of capacity of water to absorb H^+ ion (Ravi Chandra 1987) reported that pollution by sewage and its decomposition seems to be possible cause for the higher value of alkalinity in water. The mean value of alkalinity obtained in this sample is 996 mg/l. Nitrate is naturally ion that is a part of nitrogen cycle. Nitrate ion cause

Table 4 : Correlation Between Various Physico-Chemical Parameters

	Cl	*Alkalinity*	*Hardness*	*pH*	*TDS*	*Phosphate*	*Nitrate*	*C.O.D*	*Pb*	*Fe*	*Zn*	*Co*	*Cr*	*Ni*	*DO*
Cl	1														
Alkalinity	0.156	1													
Hardness	0.006	-0.62	1												
pH	-0.096	-0.49	0.892	1											
TDS	0.722	0.06	-0.077	-0.026	1										
Phosphate	-0.39	-0.23	-0.501	-0.442	-0.10569	1									
Nitrate	0.301	-0.06	0.2445	0.148	0.39173	-0.188	1								
C.O.D	0.405	-0.19	0.2631	0.2512	0.18441	-0.207	0.044	1							
Pb	0.119	0.39	-0.486	-0.721	0.22614	0.208	0.33	-0.606	1						
Fe	0.353	0.34	-0.012	0.1743	0.343	-0.048	0.234	0.2332	0.11898	1					
Zn	-0.317	-0.18	-0.314	-0.422	-0.49476	0.323	-0.508	-0.116	-0.0704	-0.85	1				
Co	0.415	0.53	0.0336	-0.189	0.3378	-0.623	0.661	-0.557	0.61237	0.076	-0.4184	1			
Cr	-0.039	0.39	0.063	0.116	0.15448	-0.349	0.747	-0.64	0.41999	0.282	-0.5711	0.9355	1		
Ni	-0.476	-0.54	-0.263	-0.067	-0.27848	0.865	-0.629	0.0434	-0.25	-0.14	0.459	-0.9258	-0.5	1	
DO	-0.468	-0.05	0.2271	0.4347	-0.30867	0.086	0.182	-0.088	-0.2099	0.061	0.0533	-0.4839	0.125	0.476	1

methaemoglobinaemia in infants less than six months old (Egereonu and Nwachuk, 2005). However, the nitrate value varies from 0.09mg/l to 1.33 mg/l. The phosphate mean value is 0.281 mg/l. The DO ranged between 3.5 to 7.2 mg/l. The average value was 5.96 mg/l. Dissolved oxygen is an important determinant for suitability of waters and survival of water organisms. The low dissolved oxygen concentration observed during the study could be attributed the lesser input of freshwater into the study area, according to Ananthan (1991 and 1995) and Sampathkumar and Kannan (1998). Karuppasamy *etal.* (1999) and Ravanaiah G. and Narasimha Murthy C.V. (2010) have stated that the low dissolved oxygen was due to the biochemical oxidation of organic matter and the combined effects of temperature, salinity and photosynthetic activity. Further the dissolved oxygen concentration showed negative correlation with COD. Micro organisms may decompose organic substances in water by means of oxygen. The mean concentration of chemical oxygen demand ranged from 36.0-180 mg/l. The mean COD concentration in the samples analyzed was 86 mg/l indicated that the solid wastes is highly polluted with both oxidizable organic and inorganic pollutants (Otukune and Biukwu, 2005). High value of COD due to accumulation of domestic sewage was also reported is given (Mohan *etal.* 2007). Iron is highly above the permissible limits (0.7-29.5 mg/l).This element can lead coloration of water thus initiating sedimentation in the system, which can result to corrosion (Egereonu, 2004). Copper and zinc are present in 0.01-0.1 mg/l and 0.1 mg/l–0.5 mg/l respectively. They affect the organoleptic quality of water. The mean value for cobalt is 0.074 mg/l during the study duration. Chromium ranged between 0.4-1.6 mg/l. The value constitutes health hazards as they are not within the permissible level of 0.05 mg/l. This must be totally removed from water because they are carcinogenic (Otukune and Biukwu, 2005: Edet and Okereke, 2001).

CONCLUSION

In India and other parts of the world human civilization was developed nearby of the banks of almost all major rivers.

Rivers and ponds provide aesthetic, cultural, religious and tourism opportunity is local inhabitants. The study area river Ahar was source of drinking water and navigation when civilization of Ahar was established here. Lateron due to growth of population and use of catchment area for agriculture this river severely polluted. Most of domestic and sewage effluents drained into this river, causes water pollution. Madri industrial area is also located near river Ahar and industrial effluents also get mix with river water. The present study reveals that the water of River Ahar is not useful for drinking, agriculture and industrial purposes. The WQI of the sample was 223.85 which exceeded 100, the upper limit for drinking water. The River Ayad application of Water Quality Index (WQI) in this study has been found useful in assessing the overall quality of water and to get right judgment on quality of the water. It is also helpful for public to understand the quality of water as well as being a useful tool in many ways in the field of water quality management opportunities.

REFERENCES

Ananthan, G. (1991) : Hydrobiology of Parangipittai and Cuddalore marine environs with special reference to heavy metal pollution. M.Phil., Thesis. Indian J. Mar. Sci., 25 : pp. 352-334.

Ananthan, G. (1995) : Plankton ecology and heavt metals studies in the marine environs of Pondicherry, India. Ph.D.., Thesis, Annamalai University, India, pp. 125.

American Public Health Association, (APHA), (1995) : Standard Method for Examination of Water and Wastewater. 19th Edn., Published by E and FNSPON, Washington DC., pp : 2-56.

Atulegwu, P.U. and J.D. Njoku, (2004) : The impact of biocides on the water quality. *Int. Res. J. Eng. Sci. Technol.*, 1 : 47-52.

Borbos, S. (1986) : Monitoring natural water for drinking water quality. *World Health Statistics Quarterly*. 39(1) : 32-45.

Braide, S.A., W.A.L. Izonfuo, P.U. Adiukwu, A.C. Chindah and C.C. Obunwo, (2004) : Water quality of Miniweja stream, a swamp forest stream receiving non-point source waste discharges in Eastern Niger Delta, Nigeria. *Scientia Africana*, 3 : 1-8.

Carpenter *etal.*, (1996) : Freshwater Ecosystems : Linkage of Complexity and Processes. In : Moony H.A. And Schulze E.D. (eds.) Functional roles of Biodiversity. A global perspective. John Wiley & sons. pp. 299-325.

Das, J. and B.C. Acharya, (2003) : Hydrology and assessment of lotic water quality in Cuttack city, India. *Water, Air, Soil Pollut.*, 150 : 163-175. DOI : 10.1023/A : 1026193514875.

Edet, A.B. and C.S. Okereke, (2001) : A regional study of saltwater intrusion in Southeastern Nigeria based on the analysis of geoelectrical and hydrochemical data. Environ. Geol., 40 : 1278-1284.

Egereonu, U.U. and U.L. Nwachukwu, (2005) : Evaluation of the surface and groundwater resources of Efuru River Catchment, Mbano, South Eastern, Nigeria. J. Assoc. Adv. Model. Simulat. Tech. Enterpr., 66: 53-71.

Egereonu, U.U., (2004) : Assessment of atmospheric aerosols from three satellite stations: Heavy metal pollutants. J. Assoc. Adv. Model. Simulat. Tech. Enterpr., 65 : 71-88.

Fotedar A. *etal.* (2009) : Anote water chemistry of Ranbir canal flowing from Akhnoor to Jammu (J & K State) *J. Curr. Sci* 14(1) : 37-46

Karuppasamy, R *etal.* (1999): Seasonal variation of nutrient in the Uppanar estuary, cuddalore, south East coast of India. In : Proc. Intl. Sem. Appl. Hydrogeachem., Annamalai University, India, pp 269-274.

Ikomi, R.B. and C.T. Emuh, (2000) : The status of the physicochemical hydrology of Upper Warri River Nigeria. J. Sci. Environ., 2 : 75-86.

La Riviere J.W.M. (1989) : Treats to the world's water. 'Sci. Am. September: pp : 80-94.

Manivasakam, N. (1996) : Physicochemical examination of water sewage and industrial effluents.

Mohan, D., Gaur, A. and Choudhary, D., (2007) : Study of limnology and microbiology of Naya talab, Joadhpur (Raj.). Proc.DAE-BRNS Nat. Symp. Limnol., 07 : 64-68.

NEERI (1988) : Water and wastewater analysis. Course Manual. National Environmental Engineering Research Institute, Nagpur, India.

Okeke, C.O. and A.H. Igboanua, (2003) : Characteristics and quality assessment of surface water and groundwater recourses of Akwa Town, Southeast, Nigeria. J. *Niger. Assoc. Hydrol. Geol.*, 14 : 71-77.

Otukune, T.V. and C.O. Biukwu, (2005) : Impact of Refinery Influent on Physico-chemical properties of a water body on Niger Delta. *J. Applied Ecol. Environ. Res.*, 3 : 61-72.

Pandey, Arun K *etal.* (2003) : Physicohemical characteristic of sewage discharge into Saryu at Faizbad- Ayodhya. *Him. J. Env. Zoo.* 17 : 85-91.

Ramakrishnaiah, C.R., C. Sadashivalah and G. Ranganna, (2009) : Assessment of water quality index for the groundwater in Tumkur Taluk, Karnataka State. *Indian J. Chem.*, 6 : 523-530.

Ravanaiah G. & Narasimha Murthy C.V. (2010) : Impact of industrial pollution on the changes of physico chemical characteristics of water of the Pulicat; lake, Nellore District A.P. *Indian J. Environ & Ecoplanning* 17(1-2)163-174.

Sampathkumar, P & Kannan L. (1998) : Seasonal variations in physico-chemical characteristics in the Tranquebar-Nagapattinam region, southeast coast of India. Poll. Res., 17(4) : 397-402.

Singh R.P. & P. Mathur (2005) : Investigation of variations of phyico chemical charteritia of a fresh water reservoir of Ajmer city, Raj. *Ind. J. Env. Sci.* 9:57-62.

12

Water Quality Status of Phalankaruh River in East Khasi Hills, Meghalaya

P.K. Bharti, Gaurav and Vijender Singh

ABSTRACT

Phalankaruh river is a minor river of East Khasi Hills district in Meghalaya along with the Umaim River near Bangladesh Border. The river ecosystem has a great importance as a natural habitat among the various ecosystems of the region, whereas there is a limestone mining area at very large scale and no more industrial and agricultural pollution. Anthropogenic factors mainly local people's activities and catchments runoff may influence the index of nutrients in river water and may alter the physico-chemical characteristics and also the whole water quality. Physico-chemical parameters play an important role into niche restoration maintenance, self-regulation of water quality. Location variation in nutrients concentration of the river was studied with special reference to physico-chemical parameters and heavy metals in the river water. Calcium and Magnesium were observed in very low range in Phalankaruh river water in upstream as well as in downstream in comparison to rivers and hill-streams of North India. Dissolved oxygen was found 9.2 to 11.0 mg/l at upstream and downstream during the study period. Heavy Metals were found almost nil or in very low concentrations at both selected site. The mean value of water quality ratings at all sites recorded under excellent (99.04) *i.e.* Water quality is protected with a virtual absence of threat or impairment; conditions very close to natural or pristine levels; these index values can only be obtained if all measurements are within objectives virtually all of the time. The present study deals with the preliminary physico-chemical characteristics of the river water and exhibits the status of natural quality of water with minimum anthropogenic activities, which contribute in pollution load of an aquatic system.

Keywords

Pollution status, Physico-chemical characteristics, Heavy Metals, Water Quality Index.

INTRODUCTION

Indian mountains are the cradle of a large number of streams and mighty rivers. In North-East region and very close to Bangladesh border, a hill-stream Phalankaruh, 90 Km. far away from Shillong city is the study site for the accounting of physico-chemical parameters and Heavy Metals in natural water and fluctuations at the different locations along with the river. Adequate understanding of the North-East regional streams is extremely important for the development of a realistic program for utilizing the potential of water that exist in the form of hidden water resource in the area. The impact of anthropogenic pollution from industrial, agricultural, sources, quarrying and tourists activity on water quality has concerned environmentalists and scientists for the past three decades. Ecological, geo-chemical and hydrological research has been carried out in various ecosystems to understand the factors controlling the chemistry of natural water (Baron & Bricker, 1990; Malik and Bharti, 2005a). Many of such studies have been under taken during the last two decades to understand the processes that control the hydrochemistry of alpine and sub-alpine systems of North America and Europe (William *etal.* 1993 and Psenner, 1989). Similar studies were also done in China and central Asia (Xue & Schnoor, 1994).

The present study reveals to characteristics the water nutrient chemistry, influenced by anthropogenic activity &

quarrying of the geologically sedimental environments and to determine the nature and degree of anthropogenic impacts on qualitative and quantitative variations occurred in nutrients in relation to physico-chemical parameters and Heavy Metals of river water.

STUDY AREA

The Meghalaya has been surrounded by the natural beauties with lovely trees and cool climate, which is not only a pleasant place to live in but relaxing for holiday also. This is not only a popular state but also valuable or important place for tourism. Shillong is the capital of Meghalaya state. Mowlong Cherra Cement Ltd and Lafarge Umaim Mining Ltd are the major industries of the region. The all seven states of North East India are quite famous as Seven Sisters. Meghalaya is very well known for the record of rainfall, the state consists of the two places namely Cherrapunjee and Mawsinram for maximum rainfall throughout the year. So, the maximum water resources are depending chiefly on total precipitations of the region.

Basically, Phalankaruh river originated from the hills of Assam-Meghalaya. Upstream of Phalankaruh is near DCCL Project office and Downstream near Village Phalankaruh. The water of Phalankaruh is very useful for the local people. Population of the region is completely depending upon the river water for drinking, bathing, and other activities. *Meteorologically*, region has a cool & pleasant climate. *Geologically*, North-east hills are enriched with various minerals and the hills near Bangladesh are rich in limestone. *Geographically*, the study area is situated in the globe on a Latitude 25° 11′ 5.8" N & Longitude 91° 37′ 26.1" E for Upstream site and Latitude 25° 10′ 51.9" N & Longitude 91° 37′ 22.5" E for Downstream site.

MATERIALS AND METHODS

The water samples were collected from Phalankaruh River upstream (U/S sampling site A) Near DCCL project office and downstream (D/S sampling site B) near Phalankaruh Village according to the analytical requirement in morning period 9:00

Hrs. to 10:00 Hrs. The samples for physico-chemical parameters and Heavy Metals were collected by using rinsed Borosil glassware, and analyzed with the help of the procedure described by APHA (1995) and Trivedi and Goel (1984). Colour, odour, turbidity, velocity, temperature and Dissolved oxygen were analyzed on sampling sites. Samples were collected from selected sites and immediately preserved in ice boxes, and transfer to the lab for further analysis. Water samples were digested and Heavy Metals were detected using Atomic absorption spectrophotometer. Water quality index (WQI) was calculated to assess the status of water quality of Phalankaruh river. The CCME Water Quality Index: The CCME-WQI produces a measure of the derivation of water quality from water quality guidelines from described in (CCME- WQI, 2001).

$$CCMEWQI = 100 - \left[\frac{\sqrt{(F_1)^2 + (F_2)^2 + (F_3)^2}}{1.732} \right]$$

Where,

F_1= Number of failed variables/Total number of variables × 100

F_2 = Number of failed test/Total number of test × 100

F_3 = (nse/001nse + 0.01)

The divisor 1.732 normalizes the resultant values range between 0 and 100, where 0 represents the "worst" water quality and 100 represents the "best" water quality.

Once WQI value has been determined, water quality is ranked by relating it to one of the following categories :

Rank	*CCME-WQI*	*Description*
Excellent	95-100	Water quality is protected with a virtual absence of threat or impairment; conditions very close to natural or pristine levels; these index values can only be obtained if all measurements are within objectives virtually all of the time.

Cont...

Good	80-94	Water quality is protected with only a minor degree of threat or impairment; conditions rarely depart from natural or desirable levels.
Fair	65-79	Water quality is usually protected but occasionally threatened or impaired; conditions sometimes depart from natural or desirable levels.
Marginal	45-64	Water quality is frequently threatened or impaired; conditions often depart from natural or desirable levels.
Poor	0-44	Water quality is almost always threatened or impaired; conditions usually depart from natural or desirable levels.

RESULTS AND DISCUSSION

Phalankaruh river is flowing throughout a valley of North-east hills chain near the Bangladesh Border, enriched with limestone and lignite rocks, which affect the water quality of river according to the locations. Nutrients concentration, heavy metals and related physico-chemical parameters from selected sites are depicted in tables. Phalankaruh river has the spatio-temporal variations of water temperature, which plays a vital role in all physico-biochemical reactions and self-purification power of aquatic system (Badola & Singh, 1981). Higher value of temperature was found 17°C in summer at downstream and minimum 12 °C in winter season at upstream. Turbidity is striking characteristic of the physical status of the water bodies. Although in Phalankaruh river water is clear because there is no more pollution, siltation was the main source of turbidity in tributaries. Detritus and other non-organic material being added to water mass due to rainfall and anthropogenic activities (Camron, 1996). Maximum turbidity was recorded 8 NTU during rainy season at downstream and minimum 1 NTU in summer season at upstream. The maximum depth of photic zone provides the better biological production for all aquatic organisms (Malik and Bharti, 2005b).

Total dissolved solids were found in the range of 118 mg/l in summer to 148 mg/l in monsoon season, due to the gradual increases in velocity of river which favoured effective sedimentation (Subramanian, 1979). Chemical oxygen demand was found 5 mg/l to 8 mg/l during the study period. Chemical oxygen demand represents chemically oxidizable organic matter load in water, while biochemical oxygen demand is only biodegradable materials (Malik and Bharti, 2005c). In the present study the values observed during monsoon months may be attributed maximum biological activities and high temperature, stimulate the growth of microorganisms (William *etal.* 1993).

The pH of natural water was controlled in a great extent by the interaction of hydroxyl ions arising from the hydrolysis of bicarbonate (Sharma, 1986). The pH of Phalankaruh river was recorded alkaline (7.5-8.4 mg/l). Total hardness is mainly due to per cent age of calcium and magnesium salts of bicarbonates, carbonates, sulphates and chlorides, while the value of alkalinity occured due to presence of bicarbonates. The concentration of hardness was analyzed 97-128 mg/l during the study. Alkalinity was also found 81-133 mg/l with a small fluctuation. A positive relationship between hardness and alkalinity was recorded in river Ganga at Rishikesh (Chopra and Patric, 1994). Maximum chloride concentration was recorded maximum (6 mg/l) in winter and minimum in summer (2 mg/l). Chloride and hardness showed a positive relationship to one another (Chopra and Patric, 1994). Chloride was found in the form of chloride ion, and one of the major inorganic anion present in natural water (Malik and Bharti, 2009).

Calcium and magnesium the dominant cations, and these represent the main weathering products, but significant hydro-chemical differences between the two sampling sites associated with the bedrock geology exist (Jenkins *etal.* 1995) Calcium is one of the essential nutrients, which plays an important role in biological system. Maximum calcium concentration was recorded (48 mg/l) in monsoon and minimum in winter (26 mg/l). Positive relationship between, calcium and temperature was also reported by Khanna and Singh (2000) in river Suswa, Dehradun. Magnesium is also an essential element but it is toxic at higher concentration. The concentration of magnesium in Phalankaruh river was found maximum (14 mg/l) and minimum (3 mg/l) and it was very low in comparison to Hill-streams of Uttarakhand (Bharti, 2004). During the summer season nutrients concentration in rivers and hill-streams became more. Miller *etal.* (1997) described the nutrients availability in selected environmental settings of the Potomac River and Cameron (1996) showed the similar type of fluctuation in Fraser river of British Columbia. Bond (1979) described similar nutrients concentration pattern in a stream draining a mountain ecosystem in Utah.

Heavy metals like cadmium, copper, iron, manganese, lead and zinc were not found in high concentration at both sites during any season. Cadmium, copper and manganese were absolutely absent in all seasons, while lead, zinz and iron concentrations were also found below detection limit in maximum samples. The concentration of iron was maximum observed 0.06 mg/l during winter season. Lead was found 0.01 mg/l in winter season at upstream site and Zinc concentration was 0.02 mg/l in summer and monsoon season. Malik *etal.* (2009) described the role of Heavy Metals in the surface water of north India. The results of Bharti *etal.* (2010) were also indicated the relation of Heavy Metals and phytoplankton in a north Indian water body.

The concentrations of Calcium were observed in similar range in Phalankaruh river water at upstream as well as downstream in comparison to rivers and hill-streams of North India, while magnesium was found in low range. Dissolved oxygen was found 8.1 to 9.8 mg/l at upstream and downstream during the study period. Heavy Metals were found almost nil or in very low concentrations at both selected site. The mean value of water quality ratings at all sites recorded under excellent (99.04) *i.e.* Water quality is protected with a virtual absence of threat or impairment; conditions very close to natural or pristine levels; these index values can only be obtained if all measurements are within objectives virtually all of the time.

The present results conclude that significant differences in river water nutrient concentrations exist among different environmental settings within the two subunits. The environmental setting with the highest potential by more soluble nutrients, fluctuations in nutrient concentrations were the land use and carbonate bedrock that was predominated in the North East valley. The spatial variations in TDS are attributed to climatic and lithological control over the ionic concentrations. Absence or low concentration of Heavy Metals shows that the water is still industrial pollution free. However, the Indian standards are not so strict (Bharti, 2007), but Heavy Metals never cross the limits during the study period. On the basis of nutrients and Heavy Metals, the water may be considered for drinking and other purposes.

Table 1 : Physical Characteristics of Phalankaruh River Water

			Phalankaruh Upstream			Phalankaruh Downstream			
S.N.	Parameters	Unit	Winter	Summer	Monsoon	Winter	Summer	Monsoon	Desirable Limit
1.	Temperature	°C	12	16	15	12	17	16	–
2.	Colour	–	Clear	Clear	Clear	Clear	Clear	Clear	–
3.	Odour	–	Nil	Nil	Nil	Nil	Nil	Nil	–
4.	Turbidity	NTU	1	4	7	1	5	8	5
5.	Velocity	m/s	0.3	0.3	0.4	0.2	0.3	0.4	–
6.	TDS	Mg/l	120	118	148	125	120	155	500

Table 2 : Chemical Characteristics of Phalankaruh River Water

			Phalankaruh Upstream			Phalankaruh Downstream			
S.N.	Parameters	Unit	Winter	Summer	Monsoon	Winter	Summer	Monsoon	Desirable Limit
1.	pH	–	7.6	8.4	7.6	7.7	8.3	7.5	6.5-8.5
2.	Alkalinity	Mg/l	92	81	133	96	85	130	200
3.	Total Hardness	Mg/l	107	97	128	109	101	128	300
4.	Calcium	Mg/l	38	34	48	26	34	38	75
5.	Magnesium	Mg/l	3	7	12	4	11	14	30
6.	Chlorides	Mg/l	5	2	5	6	4	5	250
7.	DO	Mg/l	9.6	9.8	8.1	8.6	9.0	8.6	–
8.	BOD	Mg/l	Nil	1	Nil	Nil	Nil	Nil	–
9.	COD	Mg/l	6	8	5	5	6	6	–

Table 3 : Heavy Metals in Phalankaruh River Water

			Phalankaruh Upstream			Phalankaruh Downstream			
S.N.	*Parameters*	*Unit*	*Winter*	*Summer*	*Monsoon*	*Winter*	*Summer*	*Monsoon*	*Desirable Limit*
1.	Cadmium	Mg/l	BDL	BDL	BDL	BDL	BDL	BDL	0.01
2.	Copper	Mg/l	BDL	BDL	BDL	BDL	BDL	BDL	0.05
3.	Iron	Mg/l	0.06	0.02	BDL	0.03	0.06	0.04	0.3
4.	Lead	Mg/l	0.01	BDL	BDL	BDL	BDL	BDL	0.05
5.	Manganese	Mg/l	BDL	BDL	BDL	BDL	BDL	BDL	0.1
6.	Zinc	Mg/l	BDL	0.02	BDL	BDL	0.02	0.02	5.0

Figure 1 : Showing the nutrients concentration in Phalankaruh river during the study period.

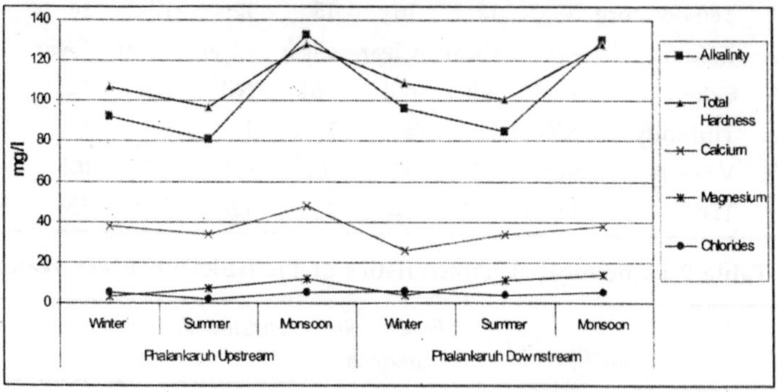

REFERENCES

APHA (1995) : Standard methods for examination of water and waste water. *American Public Health Association, 19th edition. Inc, New York.* pp : 1970.

Badola, S.P. and Singh, H.R. (1981) Hydrobiology of the river Alaknanda of Garhwal Himalaya. *Indian J. Ecol.*, 8(2) : 269-276.

Baron, J. & Bricker, O.P. (1990) : Hydrological and chemical flux in Loch Vale watershed, Rocky Mountain National Park. In: Biogeochemistry of major rivers. SCOPE 42. Wiley and sons, New York, USA.

Bharti, P.K. (2004) : Limnobiological study of Sahastradhara hill-stream at Dehradun, *M.Sc. Dissertation*, submitted to Gurukula Kangri University, Hardwar, pp : 102.

Bharti, P.K. (2007) : Why Indian standards are not so strict, *Current Science*, 93(9) : 1202.

Bharti, P.K., Malik, D.S. and Rashmi Yadav (2010) : Influence of Heavy Metals on Abundance of Cyanophyceae Members in Three Spring-fed Lake in Kempty, Dehradun, In: 'Advances in Aquatic Ecology, Vol-III' ed. by V.B. Sakhare, *Daya Publishing House, New Delhi*, pp : 107-111.

Bond, H.B. (1979) : Nutrient concentrations patterns in a stream draining a montane ecosystem in Utah. *Ecology*. 60(6) : 1184-1196.

Cameron, E.M. (1996) : Hydrogeo-chemistry of the Fraser River British Columbia: Seasonal variation in major and minor components. *J. Hydrol.* 182(1-4) : 209-255.

Chopra, A.K. and Patrick, N.J. (1994) : Effect of domestic sewage on self-purification of Ganga water at Rishikesh. *A. Bio. Science.* 13(11): 75-82.

Jenkins, A., Sloan W.T. & Cosby, B.J. (1995). Stream chemistry is the middle hills and high mountain of the Himalaya, Nepal. *Journal of Hydrology*. Vol-166, No-1-4: 61-79.

Khanna, D.R. and Singh. R.K. (2000) : Seasonal fluctuations in the plankton of Suswa River at Raiwala Dehradun. *Env. Conservations J.* 1(2 & 3) : 89-92.

Malik, D.S. and Bharti, Pawan K. (2005a): Nutrient dynamics in Rhithron zone of Shivalik Himalayan stream Sahastradhara, Dehradun (Uttaranchal), *Env. Cons. J.* 6 (2) : 63-68.

Malik, D.S. and Bharti, P.K. (2005b) : Fluctuation in planktonic population of Sahastradhara hill-stream at Dehradun (Uttaranchal), *Aquacult.* 6 (2) : 191-198.

Malik, D.S. and Bharti, P.K. (2005c) : Primary production efficiency of Sahstradhara hill-stream, Dehradun, *Env. Cons. J.* 6 (3) : 117-121.

Malik, D.S. and Bharti, P.K. (2009) : Ecology of Sahastradhara Hill-stream at Dehradun (Uttaranchal), In : 'Advances in Aquatic Ecology, Vol-I' ed. by V.B. Sakhare, *Daya Publishing House, New Delhi*, pp : 1-11.

Malik, D.S. and Bharti, P.K., Negi, K.S. and Rashmi Yadav (2009): Distribution of Metals in Water of an Artificial Lake at Mussoórie, Uttarakhand, In: 'Aquatic biology and aquaculture' edited by V.B. Sakhare, Ambajoagi, MS, *Manglam Publication, New Delhi*, pp : 77-95.

Miller, C.V.; Denis, J.M.; Ator, S.W. and Brakebill, J.W. (1997) : Nutrients in stream during baseflow in selected environmental settings of the Potomac river basin. *J. American Wat. Resources Association.* 33(6): 1155-1171.

Psenner, R. (1989) : Chemistry of high mountain lakes in siliceous catchments of central Alps. *Aquatic Sci.* 51 : 108-128.

Shrama, R.C. (1986) : Effect of physico-chemical factors on benthic fauna of Bhagirathi River Garhwal Himalaya. *Indian. J.Ecol.* 13(1) : 133-137.

Subramanian, V. (1979) : Chemical and suspended sediment characteristics of river of India. *J. Hydrol.*, 44 : 37-55.

Trivedi, R.K. and Goel, P.K. (1984) : Chemical and biological methods for water pollution studies. Karad. Environmental Publication. pp : 1-298.

William, M.W., Brown, A. & Melack, J.M. (1993) : Geochemical and hydrologic controls on the composition of surface water in the high elevation basin, *Sierra Navada. Limnol. Oceanogr*, 38 : 775-797.

Xue, H.B. and Schooner, J.L. (1994) : Acid deposition and lake chemistry in southwest China. *Wat. Air, Soil Pollut.* 75 : 61-78.

13

Non-conventional Defluoridation Technologies in Sustainable Management of Drinking Water Resources : A Review

D.S. Malik, Preetika Shukla, Anjali Goel and M.K. Sharma

ABSTRACT

The quality of drinking water is deteriorating in term of suitable parameters of physico-chemical and microbiological including accumulating heavy metals and other pesticidal residues in the ground water adjoining the vicinity of developing industrial sectors in Uttarakhand and other states. The most common water purification processes cum non-conventional technology have been evolved an economical and acceptable by the rural communities these are Bagasse, a residual cane pulp remaining after sugar extractions are being used as low cost absorbent for the removal of fluoride, heavy metal and other impurities. Nalgonda technique, an indigenous purification system and easily adopted by rural community at individual household level is very common for the removal of fluoride. A new innovative calcium removal technology is quite popularized to remove the calcium hardness from drinking water. The fly ash and brick kiln ash, red mud and activated alumina play a significant contribution in water purification and reducing the inorganic & organic pollutant compounds in the domestic waste water and can be used for decentralized application in treating waste water generated in rural areas and smaller town. The different absorbent filters have been designed and used in wastewater filtration for decolourization and treatment of microbial contamination to improve the quality of drinking water.

Keywords

Water purification, non-conventional purification, drinking water, sustainable water resource.

INTRODUCTION

Water pollution is a serious problem in India as almost 70% of surface water resources and biological, organic and inorganic pollutants already contaminate a growing number of its ground water reserves. In many cases, these sources have been rendered unsafe for human consumption as well as for other activities such as irrigation and industrial needs. This illustrate that water quality decline can in effect contribute to water scarcity as its limits the availability of for both human use and the ecosystem. Ground water with high fluoride content is found mostly in calcium-deficient ground waters in many basement aquifers, such as granite and gneiss, in geothermal waters and in some sedimentary basins. Seventeen state of India have been identified endemic to fluorosis due to abundance in natural occurring fluoride bearing minerals. Though iron content in drinking water may not content in drinking water may not affect the human system as a simple dietary overload, in long run prolonged accumulation of iron in the body may result in homochromatosis. If water containing excess amount of fluoride is consumed consistently, it is found to be highly harmful to health.

FLUORIDE POSITION

The worst affected areas in India Rajasthan, A.P., Orissa, Gujarat, M.P. and Chattisgarh (Susheela, 2001). Susheela (1999) reported that India is among 23 countries where health problems occur due to the consumption of fluoride contaminated water. Several parts of India especially Andhra Pradesh, Gujarat,

Rajasthan, Karnataka and some parts of UP (Unnao, Agra, Meerut, Aligarh, Raibareilly, and Allahabad) have high level of fluoride in ground water. The maximum range occur in Rajasthan 30 village of district Alwar is 1.5mg/L to 27.3mg/L (Tripathi and Singh (1996). The permissible level of fluoride in drinking water is 1.5 ppm as per WHO standards.

SOURCES OF FLUORIDE

The major source of fluoride in the groundwater is fluoride bearing rocks from which it get weathered and/or leached out and contaminates the water. Fluorides occur in three forms, namely, fluorospar or calcium fluoride (CaF_2), apatite or rock phosphate [$Ca_3F(PO_4)_3$] and cryolite (Na_3AlF_6). Concentration of fluorides is five times higher in granite than in basalt rock areas. Similarly, shale has a higher concentration than sandstone and limestone. Alkaline rocks contain the highest percentage of fluoride (1200 to 8500 mg/kg) (Chand, 1998; Athavale and Das, 1999).

Fluorinated hydrocarbons (refrigerants, aerosol propellants etc.), fluorinated plastics (polytetrafluoroethylene etc.), petroleum refining and hydrogen fluoride manufacturing units are mainly responsible for airborne fluoride. Since many food materials that we consume everyday, like common salts, tooth paste, etc., contain excess of fluoride, the permissible level further decreases to 1 ppm. There are lack of alternative sources of drinking water and absence of treatment of existing sources of drinking water force people of these affected areas to drink water with excess fluorides causing dental and skeletal fluorosis.

Differences in geo-chemical conditions in aquifers and differences in geo-chemical conditions in aquifers and differences in contact period between ground water and fluoride-bearing rocks, the fluoride content in ground water of Indian aquifers varies from <1ppm up to 25ppm. Yet, another factor contributing to excess fluoride in ground water in rural regions of India is the over-exploitation of ground water resources for agricultural and drinking water purposes.

HEALTH HAZARDS

Excess fluoride in drinking water can lead to fluorosis-a crippling disease manifesting as dental, skeletal and gut complaints, when the fluoride is greater than 1.5 ppm and is the maximum permissible limit prescribed by the Bureau of Indian standards in view of the health problems. This means that the body may tolerate fluoride upto a certain limit of 1.5 ppm depending upon the nutritional standards and body physiology. It is further specified that "lesser the better" as the fluoride causes where even 0.4 ppm fluoride in drinking water has caused dental fluorosis. About 96% of the fluoride in the body is found in bones and teeth. Fluoride is essential for the normal mineralization of bones and formation of dental enamel. The principle sources of fluoride to the physiology of man are water and food. Fluorine is often called a two-edge sword. Fluorine in small dosages has remarkable influence on the dental system by inhibiting dental carries, while in higher dosages cause dental and skeletal fluorosis. Fluorosis was first detected in India by Short *etal.* (1937) when the disease was prevalent in 4 states, namely Andhra Pradesh, Tamil Nadu, Punjab and Uttar Pradesh. During the period 1960-1986, 9 more states have been identified as endemic for fluorosis and during 1910-1992 two additional states, Kerala and Jammu and Kashmir have also been identified as endemic for the disease. Thus total number of states declared endemic for fluorosis at present are 17. The affected states are Jammu and Kashmir, Punjab, Haryana, Delhi, Uttar Pradesh, Rajasthan, Gujarat, Madhya Pradesh, Maharashtra, Karnataka, Andhra Pradesh, Kerala, Tamil Nadu, Orissa, Bihar, Assam and West Bengal of which Uttar Pradesh, Rajasthan, Gujarat, Andhra Pradesh and Tamil Nadu are the worst affected states.

The Indian standard specifies the desirable and permissible limits for fluoride in drinking water (>1.5ppm) lead to dental and skeletal fluorosis. The Indian standard specifies the desirable and permissible limits for fluoride in drinking water as 0.1-1.5 ppm respectively. Seventeen Indian state have been identified with the problem of excess fluoride in ground water resources (Development Alternatives, Troubled Waters) till 1999. The fluorosis problem is severe in India as almost 80% of the rural population depends on untreated ground water for potable water supplies (Ministry of Water Resources).

METHODOLOGY

Defluoridation is the process of removal of fluoride ion in drinking water. The process may be classified broadly into two categories, namely, (1) Additive methods so far tried for the removal of excess fluoride from water cab is broadly classified into four categories: (*a*) Adsorption methods, (*b*) Ion exchange methods, (*c*) Precipitation methods, and (*d*) miscellaneous methods. Some defluoridation techniques developed to control fluoride content in water are reverse osmosis, adsorption using sun flower plant dry powder, stem of phytomass, holly oak, neem bark powder, activated cotton jute carbon, bagasse ash, burnt bone powder, phosphate-treated saw dust, bone char, etc. as adsorbents, nalgonda technique, activated alumina process and ion exchange process. However, due to high cost or lower efficiency or non-applicability on mass scale these techniques are not much in use. Defluoridation is normally accomplished by precipitation (coagulation) and adsorptive process. Fluoride removal precipitation using alum, iron, lime and magnesium compound, and calcium phosphate was investigated by several researchers such as Bulusu *etal.* (1979), Nawlakhe *etal.* (1974), Parker *etal.* (1975), Bulusu and Nawlake, (1988), Boruff, C.S., (1934), Culp and Stotemberg, (1958).

Sequential addition of lime, bleaching powder (for disinfection) and alum or aluminum chloride known as the Nalgonda Technique (Rao and Mamatha (2004) is found effective, economic and is extensively used in India for defluoridation of community water supplies; but has some problems associated with operation and sludge disposal. In this respect removal of fluoride by adsorption onto activated carbon. Activated alumina (AA) and other low cost and natural materials, Alumina (Rao and Karthikeyan (2008), Flyash, red-soil and serpentine (Chidambaram *etal.* 2003) and (Bulusu *etal.* 1979), Donnan Dialysis (Durmaz, 2005), activated clay (Gogoi and Baruah, 2008), lime stone, Bentonite, Charfines, lignite and nirmali (strychnos potatorium) seed extract, bone char (Killedar, and Bhargava, 1988) serpentine and sulfonated saw-dust carbon (Bhakuni, 1970).

NON-CONVENTIONAL DEFLUORIDATION TECHNOLOGIES

Major defluoridation devices have been referred with application, significant contribution in affected society in urban, semi-urban and rural areas. Most of them are described as follows :

Activated Alumina

Activated alumina is an aluminum oxide, that is highly porous and exhibited high surface area. The crystal structure of alumina contains cation lattice discontinuities giving rise to localized area of positive charge. This makes alumina attractive adsorbent. It also does not shrink, swell, soften nor disintegrated when immersed in water. The activated alumina was proposed for the first time for defluoridation of water for domestic use in the 1930s. Since then the activated alumina has become a popular defluoridation method. The maximum adsorption capacity of activated alumina for fluoride is found to be 3.6mgF/g of alumina. Alumina is first treated with HCl to make it acidic. This acidic form of alumina when contacted with fluoride ions displaces the chloride ions and gets attached with the alumina. It has been found that maximum adsorption takes place at certain pH range, at pH > 7, silicate and hydroxide become stronger competitor of the fluoride ions for exchange sites on activated alumina and at pH > 5, activated alumina gets dissolved in acidic environment leading to loss of adsorbing media (Shrivastava *etal.* 2009). A notable feature of this study is domestic application of Activated Alumina (AA) for removal of fluorides. Studies with domestic candle type water filter by filling the hollow portion of the candles with AA were found to be effective in removal of fluoride. Therefore in endemic areas, ordinary domestic candle filters may be used for removal of fluorides as well from water especially when public supplies are non-functioning or inadequate (Srimurali *etal.* 2008).

Red Mud

It is very fine material consisting particle size is generally below 75μ and high specific surface area (around $10m^2/gm$), which is produced during the Bayer process for alumina production (Hind *etal.* 1999). It is the insoluble product after bauxite digestion with sodium hydroxide at elevated temperature and pressure. It is mainly composed of iron oxides and has a varity of elements and mineralogical phases. The removal of fluoride from aqueous solution by using the original and activated red mud forms has been studied by many researchers (Lopez *etal.* 1998). The fluoride adsorption capacity of activated form has been found to be higher than that of the original form. The adsorption is highly dependant on pH. Research has revealed that the maximum adsorption of fluoride is at pH 5.5. For $pH < 5.5$ fluoride removal decreases sharply. It was found that the sufficient time for adsorption equilibrium of fluoride ions is 2 hour. The possibility of removal of fluoride ion by using red mud is explained on the basis of the chemical nature and scientific interaction with metal oxide surfaces (Cengelo *etal.* 2002).

Nalgonda Technique

The Nalgonda Technique was developed by National Environment Engineering Research Institute (NEERI), Nagpur after extensive testing of many material and process (Nawlakhe *etal.* 1975). The Nalgonda Technique involves addition of alum (Aluminium sulphate), lime (Calcium oxide) and bleaching powder followed by rapid mixing flocculation, sedimentation, filtration and disinfection. Induced by a subsequent gentle stirring, "cotton wool" like flocks develops which is removed by simple setting (Eli *etal.* 1996). In the guidelines for household defluoridation published by NEERI in 1987, alum is to be added as a 10% solution to a 40 liter bucket equipped with a tap. This is the modification of the previously described method, where alum would be added as tablets. The dose of aluminium salt increases with increase in the fluoride and alkalinity levels of the raw water. The dose of lime is empirically $1/20^{th}$ of the dose of aluminium salt. Lime facilitates forming dense floc for rapid settling. Bleaching powder is added to the raw water at the rate of 3mg/l for disinfection. Nalgonda technique is effective even when the fluoride concentration is above 20mg/l. it is possible to lower the concentration of fluoride below 1mg/l in the treated water and adequate alkalinity is essential to reduce the fluoride level to 1mg/l or below 1 (Hind *etal.* 1999).

Bagasse Fly Ash

The bagasse fly ash, an industrial solid waste of sugar industry. The sugar industry is one of the most important agrobased industries. The material was treated with hydrogen peroxide at 60 °C for 24 hour to oxidize the adhering organic matter then washed with de-ionized water, dried at 100 °C, powdered and sieved to produce desired particle size and then batch experiments were carried out at different time of interval, different concentration, solution pH, adsorbent dose, particle size and temperature have been studied the 90% removal of cadmium and nickel is occur at 80 minute and at 6.5 pH. Bagasse fly ash was found to be stable in water, acids and bases. The composition of the adsorbent was SiO_2-60.5%, Al_2O_3-15.4%, CaO-2.90, Fe_2O_3-4.90 and MgO-0.81. Bagasse fly ash show maximum adsorption with the presence of iron oxide and magnesium oxide. It's an inexpensive and effective adsorbent for removal of nickel and cadmium from wastewater (Gupta *etal.* 2003).

Fly Ash

It's a major solid waste by-product from coal-fired power plant. The fly ash was sieved through different size of mesh. The composition of fly ash was SiO_2- 53.44, Al_2O_3-22.72, CaO-7.23, Fe_2O_3-4.48 and MgO-3.33. The removal of nickel from wastewater using flyash has been studied by batch equilibrium technique (Kannan, 1991). For each experiment, 100ml of nickel solution was taken in series in different BOD bottles and fixed doses of fly ash were added in each bottle. The pH of the solution was adjusted and kept constant. These bottles were shaken for a prescribed constant. These bottles were shaken for a prescribed contact time. Bottles were taken out and solutions from each bottle were taken out and solutions from each bottle were filtrated and filtrates were used for analysis. Fly ash shows high sorption capacity due to presence of metal oxides. The maximum adsorptions occur at pH-3. The percentage of removal of nickel depends upon the contact time and dose of adsorbent. Adsorption increases with increase in contact time and increase in amount of fly ash.

Brick Kiln Ash

It is not in use at present. It can be obtained from Agriculture Research Institute and available on free of cost but its transportation cost should be taken into account for any techno-economic study. The ash sample was washed with distilled water to their surface dust and dried in sunlight. The sample was sieving throw sieves of different size. Major component of brick kiln ash are alumina, silica, iron oxide, calcium oxide and residual carbon. All experiments were carried out at ambient temperature (25°C) in batch mode. The experiment was run in different flasks using a flask shaker at varied speed 100 to 800rpm and at different pH 83.22% COD reduction occurs. The smaller size particles with larger surface area have given better performance for the adsorption. The brick kiln ash is available as discarded material. It shows a reasonable potential to treat the domestic wastewater for decentralized applications in rural areas and small towns (Devi and Dahiya, 2006).

CONCLUSION

Nevertheless, the present indigenous water purification system have contributed by using low- cost locally available natural material for removing excess fluoride from drinking and domestic water especially for the rural population in Asian and other developing countries. In order to improve on these techniques, further detailed scientifically investigations are needed and are being undertaken. In spite of certain technical drawbacks application of the column technique is going to be very significant since most of the rural populations live below the poverty line and cannot afford to used daily treated or bottle water for daily consumption.

REFERENCES

Boruff, C.S., (1934) : Removal of fluorides from drinking water. *Industrial Engineering Chemistry, 26(1) : 69-71.*

Bulusu, K.R., Sundaresan, B.B., Pathak, B.N., Nawlake, W.G., Kulkarni, D.N. and Thergaonkar, V.P., (1979) : Fluorides in water, defluoridation methods and their limitations, *J. Institute Engineers (India)- Environmental Engineering Division 60 : 1-25.*

Culp, R.L. and Stotemberg, H.A. (1958) : Fluoride reduction at La Crosse, Kansas. *J. American Water Works Association 50(3) : 427.*

Chidambaram, S., Ramanathan, A.L. and Vasudevan, S. (2003) : Fluoride removal studies in water using natural materials.*Water SA, 29, (3), 339-343.*

Ministry of Water Resources : Problems facing the water/irrigation sector. *http://www.wrmin.nic.in/.*

Devi Rani and Dahiya, R.P., (2006) : Chemical oxygen demand (COD) reduction in domestic wastewater by fly ash and brick kiln ash. *Water, air and soil pollution, 174 : 33-46.*

Durmaz, F., Kara.H, Cengeloglu, Y., and Ersoz, M., (2005) : Fluoride removal by donnan dialysis with anion exchange membranes. *Desalination, 177, 51-57.*

Eli Dahi, Felix Matlo, Balthazar Njau, and Bregnhj Henrik (1996): Defluoridation using Nalgonda technique in Tanzania. *22nd WEDC conference. New Delhi, India. : 266-268.*

Gupta, V.K., Jain, C.K., Ali Imran, Sharma, M., and Saini, V.K., (2003): Removal of cadmium and nickel from wastewater using bagasse fly ash-a sugar industry waste. *Wat. Res, 37 : 4038-4044.*

Gogoi, Pradip K. and Baruch Ruby (2008): Fluoride removal from water by adsorption on acid activated kaolinite clay. *Indian Journal of Chemical Technology, 15, pp. 500-503.*

Hind, A.R., Bhargava S.K. and Grocott, S.C., (1999) : The surface chemistry of Bayer process : a review. *Phy. chem. Eng. aspects, 146 : 359-374.*

Indian Standards (1991) : Indian Standards drinking water specifications. pp. 65

Kannan, N. (1991) : Removal of nickel by adsorption on fly ash. *Indian. J. Environ. Protection, 11(7) : 514-518.*

Killedar, D.J. and Bhargava, D.S., (1988) : An overview of defluoridation methods (part-1), *J. Institute Public Health Engineers (India) 1 : 6-13*

Lopez E., Soto B., Arias M., Nunez A., Rubinoz D. and Barral M.T., (1998) : Adsorption properties of red mud and its use for wastewater treatment. *Colloids and surfaces,* 32(4) : 1314-1322.

Nagendra Rao, C.R. (2003) : Fluoride and environment- a review. Proceedings of third International conference on environment and health, Chennei, India. Department of geography, *University of Madras and Faculty of Environmental Studies, York University, pp. 386-399.*

Nawlakhe, W.G., Kulkarni, D.N., Pathak, B.N. and Bulusu, K.R. (1974): Defluoridation of water with alum. *Indian J. Environmental Health 16*(1) : 32-38.

Nawlakhe, W.G., Kulkarni, D.N., Pathak, B.N. and Bulusu, K.R. (1975): Defluoridation of water by Nalgonda technique. *Indian J. Environmental Health 17(1) : 26-65.*

Parker, C.L. and Fong, C.C. (1975) : Defluoridation of water with activated alumina: batch operations. *Indian J. of Environmental Health 17(3) : 262-299.*

Rao, Sudhakar M., and Mamatha, P. (2004) : Water quality in sustainable water management. *Current Science, 87(7), 942-947.*

Rao, C. R. Nagendra, and Karthikeyan, J. (2008) : Adsorption of fluoride by gamma alumina. *IWTC12.*

Rao Sudhakar M. and Mamatha, P. (2004) : Water quality in sustainable water management. *Current Science, 87(7), 942-947.*

Shrivastava, Brajesh K. and Vani, A. (2009) : Comparative study of defluoridation technologies in India. *Asian J. Exp. Sci., 23(1) : 269-274.*

Srimurali, M. and Karthikeyan, J., (2008) : Activated Alumina: defluoridation of water and household application- a study. *Twelfth International Water Technology Conference. pp-435.*

Susheela, A.K. (1999) : Fluorosis management Programme in India. *Curr. Sci, 77 (10) : 1250-1255.*

Susheela, A.K. (2001) : A treatise on fluorosis, edited. A.K. Susheela, F.R.R.D.F., *New Delhi, India pp*-654.

Tripathi, Preeti and Sultana Naheed (2007) : Fluoride content of ground water and prevalence of dental, skeletal and neurological stage of fluorosis in Tehsil Purwa of district Unnao, Uttar Pradesh. *Indian. J. Environ. Protection,* 27(8), 737-739

Yunus Cengelo Lu, Esengul Kir and Mustafa Ersoz (2002) : Removal of fluoride from aqueous solution by using red mud. *Separation and purification technology. 28(1) : 81-86.*

Piekos, R. and Paslawaska, S. (1999) : Fluoride uptake characteristics of fly ash pp-9. *International Society for Fluoride Research, 32(1), 1-6.*

14

Public Health Aspects of Wastewater Expulsion on Land

Sangeeta Madan and Akansha Bhatia

INTRODUCTION

Wastewater treatment plants generate effluents that must be disposed of safely and economically. The main objectives of land application of wastewater are further effluent treatment, groundwater recharge, and the provision of nutrients for agricultural crops. During land treatment of wastewater effluents, biological and chemical pollutants are removed by physical (settling, filtration), chemical (adsorption, precipitation) and biological (*e.g.*, plant uptake, microbial transformations, biological decay) processes.

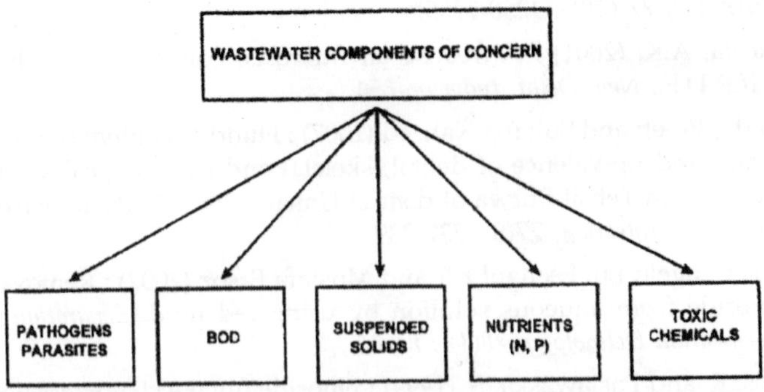

Problems associated with land application of wastewater and sludges are contamination of groundwater, soils, and crops with pathogens, heavy metals, nitrate, and toxic, mutagenic, carcinogenic, and endocrine disrupting organic compounds. Crops and grazing animals may become contaminated with heavy metals (*e.g.*, Cd, Zn, Cu, Ni) present in sludge. Nitrate from sludge and wastewater effluents may also contaminate groundwater used for drinking water supply (Bouchard *etal.* 1992). Lipophilic trace organic compounds are recalcitrant to biodegradation and may accumulate in the fatty tissues of grazing animals. Pathogenic microorganisms and parasites may survive on crops, particularly leafy vegetables (*e.g.*, spinach, lettuce), irrigated with raw wastewater or wastewater effluents. Market vegetables may become contaminated with pathogens and parasites (*e.g.*, *Salmonella*, *Giardia lamblia*, *Entamoeba histolytica*) during irrigation, transportation, and subsequent handling (Ercolani, 1976; Kowal, 1982; Pude *etal.* 1984). Survival of pathogens on crops depends on the type of pathogen, the type of crop, and environmental conditions (*e.g*, sunlight, temperature, wind, rain) and may vary from days to months (Rose, 1986). The concentration of parasite eggs applied to land may be high and may reach levels of 6000–12,000 viable eggs/m^2/year. Parasite eggs, especially those of Ascaris, may persist in soils for years (5–7 years or more) (Little, 1986). Therefore, the irrigation of processed food crops with wastewater effluents should be stopped some weeks (4–6 weeks) before harvesting to allow a sufficient inactivation of potential pathogens. Furthermore, irrigation of food crops that are eaten raw should be restricted. Crop contamination (leaves and fruits) is limited when wastewater effluents are applied via subsurface drip irrigation.

TRANSPORT OF PATHOGENS THROUGH SOILS

The detection of pathogenic microorganisms in groundwater has triggered research on their fate (transport and persistence) in the soil matrix.

BACTERIAL TRANSPORT THROUGH SOILS

Owing to their size, bacterial pathogens may be filtered out during their transport through the soil matrix. Bacterial removal by soils is inversely proportional to the particle size of soils. Furthermore, bacteria are charged biocolloids that can adsorb to soils, providing there are optimal conditions that favor their attachment to soil particles. These conditions are the presence of cations (for example, iron-coated sand adsorbs up to 6.9–108/g sand; Mills *etal.* (1994), presence of clay minerals that provide adsorption sites, low concentrations of soluble organics, and low pH conditions (Gerba and Bitton, 1984). However, heavy rainfall promotes bacterial transport through soils (Lamka *etal.* 1980; Zyman and Sorber, 1988). Laboratory experiments with sludge soil mixtures (equivalent to an average loading rate of 0.05 tons/ha) challenged with bacterial indicators have shown that only heavy rainfall (12.3 cm/day) promotes significant downward transport of the bacteria to the bottom of an 8 in. deep column. Lower rainfall did not cause significant migration of the bacterial cells (Zyman and Sorber, 1988). Under field conditions, indicator bacteria are efficiently retained by soils and are detected at low levels in groundwater. They are not suitable indicators of virus transport into groundwater (Alhajjar *etal.* 1988). After sludge application to land, most of the sludge-associated bacteria are retained at the soil surface and their transport to groundwater is unlikely (Liu, 1982).

VIRUS TRANSPORT THROUGH SOILS

The major factors controlling virus transport through soils are soil type, virus serotype, ionic strength of soil solution, pH, soluble organic compounds present in wastewater effluents, and hydraulic flow rate. Retention of virus by the soil matrix is governed primarily by adsorption to surfaces, particularly those provided by clays and other minerals such as hematite or magnetite. Both electrostatic and hydrophobic interactions are implicated in virus adsorption to soils (Bales *etal.* 1995; Bitton, 1980b; Bitton and Harvey, 1992; Gerba, 1984; Lipson and Stotzky, 1987; Shields and Farrah, 2002).

Adsorption of virus to soils is affected by soil texture. Clay soils generally have a greater virus-retaining capacity than sandy ones. Muck soils display a low affinity for viruses (Meschke and Sobsey, 2003; Scheuerman *etal.* 1979; Sobsey *etal.* 1980). Adsorption is inhibited and, thus, virus transport is promoted, by soluble organic materials found in wastewater effluents and sludges and by humic and fulvic acids (Dizer *etal.* 1984; Schaub and Sorber, 1977; Scheuerman *etal.* 1979). Domestic wastewater contains surfactants that result from the use of laundry detergents. These surfactants inhibit virus adsorption to surfaces, thus increasing transport through soils (Chattopadhyay *etal.* 2002). Virus transport is also promoted by increasing hydraulic flow rate, or by increasing pH (Bales *etal.* 1995; Lance and Gerba, 1980; Vaughn *etal.* 1981).

PERSISTENCE OF PATHOGENS IN SOILS

Persistence of Bacterial Pathogens

The main factors affecting the survival of pathogenic bacteria in soils are temperature, moisture content, sunlight, pH, organic matter, bacterial type, and antagonistic microflora, which include indigenous soil bacteria and predatory protozoa (Bitton and Harvey, 1992; Foster and Engelbrecht, 1973; Gerba and Bitton, 1984; Gerba *etal.* 1975a, b). Soil desiccation is important in the control of bacterial survival in soils. The decay rate of total and fecal coliforms in soil sludge mixtures increases as the mixture is allowed to dry naturally (Zyman and Sorber, 1988).

Persistence of Viruses in Soil

The two decisive factors controlling the persistence of viruses in soils treated with wastewater effluents or sludge are soil temperature and moisture (Bitton, 1980a). Recent studies with sludge-amended desert soils confirmed the importance of these two environmental factors (Straub *etal.* 1992).

Persistence of Pathogens in Aquifers

Groundwater monitoring for the past 20 years has shown that about half of all waterborne disease outbreaks are associated with contaminated groundwater. Enteric viruses are the pathogens most likely found in groundwater. Gastroenteritis and hepatitis A outbreaks due to the consumption of untreated groundwater have been documented (Barwick *etal.*, 2000; Beller *etal.* 1997; Bowen and McCarthy, 1983; De Serres *etal.* 1999; Ha fliger *etal.* 2000). Generally, more outbreaks occur in noncommunity than in community systems using groundwater. As discussed previously, contamination may be caused by groundwater recharge, land application of wastewater effluents, or septic tank effluents. Enteric viruses (*e.g.*, enteroviruses, hepatitis A virus, Norwalk virus, rotaviruses) have been isolated from groundwaters around the globe (Abbaszadegan *etal.* 2003; Bitton and Farrah, 1986; Farrah and Bitton, 1990).

The Groundwater Disinfection Rule was released by the U.S. EPA to address the problem of groundwater contamination by viruses and other microbial pathogens. This piece of legislation requires disinfection (*e.g.*, chlorination, ultraviolet irradiation) for all community and noncommunity public water systems using groundwater, and it sets a maximum contaminant level goal (MCLG) of zero for viruses (Grubbs and Pontius, 1992).

Persistence of viruses is generally higher in groundwater than in surface waters. Decay rates of several enteroviruses in groundwater vary between 0.0004 and 0.0037 h21 (Bitton *etal.* 1983; Jansons *etal.* 1989a). Temperature is the most decisive factor that controls virus survival in groundwater (Jansons *etal.* 1989a; Yates and Gerba, 1984; Yates *etal.* 1985).

Land Treatment Systems

Land treatment systems for wastewater effluents are capable of removing microbial pathogens and parasites, BOD, suspended solids, nutrients (nitrogen and phosphorus), toxic metals, and trace organics. Suspended solids are removed by filtration and sedimentation. Soluble organic compounds are removed by microbial action, particularly by biofilms developing on soil particles. Nitrogen is removed by sedimentation–filtration (*e.g.*, particulate-associated organic nitrogen), adsorption to soil and volatilization (*e.g.*, removal of NH_4 as NH_3), uptake by crops, and biological denitrification (Lance, 1972). Phosphorus is removed by adsorption to soil particles, chemical precipitation and uptake by vegetation. There is concern over the uptake of trace organics found in wastewaters and sludges by agricultural crops and animals (Majeti and Clark, 1981).

Slow-Rate Irrigation System : Slow-rate irrigation systems are the most frequently used land treatment systems. They provide essential nutrients to satisfy the growth requirements of agricultural crops. Pretreated wastewater is applied to land (soil texture ranges from sandy loams to clay loams), by sprinkler or surface distribution, at a relatively slow rate (0.5–6 m/year) (weekly loading rate of 1.3–10 cm) and serves as a source of nutrients for forage (*e.g.*, alfalfa, bermuda grass, ryegrass), or field crops (*e.g.*, corn, cotton, barley). (U.S. EPA, 1981). Physical, chemical, and biological processes contribute to the treatment of the incoming wastewater. This system provides the highest treatment potential and removes approximately 99 per cent of BOD, suspended solids, and coliforms. Some limitations of the slow-rate irrigation system are land cost, high operating cost, and transport of wastewater to the treatment site.

Overland Flow System : Wastewater is applied at a rate of 3–20 m/year or more and flows down a grass-covered slope (2–8 percent) with a length of 30–60 m (U.S. EPA, 1981. The most suitable soils are clay or clay-loamy soils with low permeability (equal to or less than 0.5 cm/h) to limit wastewater percolation through the soil profile. The treated effluent is captured in a collection channel. Nutrients (N, P, and BOD), suspended solids, and pathogens are removed as wastewater flows down the slope. This system achieves 95–99 per cent removal of BOD and suspended solids. Removal of nitrogen in overland flow systems is due to nitrification followed by denitrification, and uptake by agricultural crops. Phosphorus is removed via adsorption and precipitation.

Rapid Infiltration Systems : Wastewater is applied intermittently at high loading rates (6–125 m/year) onto a permeable soil (*e.g.*, sands or loamy sands) (U.S. EPA, 1981). The hydraulic pathway displayed shows that most of the applied wastewater flows to groundwater aquifers. The treated wastewater may be collected via recovery wells. The minimum depth to groundwater is 1 m during flooding periods and 1.5–3 m during dry periods. The treatment potential of rapid infiltration systems is lower than in slow-rate systems. Removal of nitrogen is generally low, but may be increased by encouraging denitrification. Denitrification requires adequate carbon levels (as found in primary effluents) and low oxygen levels, necessitating flooding periods as long as 9 days followed by drying periods of about 2 weeks.

Subsurface Drip Irrigation : Subsurface drip irrigation is practiced in arid areas to save water. It consists of dripping water near the root zone (30–50 cm below the soil surface) at a rate that depends on the plant water requirements. This practice helps avoid or minimize biological aerosol production, contamination of aerial parts of crops and wastewater flow to groundwater (Oron *etal.* 1991a, b). Using PRD1 and MS2 phages as model viruses, it was shown that subsurface drip irrigation of turfgrass can reduce the risk of contamination by potential viral pathogens when compared to sprinkler irrigation (Enriquez *etal.* 2003). Similarly, as compared to sprinkler irrigation, surface and subsurface drip irrigation were found to reduce vegetable contamination with Cryptosporidium oocysts or Giardia cysts (Armon *etal.* 2002).

BIODEGRADATION IN SOILS AND AQUIFERS

An Introduction to Bioremediation

There is a growing concern over soil and groundwater contamination with effluents from domestic and industrial wastewater treatment plants. Environmental biotechnologists are working at finding ways to reduce environmental contamination by using modern tools in microbiology, molecular ecology, chemistry, and environmental and engineering sciences (Sayler *etal.* 1991).

Bioremediation of Soils

There are four approaches to microbiological soil decontamination (Beaudette *etal.* 2002; Compeau *etal.* 1991;

Hanstveit *etal.* 1988): (1) *in situ* techniques, (2) addition of microorganisms or enzymes, (3) bioreactors for treatment of excavated soils, and (4) natural attenuation.

In Situ Decontamination Techniques

The contaminated soil is amended with nutrients (N and P) and plowed to provide oxygen. For example, the biodegradation of diesel oil in soil can be stimulated by remediation measures consisting of liming, addition of nitrogen and phosphorus, and tilling. This approach was found to reduce total hydrocarbons by 95 per cent, eliminate polycyclic aromatic hydrocarbons, resulting in complete detoxification in 20 weeks, as measured with Microtox (a microbial toxicity test) and the Ames test (a mutagenicity test) (Wang *etal.* 1990). In situ biorestoration, using above-ground biodegradation cells, resulted in 95 per cent removal of oil and grease in a clay soil (Vance, 1991). Some have proposed the stimulation of soil indigenous microflora by treating the soil with a structural analog of the chemical to be removed. For example, removal of 3, 4-dichloroaniline can be enhanced by stimulation of aniline-degrading microorganisms (You and Bartha, 1982).

Addition of Microorganisms or Enzymes to Soils (Bioaugmentation)

Addition of microorganisms and enzymes stimulates the biodegradation of xenobiotics in soils (Crawford and O'Reilly, 1989). Fungal laccases are polyphenol oxidases that catalyze the binding, by oxidative cross-coupling, of phenolic compounds to the humic fraction of soils and their subsequent immobilization and detoxification (Bollag, 1992). Fungal spores or mycelial fragments suspended in a sodium alginate hydrogel can be used to coat pellets of solid substrates for bioaugmentation of pentachlorophenol degradation in contaminated soils (Lestan and Lamar, 1996). Most efforts have been concentrated on the use of bacterial inoculants grown in large fermenters (*e.g.*, Arthrobacter, Rhodococcus chlorophenolicus, Flavobacterium) for the bioremediation of pentachlorophenol-contaminated soils. Under laboratory conditions at 308 °C, soil inoculated with 106 PCP-degrading Arthrobacter per gram dry soil, reduced the half-life of PCP from two weeks to less than one day (Edgehill and Finn, 1983b). Microbial immobilization on bark chips or their encapsulation in polyurethane or alginate enhances their PCP-degrading ability, as well as their resistance to PCP toxicity (Crawford *etal.* 1989; Salkinoja-Salonen *etal.* 1989). Microcosm studies have shown that bacterial cells immobilized in gellan gum microbeads (16–53 mm diameter) can degrade gasoline, leading to 30–50 per cent removal of the contaminant (Moslemy *etal.* 2002). These results can be applied for the in situ bioaugmentation of contaminated aquifers. The use of commercial bacterial inoculants, however, did not enhance the biodegradation of hydrocarbons in a site contaminated with bunker Carbon fuel (Compeau *etal.* 1991). Successful soil bioremediation was obtained by amending a petroleum hydrocarbon-polluted soil with an enriched culture of indigenous hydrocarbon-degrading bacteria. A biological activated carbon (BAC) system, seeded with indigenous bacteria and fed hydrocarbons and nutrients, was used to enrich for the indigenous bacteria. Natural attenuation resulted in 1.7 per cent removal of hydrocarbons in 32 days as compared to 42 per cent for enhanced bioremediation (Fig. 21.10; Li *etal.* 2000). In a similar study, bioaugmentation of soils with an enriched culture of indigenous microorganisms increased the biodegradation of petroleum hydrocarbons in by 20.4 to 49.2 per cent (Yerushalmi *etal.* 2003). Genetically modified microorganisms (GMM) are being considered for use in environmental bioremediation for cleanup of hazardous chemicals, biomining, and as biofertilizers and biopesticides in agriculture. Agricultural applications of GMMs include their use as biological control agents (*e.g.*, *Bacillus thuringiensis*), frost control, bioremediation of hydrocarbons and other xenobiotics, or for increasing nitrogen fixation in soils (*e.g.*, use of nitrogen-fixing Rhizobium strains) (van Overbeek and van Elsas, 2002; Ripp and Sayler, 2002). Such GMMs are applied to soil via injection, spraying, or as immobilized cells within carriers such as alginate. They can be transported vertically or laterally following rain events. Their transport from the site of application can be tracked, using cultivation-based (*e.g.*, plating on selective growth media) or cultivation-independent (*e.g.*, use of molecular probes) techniques. There are three main concerns over the application of GMMs to soils :

1. Fate of GMMs in Soils (Survival and Transport) : The GMMs must survive in sufficient numbers after soil inoculation in order to accomplish the task at hand. Experience from soil microbiology has taught us that non-GMMs microorganisms (*e.g.*, nitrogen-fixing bacteria such as *Rhizobium*, *Azotobacter*, or *Azospirillum*) released in soils for bioremediation or biocontrol purposes do not persist well, due to adverse conditions in soils (*e.g.*, limiting nutrients in soils, harsh environmental conditions) and competition with indigenous soil microorganisms. The GMMs may also display a reduced level of fitness due to the additional energy necessary for carrying the recombinant genes. Culture-based and molecular-based techniques have shown that GMMs or their genetic constructs can, however, persist in soils for years after field release. Another concern is the transport of GMMs from the application site. Transport of GMMs may be due to aerosolization during application to soils, water, or animal vectors.

2. Ecosystem Alteration : Introduction of GMMs into soils may affect the diversity of indigenous microbial populations, possibly leading to changes in primary production and nutrient cycling in soils. Owing to conflicting results obtained so far, more research is needed on this topic.

3. Containment of the Released GMMs : Another important concern is the transfer of recombinant DNA in GMMs to indigenous microbial species in soils, possibly leading to dire ecological consequences. The possibility of gene transfer in soils mostly by processes such as conjugation and transformation has been demonstrated (Nielsen *etal.* 1997; van Elsas *etal.* 1998). To reduce the chances of gene transfer, the foreign genes should be inserted in the chromosome rather than plasmids. Culture-based and molecular techniques can be used to assess the changes in indigenous microbial populations following the introduction of GMMs. From a public safety viewpoint, biocontainment genes (*i.e.*, "suicide genes") can be inserted into

GMMs to ensure that they are killed when they spread out from the application site (*e.g.*, rhizosphere). Examples of biocontainment genes are the gef, hok, relf genes derived from *E. coli*, and the *nuc* gene derived from Serratia marcescens. These genes can be placed under the control of a nutritional or environmental factor such as carbon limitation or temperature. This will ensure that the GMM will not escape from the application site (Ahrenholtz *etal.* 1994; Knudsen *etal.* 1995; van Overbeek, 1998; van Overbeek and Elsas, 2002).

Improvement of molecular techniques to monitor relatively small numbers of GMMs or their constructs in soils will help in enhancing our knowledge of their ultimate fate in soils and in predicting the risks associated with their release into the environment.

Bioreactors for Treatment of Excavated Soils

These bioreactors include soil slurry reactors, composting, and land treatment units. The removal of PCP and hydrocarbons, using land treatment units and soil slurries, has been demonstrated. Biodegradation of PCP in soil slurries was enhanced by the addition of a PCP microbial consortium (Compeau *etal.* 1991). Composting was also explored as a technology for bioremediation of soils contaminated with explosives such as trinitrotoluene (Myler and Sysk, 1991). More than 90 per cent of the explosives were removed from contaminated soils within 80 days (Williams *etal.* 1989).

Natural Attenuation

Natural attenuation relies on natural processes to bring about the biodegradation of xenobiotics in soils. This includes physical processes (*e.g.*, volatilization, dilution, sorption), abiotic and microbial degradation of the contaminants.

Treatment Strategies for Aquifers

There are three basic approaches to the treatment of contaminated aquifers (Bouwer *etal.* 1988) : physical containment, above-ground treatment, and *in situ* bioremediation. A new approach that is being considered is the natural attenuation of contaminants in aquifers. This approach requires no intervention and consists of monitoring the change in contaminant concentration with time.

Physical Containment

Physical containment includes the use of temporary physical barriers to slow down or halt contaminant movement. This approach has been adopted with some success in hazardous waste sites.

Above-Ground Treatment (Pump and Treat Technology)

The contaminated water is pumped out from the aquifer via extraction wells, and treated above-ground, using one of several treatment processes. However, it is difficult to extract chemical contaminants adsorbed to the aquifer matrix. Since the removal of organic contaminants by this technology is relatively slow, this approach may be regarded as a means for preventing further migration of the contaminant in the aquifer (Mackay and Cherry, 1989). The major available treatment technologies are air stripping to remove volatile organic compounds, adsorption to granular activated carbon (GAC), ultrafiltration, oxidation with ozone/UV or ozone/H_2O_2, activated sludge, and fixed-film biological reactors (Bouwer *etal.* 1988). For example, fixed-film bioreactors, when using sand as the matrix and methane or natural gas as the primary substrate, are capable of removing up to at least 60 per cent of trichloroethylene from polluted water. They also remove more than 90 per cent of TCE and TCA from vapor streams generated by air stripping of polluted groundwater (Canter *etal.* 1990).

In Situ Bioremediation

In situ bioremediation is the enhancement of the catabolic activity of indigenous microorganisms by adding nutrients and, if necessary, oxygen (added as air, pure oxygen, or hydrogen peroxide). This approach allows the degradation of the target compound *in situ*, avoiding the need for excavation and treatment of the contaminated material. *In situ* treatment depends on aquifer characteristics (*e.g.*, permeability as measured by hydraulic conductivity), contaminant characteristics, oxygen level, pH, availability of nutrients, redox conditions, and the presence of microorganisms able to degrade the contaminant under consideration (Alexander, 1985; Bedient and Rifai, 1992; McCarty *etal.* 1984; Rittmann, 1987; Thomas and Ward, 1989; Wilson *etal.* 1986). This approach has been used mostly for gasoline spills. Indigenous subsurface bacteria are able to grow on aromatic compounds such as naphthalene, toluene, benzene, ethylbenzene, p-cresol, xylene, phenol, and cresol, which are used as the sole source of carbon and energy (Brockman *etal.* 1989; Frederickson *etal.* 1991; Glynn *etal.* 1987).

Bioaugmentation

Bioaugmentation consists of adding specific microorganisms or mixed bacterial cultures to polluted aquifers to enhance the biodegradation of target contaminants (DeFlaun, 2002; Thomas and Ward, 1989). These cultures are obtained via traditional enrichment techniques, chemical mutagenesis or molecular based techniques. However, the use of the latter approach is still being debated. The added bacteria should have some or all the following desirable characteristics (DeFlaun, 2002). Ability to degrade the target contaminant, depending on the aquifer and type of contaminant, the organism should be chosen based on its ability to degrade the contaminant under aerobic or anaerobic conditions. Ability to produce constitutive degradative enzymes. Ability to migrate in the aquifer material, Adhesion-deficient bacteria should be selected because they can migrate further in the aquifer and help avoid or reduce plugging of the injection well. Other means to enhance bacterial transport in aquifers include the use of surfactant foams and liquid surfactants or the use of "ultramicro" bacteria, which travel more easily than full-size bacteria (starvation reduces bacterial size). Ability of the cells to store energy-rich materials such as poly-b-hydroxybutyric acid (PHB). This storage polymer will help in the survival of the added bacteria in the aquifer.

As an example, a bioaugmentation study was carried out in California, where commercial blends of hydrocarbon-degrading bacteria were used for the bioreclamation of soils and aquifers contaminated with petroleum hydrocarbons (von Wedel *etal.* 1988). The process consisted of the following :

1. Contaminated groundwater was treated above ground in chemostat bioreactors in the presence of the commercial bacterial cultures.

2. The treated groundwater was further amended with additional bacteria and was reinfiltrated into the contaminated soil for *in situ* treatment of both contaminated soil and aquifer. After several months of operation, this augmented biorestoration scheme resulted in 93 per cent removal of total hydrocarbons. For benzene, toluene, and xylene, the per cent removal rates were 85 per cent, 98 per cent, and 33 per cent, respectively (von Wedel *etal.* 1988). Bioaugmentation of an aquifer with a MTBE-degrading bacterial consortium and oxygen addition was successful in reducing MTBE concentrations to less than 0.001–0.01 mg/L (Salanitro *etal.* 2000). Other examples of field-scale demonstrations of bioaugmentation are given by DeFlaun (2002).

Factors Affecting Bioremediation of Soils and Aquifers

These factors include those pertaining to environmental conditions in soils or aquifers, contaminant characteristics, and the microorganisms.

1. Suitable hydrogeological features must allow extraction of contaminated groundwater (*i.e.*, recovery of the free contaminant phase) and reinjection and circulation of the treated water. Other hydrogeological characteristics that must be evaluated include depth and specific yield of the aquifer, permeability, and direction of groundwater flow.

2. Oxygen can be added to the aquifer by injecting pure oxygen, air, or hydrogen peroxide. Hydrogen peroxide, at a concentration of up to 750 mg/L, was used as a source of oxygen

for the bioremediation of a shallow sandy aquifer contaminated with fuel spills in Michigan (Wilson *etal.*, 1989). Since H_2O_2 may be toxic to aquifer microflora, some recommend a gradual increase in its concentration to avoid toxicity. Solid peroxides (*e.g.*, Na percarbonate, Ca peroxide) have also been considered as sources of oxygen for *in situ* subsurface bioremediation. Toxicity is reduced when these solid peroxides are encapsulated in polyvinylidene chloride (Vesper *etal.* 1992). Nitrate, and perhaps nitrous oxide, can also serve as electron acceptors. Aromatic hydrocarbons (*e.g.*, toluene, xylene, ethylbenzene) are biodegraded under denitrifying conditions in aquifer materials (Evans *etal.* 1991; Hutchins, 1989; 1991; Hutchins *etal.* 1991). Sulfate is another useful electron acceptor for the biotransformation of halogenated organic compounds (Cobb and Bouwer, 1991). Benzene bioremediation by sulfate-reducing bacteria has been demonstrated in groundwater under field conditions (Anderson and Lovley, 2000). In a petroleum-contaminated aquifer, the *in situ* anaerobic biodegradation of BTEX (benzene, toluene, ethylbenzene and xylene) compounds was enhanced by the combined injection of nitrate and sulfate into the aquifer. The use of oxygen in conjunction with other electron acceptors (*e.g.*, nitrate, sulfate) for bioremediation of aquifers could improve the biodegradation of hydrocarbons in aquifers (Cunningham *etal.* 2001). Addition of inorganic and organic nutrients. Sometimes, it is necessary to add nitrogen and phosphorus to enhance biodegradation in soils and aquifers (Thomas and Ward, 1989). Methane and air can enhance the biodegradation of chlorinated hydrocarbons (*e.g.*, trichloroethylene) by methanotrophs.

Contaminant Characteristics

Parameters influencing the biodegradation in soils and aquifers include xenobiotic concentration, solubility, hydrophobicity, toxicity to indigenous or added microorganisms, sorption to inorganic or organic solids, and bioavailability.

Microbiological Considerations

Bioremediation is affected by the presence in soils or

aquifers of sufficient numbers of microorganisms capable of degrading the xenobiotic. Molecular techniques played for assessing microbial community structure and activity in Bioremediation. Several methods are available for assessing the impact of bioremediation and are as follows (Beaudette *etal.* 2002; Widada *etal.* 2002) :

1. DNA-Based Methods : These methods allow the detection of genes or gene sequences in DNA extracted from environmental samples, using specific DNA probes. Greater sensitivity of these probes can be obtained by PCR amplification. These probes can be used in combination with Most Probable Number (MPN) techniques to follow the biodegradation of a given xenobiotic in soils or water.

2. RNA-Based Methods : rRNA probes are hybridized with extracted target RNA and give information on community activity and are used for the identification of potential degraders of a given contaminant. Furthermore, the detection of mRNA gives information about the expression of the gene involved in the biodegradation of a given xenobiotic.

3. Genetic Fingerprinting Methods : Following PCR amplification, products with the same length but slightly different sequences are identified. The DNA fragments can be resolved by using denaturing gradient gel electrophoresis (DGGE).

4. Reporter Gene : (*i.e.,* marker genes) systems and immunological techniques for tracking degrading bacteria in bioaugmentation projects. In bioaugmentation operations one needs to track the persistence, transport pattern, and activity of the added bacteria.

REFERENCES

Bitton, G. (2010) : Public Health Aspects of Wastewater and Biosolids Disposal on Land. In : Wastewater microbiology. John Wiley & Sons, Inc., Hoboken, New Jersey.

Rose, J.B. (1986) : Microbial aspects of wastewater reuse for irrigation. *CRC Critical Reviews in Environmental Control* 16(3) : 231-256.

Shuval, H.I. (1991) : Health guidelines and standards for wastewater reuse in agriculture : historical perspectives. *Water Science and Technology* 23(10/12) : 2037-2080.

Web resources

http://www.soil.ncsu.edu/publications/Soilfacts/AG-439-03/ (land application of sludge from North Carolina State University)

http://www.brownfieldstech.org/technology/bioremediation/ (lists bioremediation

technologies) http://www.clu-in.org/remed1.cfm (EPA remediation site)

15

Biological Indicators of Water Pollution

B.B. Sandwar and Geetika

ABSTRACT

As the human population, their demand and their ability to fulfill the demand is increasing, causing industrialization, urbanization, deforestation etc. All these factors are responsible for water pollution. This water becomes unfit for all living organisms. But this polluted water first affects badly to the aquatic flora and fauna. Sometimes these aquatic organisms escape from polluted area for their survival or sometimes they can't resist the polluted condition and die but sometimes they adapt the polluted condition. In any case, being dumb they indicate about the condition of water bodies. Some pollutants get accumulated in the body of aquatic organisms and pass through the food-chain and food web causing unbalance to the whole water ecosystem and ultimately affects badly to human beings. But due to these sensitive aquatic organisms we are able to know the nature of polluted water.

Keywords

Bio-indicators, Water Pollution, Aquatic Organisms, Ecosystem, Biomonitoring Techniques.

INTRODUCTION

Biological indicators or Bioindicators of pollution are the organisms which indicate the specific types of pollutants in the aquatic environment by their presence or complete absence or few in number. Presence of a population of a single species is more reliable bioindicator than few individuals. Sometimes some indigenous organisms are completely absent from particular ecosystem, this also indicates the presence of pollution in that aquatic ecosystem. When plants are used as indicator of pollution in any aquatic environment it is called as Plant Indicators and when animals are used it is called as Animal Indicators. Biological indicators provide informations about the quality of particular aquatic environment in which pollutants are present by following ways :

(*i*) Metabolically

(*ii*) Morphologically, like chlorosis, stunted growth etc.

(*iii*) Cytologically

(*iv*) Physiologically

(*v*) By complete absence of certain species.

Levels of pollution vary in a stream, example from source of a stream to its mouth. This variation is due to temperature of water, types of pollutants, volume of pollutants, velocity of stream and also on the area whether it is plain or hilly. All these factors decide the presence or absence of specific types of species. It is a fact that pure or unpolluted water has great biodiversity but polluted water bodies contain less biodiversity and most pollution tolerant species. Aquatic organisms like macroinvertebrates, insects or phytoplankton's are the most reliable indicators because they are very sensitive to pollutants, they live in water for longer period if they are living in pond or lake and they do not have option to leave that area and they can easily monitored. They are the integrators of sum of all condition of water body. Thus they provide all the information's about the health of the specific water ecosystem.

PLANT INDICATORS IN AQUATIC ENVIRONMENT

As the plants are static they give right information about the past and present state of aquatic environment due to presence of pollutants. Problem of water pollution is increasing tremendously due to population growth, industrialization, agricultural advancement and urbanization. Pollutants from various kinds of industries may be heavy metals, PCB's and other organic and inorganic substances. Different kinds of fertilizers and organic supplements are used in nearby agricultural field surrounding any pond or lake and are carried into it by runoff water causing luxuriant growth of algae. Thus presence of algal bloom in any water body indicates organic pollution or may be presence of nitrogen and phosphorous. This algal bloom causes suffocation to other organisms living in the same water body. Brown algae are the indicator of heavy metal pollution.

In alga like *Anabaena cylindrica* it was reported that there is significant increase in heterocyst frequency under cadmium stress. Also, in *Anabaena enequalis*, there is induction of abnormally long filaments and loss of cellular content due to cadmium pollution. Cains and Cains (1983, 1984) have been reported remarkable decrease in zygospore germination in Chlamydomonas in the presence of certain herbicides and insecticides. Some vascular plants are also good indicator of pollution and mostly they provide information about heavy metals. As the heavy metals do not degrade but accumulate in living bodies and some vascular plants can accumulate them. Panda and co-workers have reported formation of micronuclei in dividing root cells of *Eichhornia* plants growing in mercury polluted sites (Science Reporter, April-May 1989). Similarly in the leaves of *Elodea densa*, surface membrane and chloroplast structure were found damaged due to excessive accumulation of methyl mercury.

Various kinds of plants grow in polluted water due to their ability to withstand the unnatural condition or sometimes the pollutants become nutrient for them, like deposition of organic substances from dead animal bodies, fertilizers from nearby agricultural fields etc. support the luxuriant growth of some algae like *Utricularia, Chara, Wolffia* etc. Water polluted by sewage is indicated by presence of Diatoms. When there is massive growth of lower plants in any aquatic body occurs, after sometimes death of flora and fauna takes because of the deficiency of oxygen in the water and then there is growth of decomposers like bacteria and fungi takes place. Thus, bacteria like *Escherichia coli* are also bio-indicator of water pollution.

Animal Indicators of Pollution in Aquatic Environment

Some invertebrates accumulate trace elements efficiently by detoxification mechanism, such as intracellular compartmentalization or metal inactivation by binding to metallothioneins. Thus metal concentration in the animal's body indicates types of pollution in that water body. Sometimes due to metal pollution stress they start synthesizing metallothioneins. In addition to this, metallothioneins are present in different isoforms in the metal loaded organisms due to presence of different metals. Thus Metallothioneins of invertebrates of polluted water become indicator of water pollution caused by metals.

Some fishes like *Catla catla*, *Labeo bata*, *L. rohita* and *Natopterus natopterus* indicate presence of industrial pollution due to their complete absence in that water ecosystem. Thus they are pollution-intolerant organisms. Some pollution-tolerant species are leeches, midge larva, bullheads and carp. Some aquatic animals provide information about presence of heavy metals which are discharged by nearby industries in the water bodies. Because heavy metals do not degrade rather they accumulate in the aquatic plants and animals then pass through the food chain or food-web also. Thus, the whole food chain or food web becomes able to provide the presence of heavy metals or in other words they themselves become bio-indicator. In addition to this when industries nearby a water body release hot water into it temperature of whole water body increases thus suspended solid particles settle down and cover the eggs of fishes thus they get suffocated and die and population of fish becomes thin in the water body. Some amphibians are also sensitive to pollutants like pesticides, which are used in agricultural land close to water body and are reached into the water with flow of irrigated water and affects badly to these organisms. Thus their population starts decreasing and sometimes they give birth to deformed ones.

Biomonitoring Techniques

Techniques of Biomonitoring is based on the fact that environmental stress will cause significant change in the physical & chemical environment of the water and that changes are bound to cause ecological imbalance in the system. Thus, by measuring the extent of ecological imbalance, the severity of the adverse impact can be easily assessed. In fact, there are several methods available for assessment of Water quality using biological structures (Trivedi, 1981). Each method has it own application, advantages & disadvantages, which are summarized in the Table 1, from which it can be observed that a combination of both techniques is definitely preferable to either single one. The Biological methods could be useful for assessing the overall impact of of pollutants on the ecosystems. Such impact are generally independent of the time (s) during which the pollutants were discharged and this approach could help a lot in integral assessment of the water characteristics. This is the most significant advantage of biological techniques. They do however, suffer from some limitations also.

Table 1 : Advantage & Disadvantage of Different Water Quality Monitoring Techniques

Realm	*Performance of Chemical Monitoring*	*Performance of Biological Monitoring*
Precision (*i.e.* pollutant concentration Assessment)	Good	Poor
Discrimination (*i.e.* what kind of pollution?)	Good	Poor
Reliability	Poor	Good
Measure of ecological effects	No	Yes
Cost involved	Relatively High	Relatively Low

BIOLOGICAL MAPPING OF WATER BODIES

In fact, water quality mapping is the most important exercise for the various state pollution control boards and non-government agencies involved in these areas to establish magnitude of water quality degradation in different water bodies in the country. Central Pollution Control Board, Delhi has taken serious efforts in the direction of mapping of some of the Indian Rivers using biological information's.

CONCLUSION

Biological indicators of pollution leasing to biomonitoring has numerous advantages. It is highly desirable to use biomonitoring techniques in environmental impact assessment (EIA. Biomonitoring results can be successfully used in indentifing environmental problems, establishing priority for pollution control, efforts, identification and implementation of appropriate pollution control measures). In fact, the primary aims & mission of indicators are to characterize the current environmental status and to track or predict any significant change with a strong foundation of diagnostic research, and ecological indicator may also be used to identify major ecosystem stress. Bioindicators generally respond in a predictable manner to sressors from human disturbances. Assessing the condition of biological communities provides a basis both to determine biological potential as well as to measure success in achieving that potential biological indicators and related data are the basis for setting protection goals as well as reporting the effectiveness of management tools adopted.

REFERENCES

Trivedi R.C. (1981) : Use of Diversity, Index in Evaluation of water quality, proceedings of the WHO workshop on Biological indicators & environmental pollution. PROBES/6/1980-81. C.P.C.B., New Delhi.

CPCB, 1978-79 : Scheme for zoning & classification of Indian Rivers, Estuaries and coastal waters (Part one: Sweet water) ADSORBS/3/78-79, Central Pollution Control Board, Delhi.

Trainor, F.R., (1984) : Indicator Algal Assay Laboratory & fields approaches. In "Algae as Ecological Indicators, Ed. Shubert, Academic press, London PP 3-14.

Venkatraman, L.V., M.K. Krishi and Suvarnalatha, G (1994) : Algae as tool for bio monitoring and abatement of Pesticide pollution in aquatic system. J. Phykos 33 (1 & 2) 171-193.

Hosmani, P.S. (2002) : Biological Indicators of Water quality, Lake symp. on conservation, Restomanagement of aquatic ecosystems, (Dec., 9-13) org. by CES, IISC, Bangalore, pp. 73-76.

Trivedi, R.C., Bhardwaj, R.M. and Agrawal, Sanjeev (2008) : Biological, Monitoring of water quality in India Needs and constraints, Proc. of Tall 2007 : The 12th world has conference, pp. 1-6.

Trivedy, Trivedy, R.K. and Goel, P.K. (1986) : Chemical & biological methods of water pollution studies, Enviromedia publications, Karad (M.S.), India.

Trivedy, R.K. (1986) : Role of Algae in Biomonitoring of water pollution, Asian Environment, Phillipines.

Hosmani, S.P. and Bharti, S.G. (1982) : Use of Algae in Classification of Water bodies, J. Phykos, 21 PP. 48-51.

Palmer, C.M. (1980) : Algae and water Pollution, Castle House Publication Ltd. U.S.A.

The National Environmental Policy (2006) : Ministry of Environment & Forest, Govt. of India. Environment (Protection) Act, (1986), Gazette of India.

16

Spatial Distribution of Heavy Metals in Ground Water Regime of Panipat Industrial Area, India

D.S. Malik and P.K. Bharti

ABSTRACT

The present study is aimed towards an assessment of ground water quality at different place and different depth of water table, directly or indirectly affected by textile industrial effluents in industrial area of Panipat region in Haryana state of India. The ground water samples were collected from different depths and analyzed pH, TDS, BOD, COD and heavy metals. The ground water of shallow aquifer was found highly polluted in comparison to deep aquifer. Concentrations of trace metals of Fe, Mn, Pb, Cu and total dissolved solids of shallow ground water were recorded above the level of standard of WHO and BIS. Total toxic metals load in all shallow ground water samples was observed as exceed the limit *i.e.* 2.0 mg/l. Cu concentrations in all ground water samples and Fe contents in all samples were higher than WHO, USPHS, BIS. However, BOD and COD was found low in ground water for irrigation purposes but TDS load was found high for irrigation uses. The present observation reveals that sallow ground waters cannot be recommend for drinking purpose as well as irrigation, otherwise it may cause toxic effect on human beings, agricultural soil, plants, and animals.

Keywords

Ground water, textile effluent, heavy metals, vertical distribution, spatial variation.

INTORDUCTION

Environmental pollution by heavy metals such as cadmium, nickel, zinc, lead, copper etc., is accelerated dramatically during the last two decades due to mining, use of agricultural fertilizers, pesticides, traffic emissions, smelting, manufacturing, municipal wastes, industrial effluents and industrial chemicals etc. The air, soil and water are increasingly polluted by industrial pollutants such as organic and inorganic chemicals and toxic metals (Irshad *etal.* 1997).

Heavy metals are mainly responsible to induce tumor and mutations in animals at higher concentrations (Degraeve, 1981) and capable of causing genetic damage to germ cells of male and female animals including humans (Groten and Vanbladeren, 1994; Wagner, 1993). They are regarded as cumulative toxins, which through bio-magnification in plants affect human health (Groten and Vanbladeren, 1994). Heavy metals persist in soils and can be adsorbed in soil particles or leached into ground water. Human exposure to these metals through ingestion of contaminated food or uptake of drinking water can lead to their accumulation in humans, animals and plants (Malik *etal.* 2004).

In the race towards industrialization, a number of industrial estates were established near G.T. Road at Panipat, along with Jatal road in India. The major industries include textile spinning, textile dying and printing, carpet, blanket, etc., which are spread all over the Panipat and also in other cities of India. Effluents from these industries consist of toxic heavy metals, which further disperse in surroundings. The textile effluents have consisting high concentrations of heavy metals, organic pollutants and toxic colours, which may alter the surface water quality of the region. Toxic pollutants may percolate down via soil profile and reach in ground water, which ultimately cause the health hazards among human being and livestock after consumption as daily drinking requirements. The waste-water without any treatment may cause adverse effect on the health of human, domestic animals, wildlife and environment (Sharma *etal.* 1999). Thus, contaminated ground water has deteriorated immensely the drinking utilities, post agriculture irrigation and impacts on soil systems and crop productivity.

The quality of industrial effluents may vary with industrial processes in different industrial regions. The effluents discharged by different industries contain a high range of physico-chemical parameters like temperature, pH, conductivity, hardness, alkalinity, COD, TDS, cations (Ca^{2+} and Mg^{2+}) and anions (Cl''). These effluents from different industries also contain heavy metals and trace metals including chromium, cadmium, copper, lead, nickel, zinc, cobalt, magnesium, iron and arsenic (Malik and Bharti, 2007). Although India is blessed with one of the best canal irrigation systems of the world, there is a shortage of good quality surface irrigation water with this shortage being met by the conjunctive use of good quality irrigation water. So there is urgent need to quantify these toxic elements in the ground water prior to their use for irrigation, drinking and other domestic purposes.

LOCATION AND METHODS

Location

In north India, Panipat is situated between 29' 09' 50" and 29´ 50´ North latitude & 76°31' 15" and 77°12' 45" East longitude on globe along with the bank of river Yamuna with a height of 255 meters above sea level. Karnal, Sonipat, Jind, Kaithal and Uttar Pradesh bordered this district. Average population density of the urban area of Panipat city is 25, 278 per sq. km. Panipat town is located on the national highway no.1 about 90 km toward north of Delhi.

Dye houses in Panipat city are situated on Jatal road in industrial area and quite famous for handloom business in India and abroad. Dyeing units of textile industries consume a huge quantity of fresh water and discharged huge volume of their effluents compositely into a common effluent drain, ultimately which goes to a large pond of village Binjhole situated in the vicinity of industrial area (Malik *etal.* 2006). The ground water samples were collected from hand pumps, tube wells and deep bored wells of different depth of water tables at village Binjhole and adjoining agriculture fields **(Fig.1)**, where the effluent drain ends up into a large surface water body and percolate down into shallow aquifers, and divided into three categories viz. Shallow ground water (SGW) below 15-40 ft, middle ground water (MGW) from 41-80 ft and deep ground water (DGW) above 80 ft. Ground water samples were also collected from different spatial distance from pollutant point source (1 km and 2 km far from disposal point of dye houses).

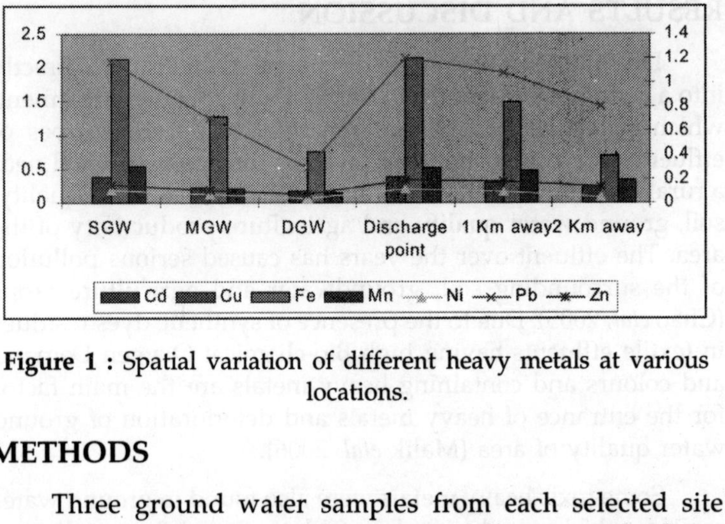

Figure 1 : Spatial variation of different heavy metals at various locations.

METHODS

Three ground water samples from each selected site including discharge outlets points of different textile dying industrial units from Panipat industrial area were collected and brought to the laboratory for various chemical parameters analysis. The pH of the water samples was determined on the spot by the digital pH meter. Methods for the determination of TDS, chlorides were as per methods of Trivedi and Goel (1984). BOD and COD were also determined (APHA, 1995) and the contents were compared with the standard of WHO, BIS, USPHS. Methods for estimation of heavy metals such as Fe, Zn,

Cu, Pb, Ni, and Mn in the ground water used for drinking and irrigation in the Panipat area were as per standard methods from APHA (1995).

Water samples were acidified in the field with concentrated HNO_3 (5 ml/l of water sample, to reduce the pH of the sample, pH > 2.0) for the total metal estimation. The total metal content was determined by digesting 200 ml of sample with a mixture of concentrated HNO_3 and $HClO_4$ acid (10 ml + 2 ml) by Singh *etal.* (2005). The digested samples were filtered through Whatman filter No. 42 and finally volume were made 10 ml with 0.1 N HNO_3 and analyzed for heavy metals using AAS (Model ECIL-4129).

RESULTS AND DISCUSSION

Dye houses of the region discharged their effluent directly into a common effluent drain without any adequate treatment, which collectively carry the sum total of different types of effluents in a composite form towards the pond, situated near a rural community, which has altered the surface water quality, soil, ground water quality and agriculture productivity of the area. The effluent over the years has caused serious pollution of the surrounding soil, groundwater and agriculture crops (Chao *etal.* 2005). Due to the presence of synthetic dyes residues in textile effluents having high Bio-chemical Oxygen Demand and colours and containing heavy metals are the main factor for the entrance of heavy metals and deterioration of ground water quality of area (Malik *etal.* 2006).

Some toxic heavy metals were also found in ground water regime, which are depicted in table 1, 2 and 3 according to spatial differences, while the physico-chemical parameters of ground water of industrial area of Panipat are given in Table 4. The mean values of ground water parameters are given with the comparison to drinking water quality standards of WHO, BIS, USPHS along with each table. Cadmium (0.02 mg/l) and Lead (0.31 mg/l) concentrations in ground water samples were found very near to the permissible limits of standards in some individual samples, while copper concentration (0.03 mg/l) were found in level to standards mostly.

Total dissolved solids (TDS) in ground water varied from 0.80 to 1. 32 gram per liter, including the high amount in hand pumps (56-96 mg/l) of the area, due to shallow ground water level occurs in area. The average value of TDS was calculated similar to the WHO standards. BOD was found more in hand pumps samples in comparison to tube wells (29-34 mg/l). Nutrients like Na, Cl, K, Ca, Mg were found comparatively high, and heavy metals were also present in high concentration in the SGW obtained from hand pumps in comparison to DGW. Vertically metals were found maximum in SGW > MGW > DGW of Panipat industrial area.

High concentrations of heavy metals in ground water like Pb (0.4 mg/l), Mn (0.18 mg/l), Fe (0.8 mg/l), Ni (0.1 mg/l) were found in the point source of textile industrial area. Rahman (1996) also stated the high level of Pb in ground water of some megacities in Karachi, Pakistan. This trend was also supported by Malik *etal.* (2004) when they found heavy metals in crop plant samples from agricultural field adjoining of textile industrial effluent flowing channels contained heavy metals in high amounts *e.g.* Pb, Cu, Mn, Ni, Fe, Cd and Zn, indicate the poor water quality of ground water used in irrigation. During the percolation of heavy metals from effluent drain or adjoining pond, some concentrations of heavy metals remained in soil environment by soil adsorption. This is due to adsorptive nature of soil particles to the heavy metals (Mido and Satake, 2003). During the heavy metals passed through soil profile with a long period of time, some concentration of heavy metals was absorbed and remains in soil, while rest goes to deep aquifers. So, the concentrations of heavy metals in deep ground water remained always low than the shallow ground water. This repeated and continuous process of percolation may accumulate the enough amounts of heavy metals in ground water and transfer it to other trophic levels through food chain.

Heavy metals concentrations in shallow ground water of hand pumps were found higher than deep ground water. In the samples collected from hand pumps, Mn (0.1 mg/l), Ni (0.03 mg/l), Fe (0.76 mg/l), Cu (0.17 mg/l), Cd (below detection limit), Pb (1.1 mg/l), Zn (0.2 mg/l) were found higher than the samples collected from deep ground water of tube wells. In tube wells samples the maximum concentration of heavy metals were found Mn (0.06 mg/l), Ni (0.04 mg/l), Fe (0.05 mg/l), Cu (0.01 mg/l), Cd (0.02 mg/l), Pb (0.29 mg/l), Zn (0.07 mg/l) respectively, because in deep aquifers heavy metals always remained in minor quantities due to leaching rate and absorption of heavy metals by soil system (Lokeshwari and Chandrappa, 2006). It is concluded that the ground water quality became deteriorated in the adjoining area of industrial sector. On the behalf of different spatial variations the heavy metals were found in maximum quantities at point source of pollution, while it was observed comparatively less at 1 km ahead from pollution source and so on. High concentrations of cadmium, lead and copper were found in most samples of ground water in point source of textile industrial area, while rest metal concentrations were found within the level of drinking water quality criteria.

Heavy metals like cadmium, lead and nickel are highly toxic at relatively low concentrations to plants. Heavy metal toxicity is the result of complex interaction of major toxic ions with other essential or non-essential ions. The metals can cause a reduction in the hydrolysis products *viz.*, α-amylase, RNAs, proteins and phosphatase. They interfere in the enzyme action by replacing metal ions from the metalo-enzymes and inhibit different physiological processes of plants (Agarwal, 1999). On the other hand, heavy metals are highly toxic at relatively low concentrations to animals and human beings. Eating food or drinking water with very high levels of heavy metals severely irritates the stomach leading to vomiting and diarrhea. Similarly high levels of lead may decrease reaction time, and cause anemia, a disorder of blood in human beings (De, 2002). Now it is our responsibility to advocate the proper use of different contaminated ground water in order to avoid surface water and soil contamination because soil is relatively valuable natural resource for mankind. On the whole, these ground waters cannot be used for drinking as well as irrigation otherwise that may cause toxicity to human beings, agricultural soil, plants, and animals as well add to the problems of salinity and sodocity. Sodocity is caused by the presence of sodium attached to clay in soil. A soil is considered sodic, when the sodium reaches a concentration, where it starts to affect soil structure. The sodium weakens the bonds between soil particles when wetted resulting in the clay swelling and often becoming detached.

The work presented here only studied the chemical quality of the ground water, it is equally important to extend this study to include (1) measurement of a range of biological as well as physico-chemical properties of the soil which receive the polluted ground water for irrigation purposes; (2) identification and chemical analyses of the plants growing on soils receiving ground water by irrigation; and (3) microbial characterization of the soil (Khan *etal.* 1998). Its understanding may be a powerful tool for managing toxic wastes through the use of adapted plant species growing on heavy metal contaminated area and industrial sites, *i.e.* phytoremediation (Chaudhry *etal.* 1998; Hayes *etal.* 2003; Khan, 2006b). Plants growing on such sites need to be studied for heavy metal contents and their role in phytoremediation.

Table 1 : Vertical Distribution of Heavy Metals in Ground Water at Various Depths (Mean ± S.D.)

Heavy Metals (ppm)	*SGW*	*MGW*	*DGW*	*WHO Standards*
Cd	0.005 ± 0.004	0.003 ± 0.002	BDL	0.01
Cu	0.373 ± 0.22	0.223 ± 0.11	0.157 ± 0.06	–
Fe	2.12 ± 2.18	1.26 ± 0.96	0.76 ± 0.50	1.0
Mn	0.54 ± 0.09	0.23 ± 0.11	0.13 ± 0.14	0.1
Ni	0.099 ± 0.08	0.068 ± 0.05	0.037 ± 0.03	0.2
Pb	1.163 ± 0.82	0.073 ± 0.32	0.029 ± 0.17	0.1
Zn	0.15 ± 0.095	0.10 ± 0.08	0.07 ± 0.05	5.0

Table 2 : Horizontal Distribution of Heavy Metals (Mean ± SD) in SGW

Heavy Metals (ppm)	*Discharge Point*	*1 Km Away*	*2 Km Away*	*ISPHS Standards*
Cd	0.007 ± 0.001	0.006 ± 0.001	0.005 ± 0.001	0.01
Cu	0.38 ± 0.24	0.29 ± 0.19	0.23 ± 0.15	0.05
Fe	2.133 ± 2.19	1.486 ± 1.52	0.696 ± 0.45	0.1
Mn	0.517 ± 0.11	0.483 ± 0.09	0.333 ± 0.07	0.05
Ni	0.106 ± 0.086	0.092 ± 0.07	0.041 ± 0.059	-
Pb	1.2 ± 0.88	1.077 ± 0.66	0.800 ± 0.35	0.1
Zn	0.193 ± 0.13	0.173 ± 0.12	0.136 ± 0.094	5.0

Table 3 : Horizontal Distribution of Heavy Metals (Mean±SD) in DGW

Heavy Metals (ppm)	*Discharge Point*	*1 Km Away*	*2 Km Away*	*BIS Standards*
Cd	0.001 ± 0.002	BDL	BDL	0.01
Cu	0.187 ± 0.07	0.160 ± 0.072	0.117 ± 0.06	0.05
Fe	0.800 ± 0.49	0.793 ± 0.44	0.640 ± 0.40	0.3
Mn	0.147 ± 0.15	0.127 ± 0.14	0.097 ± 0.08	0.10
Ni	0.044 ± 0.04	0.028 ± 0.27	0.017 ± 0.01	-
Pb	0.403 ± 0.12	0.317 ± 0.11	0.230 ± 0.09	0.05
Zn	0.097 ± 0.02	0.070 ± 0.03	0.047 ± 0.03	5.0

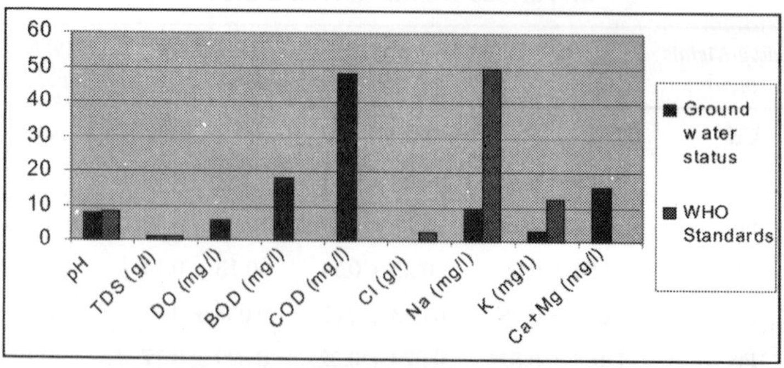

Figure 2 : Comparison of different ordinary parameters of ground water with WHO standards.

REFERENCES

APHA, (1995) : Standard methods for examination of water and waste water. *American Public Health Association*, 19th edition. Inc, New York. pp : 1170.

Chao, Keh-Ping; Tsai, Ching-Tsan; Wang, John H.C.; Lin, Shaw-Tao and Chiang, Chow-Feng, (2005) : Health risk assessment of a heavy metal contaminated site in Taiwan. *Jr. Practice Periodical of Hazardous, Toxic, and Radioactive Waste Management*, pp : 167-171.

Chaudhry, T.M., Hayes, W.J., Khan, A.G. and Khoo, C.H., (1998): Phytoremediation-focusing on accumulator plants that remediate metal-contaminated soils. *Australas. J. Ecotoxicol.*, 4 : 37-51.

Hayes, W.J., Chaudhry, T.M., Buckney, R.T. and Khan, A.G., (2003): Phytoaccumulation of trace metals at the Sunny Corner Mine, New South Wales, with suggestions for a possible remediation strategy. *Australas. J. Ecotoxicol.*, 9 : 69-82.

Irshad, A., Ali, S., and Jan, M.R., (1997) : Physico-Chemical Studies of Industrial Pollutants. NSMTCC'97 on Environment Pollution. Islamabad, Pakistan.

Khan, A.G., Chaudhry, T.M., Hayes, W.J., Khoo, C.H., Hill, L., Fernandez, R. and Gallardo, P., (1998): Physical, chemical and biological characteristics of a steelworks waste site at Port Kembla, NSW, Australia. *Water, Air Soil Pollut.*, 104(3/4) : 389-402.

Løbersli, E.M. and Steinnes, E., (1988) : Metal uptake in plants from a birch forest area near a copper smelter in Norway. *Water, Air Soil Pollut.*, 37(1-2) : 25.

Lokeshwari, H. and Chandrappa, G.T., (2006) : Impact of heavy metal contamination of Bellandur Lake on soil and cultivated vegetation. *Current Science*, 91(9) : 622-627.

Malik, D.S.; Bharti, P.K. and Sumit Grover, (2006) : Alteration in surface water quality near textile industries at Panipat (Haryana). *Environment Conservation J.*, 7(2) : 65-68.

Malik, D.S.; Yadav, Rashmi and Bharti, P.K., (2004) : Accumulation of heavy metals in crop plants through irrigation of contaminated ground water in Panipat region. *Environmental Conservation Journal* 5(3) : 101-104.

Malik, D.S. and Bharti, P.K., (2007) : Soil quality of irrigated agricultural fields in textile industrial area of Panipat city. *Asian Journal of Experimental Sciences*, 21 (2) : 445-451.

Mido, Y. and Satake, M., (2003) : Chemicals in the environment. In: *Toxic Metals* (eds. Sethi, M. S. and Iqbal, S. A.), Discovery Publishing House, New Delhi, pp : 45–68.

Rahman, A., (1996) : Ground water as resources of contaminations for water supply in rapidly growing Megacities of Asia : Case of Karachi, Pakistan, In: Proceeding of 18th biennial conference of international association of water quality, 34 (7-8) : 285.

Sharma, K.P.; Sharma, K.; Bharadwaj, S.M. and Chaturvedi, R.K., (1999): Environmental impact assessment of textile printing industries in Sanagar, Jaipur: A case study. *J. Int. Bot. Soc.* 78 : 71-74.

Singh, V.K., Singh, K.P., and Mohan, D. (2005) : Status of heavy metals in water and bed sediments of river Gomti-A tributary of the Ganga river, India. *Environ. Monit. Assess.* 105 : 43-67.

17

A Capability Approach Perspective of Health Poverty Implications Following Non Point Water Pollution : The Case of Grand River North West in Mauritius

Shalini T. Ramessur, Roshan T. Ramessur and Kishore Boodhoo

ABSTRACT

Though two-thirds of the earth surface is covered with water, only 1% of the water resource is available as fresh water and even this source is under threat due to high levels of pollution occurring from agricultural activities, urbanization and industrialization, thereby menacing human health. For instance according to the World Health Organization (WHO) (2002), as many as 4 million children die every year as a result of diarrhoea caused by water-borne infection. In Mauritius, an area suffering from acute water pollution is the Grand River North West where computed ratio of dissolved [nitrate]: [phosphate] in lagoon waters was less than 2 due to denitrification or possible high phosphate input attributed to road runoff resulting from increasing urbanization. Existing literature which have assessed the impact of water pollution have focused on the economic costs and in some cases the social effects through the application Livelihood Approaches. However these approaches only take into account impact on resources. This study contributes to the literature in two ways first by using the capability approach in the context of GRNW area, which looks at how water pollution affects the capacity of people to convert resources into well-being. Secondly by acknowledging that poverty is multidimensional and any study on overall poverty will be misleading for policy implications, it focuses only on the health dimension of poverty, in other words on health-poor.

Keywords

Water Pollution, Dissolved Nutrients, Lead Contamination, Estuaries, Urbanization, Health and Capability Approach.

INTRODUCTION

Out of the many activities leading to water pollution it is argued that agriculture is the main cause of water pollution first through its release of pollutants and deposits to surface and/or groundwater, second through poor agricultural practices resulting in net loss of soil, and thirdly through salinisation and water-logging of irrigated land. In the case of agriculture based pollution the most affected region in the world is sub Saharan Africa where agriculture accounts for about 35% of GNP, 40% of exports, 70% of employment and nearly 90% of water withdrawal. The UN Environment Program (UNEP) (2004) discloses that only 22 per cent to 34 per cent of populations in at least eight sub-Saharan countries have access to safe water and hence the region is the worse in terms of water scarcity on the continent. Africa ranks worst in terms of water availability per capita and within the region sub Saharan Africa accounts for the least percentage (3.5%) of irrigated cropped land (Food and Agriculture Organization (FAO) 1986a, b, c). Alongside around 40-60 % of irrigation and water is lost through seepage and evaporation, which result in severe soil salinisation and waterlogging problems. Poor agricultural land use coupled with poor management lead to high levels of eutrophication due to increased deposition of nitrogen and phosphorous into water bodies.

The call for food security and in turn demand for higher food production in the region is causing higher pressures on water supplies and also the increased use of fertilizers and pesticides to expand yields, thereby resulting in higher water pollution. For instance the findings of GEM (2004) show that since 1990, nitrate contamination resulting from agriculture runoff has more than doubled in the Eastern Mediterranean and Africa. There is a widespread consensus that regions that suffer from water stress and pollution serve as catalysts for the spread of disease which consequently is a major threat to public health. For example in a global study conducted by the United Nations, it was found that unsafe water and water stress is indirectly responsible for delaying the recovery process of those suffering from malaria, in Africa, which accounts for 90 per cent of global cases of malaria. Alongside among the major causes of mortality and morbidity in Africa, appear water-borne diseases like typhoid, cholera and dysentery, partly due to the fact that there is little or no treatment of wastewater which is increasingly being used directly or recycled from receiving waters, into irrigated agriculture.

This a major concern for mainly the poor since 70 % of the poor in sub Saharan Africa live in rural areas, where agriculture is their mainstay and hence are extremely vulnerable to health problems associated with water pollution arising from agricultural activities. Given they have no cash reserves to pay for a sudden illness, being exposed to increasing contamination and transmission of diseases, reduces their productivity, resulting in loss of income and thereby further push them into poverty and debt. It must be noted that agriculture itself can be a victim of water pollution through use of wastewater and polluted surface and groundwater which contaminate crops, but this is out of the focus of the study. The accumulation of toxic and persistent substances in the coastal environment continuously increases owing to anthropogenic activities as the estuaries and their adjacent coasts are the focus for many economic activities (Singh and Singh, 2006). The coastal zone of Mauritius, (1850 km^2, 20 °S and 57 °E, Western Indian ocean) is made up of terrestrial and marine interfaces where industrial and urban activities and baseline investigations carried out on the west coast of Mauritius have highlighted the susceptibility of local estuaries to metal and nutrient inputs and the threats posed by rapidly increasing residential and industrial development (Murugan 2004; Burnett *etal.* 2006; Balgobin 2009, Ramessur 2010; Ramessur *etal.* 2010). Sources of input of trace metals in particular, lead in Mauritius include industrial effluents, urban and road runoff, sewage sludge and landfill leachates. The aim of the paper is to investigate dissolved nutrients and lead contamination in sediments in an estuary in Mauritius highlighting the empirical evidence on how the polluted water in turn can affect the health of individuals, mainly the poor as they are more exposed to agricultural activities and urbanization. The next part introduces the Capability Approach emphasising on capabilities and valued functionings in the context of health and is followed by assessing the impact of water on health dimension of poverty from a capability perspective.

POLLUTION REVIEW

Sources of water pollution can be grouped under point source water pollution and non-point source water pollution. The former can be easily measured and controlled since it represents those activities where wastewater is routed directly into receiving water bodies by, for example, discharge pipes, while the latter emerges from a wide range of human activities for which the pollutants have no obvious point of entry into receiving watercourses. Their dispersed characteristic makes it harder to design a system of controls for pollutants.

As far as agricultural activities are concerned, inclusive of land use and animal feeding operations, they are regarded as non-point sources. Figure 1 adapted from Rickert (1993) shows the relative complexity of agricultural non-point source pollution. According to the FAO, agricultural activities can lead to both surface water and groundwater pollution as summarised in Table 1.

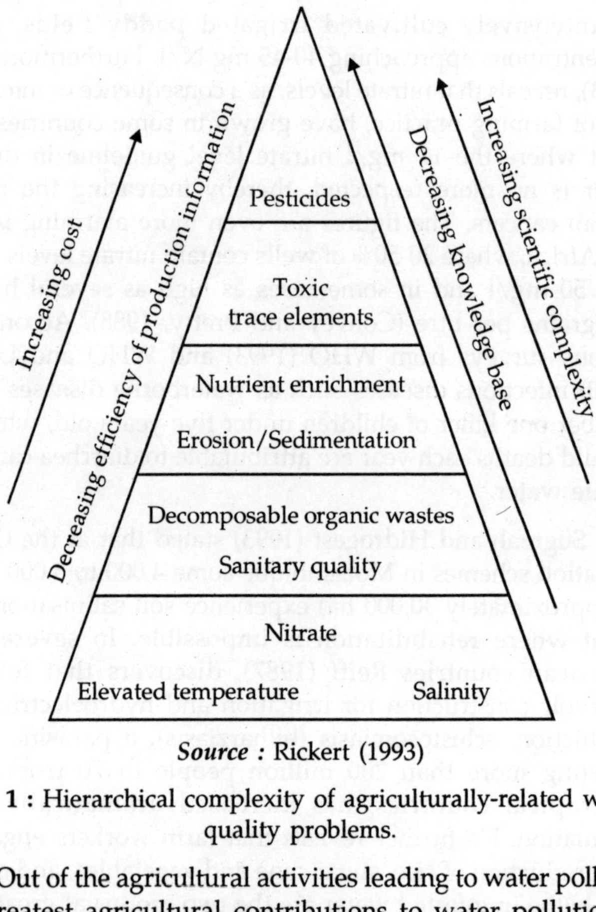

Source : Rickert (1993)

Fig. 1 : Hierarchical complexity of agriculturally-related water quality problems.

Out of the agricultural activities leading to water pollution, the greatest agricultural contributions to water pollution are through nutrient and sediment pollution. Such agricultural-led water pollution is both a direct and indirect cause of human health impacts. For instance according to Kamrin (1995) the toxicological effects of pesticides on the endocrine system of the human body have impacts on the reproductive system. From an empirical point of view, various studies have found evidence for significant negative impact of water pollution caused by agricultural activities on human health. For instance according to the WHO, water-borne infection results in as many as 4 million deaths of children every year. Lawrence and Kumppnarachi (1986) testified that in irrigation wells that are located close to the intensively cultivated irrigated paddy fields, nitrate concentrations approaching 40-45 mg N/l. Furthermore WHO (1993), reveals that nitrate levels, as a consequence of intensification of farming practice, have grown in some countries to the point where the 10 mg/l nitrate level guideline in drinking water is no more respected, thereby increasing the risks of human cancers. The figures are even more alarming in India and Africa, where 20-50% of wells contain nitrate levels greater than 50 mg/l and in some cases as high as several hundred milligrams per litre (Convey and Pretty, 1988). According to various surveys from WHO (1993) and WHO and UNICEF (2000) infectious diseases such as waterborne diseases are the number one kiiler of children under five years old, while 15% of child deaths each year are attributable to diarrhea caused by unsafe water.

Sogreah and Hidrogest (1993) stated that at the Chokwe irrigation schemes in Mozambique some 4,000 to 5,000 ha (out of approximately 30,000 ha) experience soil salinisation to the point where rehabilitation is impossible. In several Latin American countries Reiff (1987), discovers that following reservoir construction for irrigation and hydroelectric power production, schistosomiasis (bilharziasis), a parasitic disease affecting more than 200 million people in 70 tropical and subtropical countries, has increased dramatically in the popuiation. He further reveals that farm workers engaged in the production of rice, sugar cane and vegetables, and children that bathe in infested water are the two groups at greatest risk of infection.

Table 1 : Agricultural Impacts on Water Quality

Agricultural Activity	*Impacts Surface Water*	*Groundwater*
Tillage/ploughing	**Sediment/Turbidity** : Sediments carry phosphorus and pesticides adsorbed to sediment particles; siltation of river beds and loss of habitat, spawning ground, etc.	

Cont...

Fertilizing	Runoff of nutrients, especially phosphorus, leading to eutrophication causing taste and odour in public water supply, excess algae growth leading to deoxygenation of water and fish kills.	Leaching of nitrate to groundwater; excessive levels are a threat to public health.
Manure spreading	Carried out as a fertilizer activity; spreading on frozen ground results in high levels of contamination of receiving waters by pathogens, metals, phosphorus and nitrogen leading to eutrophication and potential contamination.	Contamination of ground-water, especially by nitrogen.
Pesticides	Runoff of pesticides leads to contamination of surface water and biota; dysfunction of ecological system in surface waters by loss of top predators due to growth inhibition and reproductive failure; public health impacts from eating contaminated fish. Pesticides are carried as dust by wind over very long distances and contaminate aquatic systems 1000s of miles away (*e.g.* tropical/subtropical pesticides found in Arctic mammals).	Some pesticides may leach into ground water causing human health problems from contaminated wells.
Feedlots/animal corrals	Contamination of surface water with many pathogens (bacteria, viruses, etc.) leading to chronic public health problems. Also contamination by metals contained in urine and faeces.	Potential leaching of nitrogen, metals, etc. to groundwater.
Irrigation	Runoff of salts leading to salinization of surface waters; runoff of fertilizers and pesticides to surface	Enrichment of groundwater with salts, nutrients (especially nitrate).

Cont...

Agricultural Activity	*Impacts Surface Water*	*Groundwater*
	waters with ecological damage, bioaccumulation in edible fish species, etc. High levels of trace elements such as selenium can occur with serious ecological damage and potential human health impacts.	
Clear cutting	Erosion of land, leading to high levels of turbidity in rivers, siltation of bottom habitat, etc. Disruption and change of hydrologic regime, often with loss of perennial streams; causes public health problems due to loss of potable water.	Disruption of hydrologic regime, often with increased surface runoff and decreased ground-water recharge; affects surface water by decreasing flow in dry dry periods and con-centrating nutrients and contaminants in surface water.
Silviculture	Broad range of effects : pesticide runoff and con-tamination of surface water and fish; erosion and sedimentation problems.	
Aquaculture	Release of pesticides (*e.g.* Tributyltin TBT) and high levels of nutrients to surface water and groundwater through feed and faeces, leading to serious eutrophication.	

Source : http://www.fao.org/docrep/w2598e/w2598e04.htm

MATERIALS AND METHODS

Study Sites

The Grand River North West estuary along the western coast of Mauritius is shown in Figure 2.

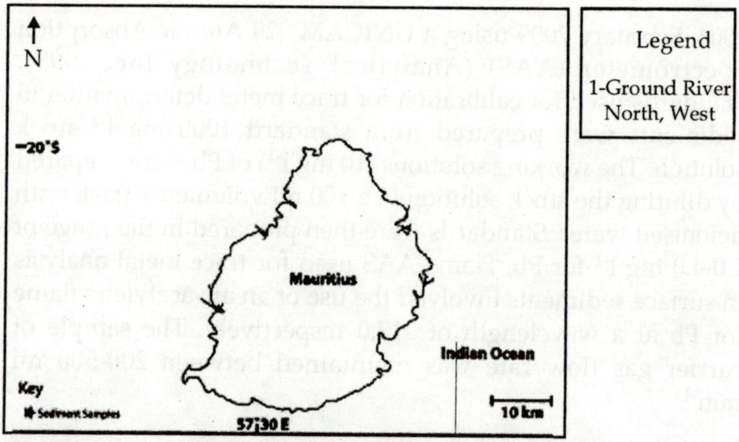

Fig. 2 : Sampling station along GRNW estuary on the western coast of Mauritius. (Urban estuary at GRNW, station 1).

The Grand River North West (GRNW) is situated south of Port Louis with a catchment area of 116 km^2. The GRNW area receives land and road runoff in Coromandel and from St. Louis river which flows through Pailles and Plaine Lauzun industrial area, Fig 2.

Dissolved Nutrient Analysis and Lead Analysis in Surface Sediments

Replicate samples of water were collected in 200ml plastic bottles (dissolved nitrate) and glass bottles (reactive phosphate) in the coastal area in GRNW. Samples were stored at 4°C and analysed within 24h. The concentration of dissolved nitrite, dissolved nitrate, and dissolved phosphate were determined using standard spectrophotometric methods (Parsons *etal.* 1984) at 543nm and 882nm respectively using a PU 8710 spectrophotometer and a UNICAM 8700 Series UV/VIS spectrometer following calibration using known standard solutions. Quality control was achieved by analysing an internal reference independently prepared from the standard and the standard curves were verified after 10 successive runs by analysis of one standard solution within the linear range for each nutrient. Pb were determined in the extracted solution from surface sediments collected along the estuaries during the period August

2004–February 2009 using a UNICAM 929 Atomic Absorption Spectrometer (AAS) (Analytical Technology Inc. 1993). Standards used for calibration for trace metal determination in sediments were prepared from standard 1000 mg l^{-1} stock solution. The working solutions (10 mg l^{-1}) of Pb were prepared by diluting the stock solution in a 100 ml volumetric flask with deionised water. Standards were then prepared in the range of 1.0-4.0 mg l^{-1} for Pb. Flame AAS used for trace metal analysis in surface sediments involved the use of an air-acetylene flame for Pb at a wavelength of 217.0 respectively. The sample of carrier gas flow rate was maintained between 200-500 ml min^{-1}.

Quality Control for Lead

The accuracy and precision of the method was evaluated using three replicate determination of Standard Reference Material SRM 1646a Trace Elements in Estuarine Sediments from National Institute for Science and Technology (USGS 1884, Colorado, U.S.A) which yielded 8.7 ± 1.3 mg kg^{-1} (Pb) compared to certified values of 11.7 ± 1.2 mg kg^{-1} (Pb) respectively (mean and standard deviation) as determined using ICP MS by NIST with % recovery as 74.4% for Pb. The main causes of losses were during the digestion procedure and also included random

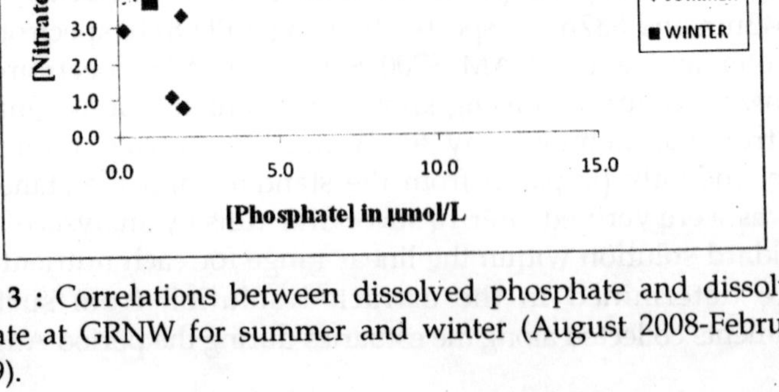

Fig. 3 : Correlations between dissolved phosphate and dissolved nitrate at GRNW for summer and winter (August 2008-February 2009).

and systematic sources of uncertainty during analysis using the atomic absorption spectrometer. Necessary corrections were made accordingly. The detection limits was 1.2 mg kg^{-1} Pb respectively. (The limits of detection were taken as 3s of three replicates of the procedural blank digested filter paper (3x standard deviation about the mean).

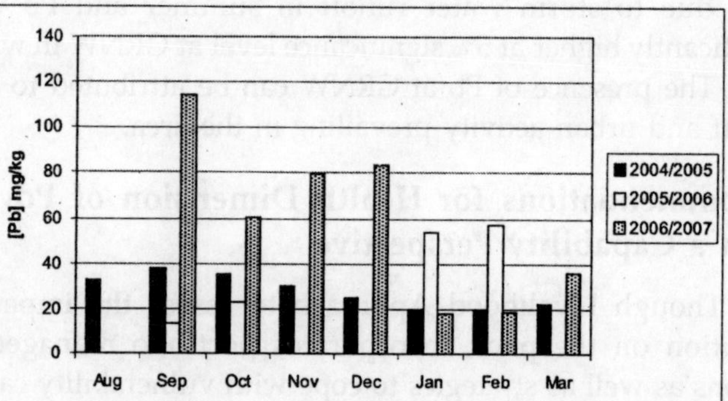

Fig. 4 : Lead concentration in estuarine sediments at GRNW from 2004-2007.

RESULTS AND DISCUSSION

Dissolved Nutrients and Lead Contamination in Estuarine Sediments in Mauritius

The ratios of dissolved [nitrate] : [phosphate] in coastal waters computed for GRNW situated in urban areas were less than 2 and were attributed to denitrification or high phosphate input. A strong positive correlation between nitrate and phosphate ($R^2 = 0.78$) was obtained at GRNW during summer as shown in Figure 3. This indicated that in summer biological activities in water for phosphate and nitrate in summer are linked. The low ratio of [nitrate] : [phosphate] is associated with low oxygen concentrations and is probably caused by denitrification as reported by Tyrrell and Law (1997) with algal bloom observed during this period. Monthly variations of Pb in estuarine sediments from August 2004–March 2007 are shown in Fig. 4. The mean concentration of Pb (27.0 mg kg^{-1}) in sediments along the GRNW estuaries can be considered below those from contaminated estuarine sediments. Pb was however significantly higher at 5% significance level at GRNW in summer 2005 compared to winter 2005 and significantly increased from summer 2005 to summer 2006. Pb again significantly decreased from winter 2006 to summer 2007 at GRNW at 5% significance level due to storm water runoff in summer and Pb were significantly higher at 5% significance level at GRNW in winter 2004. The presence of Pb at GRNW can be attributed to road runoff and urban activity prevailing in the area.

Recommendations for Health Dimension of Poverty from a Capability Perspective

Though Livelihood Approach to assess the impact of pollution on the poor, incorporates portfolio management options as well as strategies to cope with vulnerability caused by the volatile returns of the poor's assets and promotes the use of multidimensionality analyses of poverty-environment links, Sen claims that given people differ in their capacity to convert resources into well-being and also live under different cultural and social constraints, resources tend to be imperfect measures of well-being. Consequently he advocates for the Capability Approach which focuses on the question of what people are able to do and hence promotes equality in the space of capabilities. This approach considers indicators that reflect the importance of assessing the quality of processes rather than simply outcomes of policies and also stresses on the significance of measures to enhance the ownership and participation of local communities in the management of their natural resources. The main notion of the Capability Approach is that social arrangements should target at increasing people's capabilities – their freedom to upgrade or accomplish what they value doing and being. Two important elements of the approach are: functionings and capabilities. The former refers to being and doing activities that people value and have reason to value, while the latter denote the freedom to enjoy valuable functionings and thus convey some kind of opportunity freedom.

As far as health is concerned, the approach stresses on the process of generating health and acknowledges that due to human diversity and agency, different people may have different values in terms of health and at the same time there are many factors, at the individual, social, institutional and environmental level, affecting how different individuals convert resource inputs into valued capabilities and in turn which functionings are achieved. In the context of health, inputs can be grouped into food or food supplements, preventive interventions (e.g. immunization), access to medicines and health-care personnel and provision of health education or information on particular aspects of health. How these health inputs are converted into valued health outputs depends on conversion factors which may have individual, social, occupational, and authority dimensions (Deneulin and Shahani 2009). In the case of individual conversion factors, physiology plays a major role, since for instance an individual, who is suffering from a disease disrupting one's absorptive capacity, will receive less nutritional value from the same amount and quality of food consumed by an individual who is not infected (Deneulin and Shahani, 2009). In the context of social conversion factors as the WHO Commission on Social Determinants of Health has pointed out that 'inequities in health arise because of the circumstances in which people grow, live, work and age, and the systems put in place to deal with illness. The conditions in which people live and die are, in turn, shaped by political, social, and economic forces' (Commission on Social Determinants of Health, 2008, pi.).

As far as occupational conversion factors are concerned it is argued that those in higher occupational hierarchy faces less mortality rate, since though nearly all civil servants have equal access to resource inputs, their ranks in the occupational hierarchy provide them with different conversion factors, affecting their respective health functionings in the long term (see, Marmot, 2006 for a more comprehensive study on this issue). Finally, the extent of authority an individual possesses within their household or community also affects one's ability to access or convert a particular resource into a valued health outcome. Another important consideration from a capability perspective is the magnitude of choice which is influenced by the set of available information which can be incomplete and/or modified and individual values, social, cultural and religious norms. In respect to health issues people choices become more complex given the multidimensional nature of health and also the fact that health values are weighted against other values. The capability approach further recognizes that health capabilities have both an intrinsic and instrumental function in promoting human well-being since being deprived in health dimension may also imply being deprived in other dimensions such as education, participation and employment.

Impact of Water Pollution on the Health of the Poor–A Capability Approach

To show how non-point water pollution, namely through sediments, use of fertilizers, use of pesticides and use of lead can affect the poor through health deprivation in the area of Grand River North West in Mauritius, we make use of Nussbaum's (1997) list of fundamental capabilities namely: life, bodily health, bodily integrity, senses, imagination and thought, emotions, affiliation and play. Pollution by sediment have both a physical dimension which leads to high levels of turbidity in receiving waters and also a chemical dimension as a carrier of absorbed chemicals like phosphorus, chlorinated pesticides and other metals. High levels of turbidity mainly affect the quality of water as it enhances the growth of microbes and bacteria and hence leads to bodily health deprivation in the form of lack of health which greatly reduces all aspects of bodily health capability.

Deprivation in bodily health can be a serious problem for people suffering from HIV/AIDS given lack of health implies greater vulnerability for those infected by HIV/AIDS which in turn may lead to lower life expectancy and hence life deprivation. At the same time if highly toxic metals are leaked into drinking water sources which then these metals may affect the liver and kidneys and can cause miscarriage in pregnant women, thereby affecting health capabilities in terms of life deprivation, as they reduce life expectancy and number of births. In the case of fertilizers, the major type of water pollution they lead to is known as "eutrophication", that is the enrichment of water surface with plant nutrients which leads to the excessive growth of blue-green algal (Cyanobacteria) and aquatic macrophytes. In the case of blue-green algal, when they grow out of proportion, they can lead to high concentration of cyanobacteria which affect water quality in terms of taste, odour and they can even interfere with certain water treatment processes and in extreme cases can produce toxins. These toxins act on the molecular system of human beings and can lead to disastrous impacts on the nervous, digestive, respiratory and cutaneous systems, though the age and physiological conditions of those affected will determine the severity of the symptoms. Nevertheless, such toxins through use of fertilizers influence various dimensions of health capabilities such as firstly, by affecting the nervous system they lead to senses, imagination and thought deprivations since mental impairment may affect the ability to use sense, imagination and thought and in the case of children this may reduce their chances of acquiring education and as such deprivation in health capability spills over to deprivation in education capability and perpetuates the vicious poverty cycle. Negative impacts on the nervous system can also lead to emotions deprivation as mental and psychological sickness can affect the emotional health of individuals thereby impeding their emotional development.

At the same time, negative influences on the nervous system leading to mental disorders can as well be the foundation of affiliation and play deprivations since mental disorders affect an individual's sense of self-esteem, self-respect and professional opportunities coupled with the fact that one may not be able to play and participate in recreational/leisure activities in the same ways as healthy individuals do. While negative impacts on the respiratory and cutaneous system will influence life and bodily health deprivations. As far as pesticides are concerned, their use can be in the form of herbicides (weeds), insecticides (insects), fungicides (fungi), nematocides (nematodes), and rodenticides (vertebrate poisons), all of which are chemicals.

Health capabilities are affected in various ways through the use of pesticides. In the first instance while handling, preparing and applying pesticides, individuals may end up with skin and inhalation problems- like asthma attacks (mainly farmers) as well as ingestion which in turn may head towards bodily health. Some more acute health problems following exposure to pesticides range from chronic diseases like cancer to developmental disorders which then have impact on life, bodily health and even bodily integrity, affiliation, play, emotions, senses, imagination and thought if developmental disorders also imply nervous disorders.

CONCLUSION

Excessive nutrients at GRNW can promote algal blooms, leading to oxygen depletion and severe deterioration of water quality as well as fish mortality. It can be argued that agricultural, urbanization and tourism activities have contributed to an increase in anthropogenic activities and hence have increased the potential for increased sewage effluent and urban runoff in this area. The experimental data also demonstrate the potential for Pb concentrations in estuarine sediments to exceed contamination limits. The contamination of sediments with Pb at GRNW was significant showing an increasing trend over four years which could be considered to arise from accumulation and the heavy use of leaded petrol and potentially constitute a health hazard. However, storm runoff during flash floods in summer also caused a significant decrease in the levels of Pb in the urban estuary due to dilution with cleaner background sediments comparable to levels at *Flic en Flac*.

The use of capability approach, recognizing that poverty is multidimensional, also shows that non-point water pollution leads to poor health, but poor health does not only imply suffering from any form of diseases. Poor health is much beyond that if we make use of health capabilities and deprivations dimensions in terms of achieved valued functionings, for those living in contexts of poverty in Mauritius. Being deprived in health capabilities lead to productivity loss, reduction in income, loss of employment, deprivation in education thereby pushing individuals and households further down the poverty ladder. Given the feedback effect from health deprivations into deprivations in other dimensions, the poverty cycle is further reinforced. Alongside deprivations in health may be irreversible compared to deprivations in other dimensions such as education which can be offset at later stages of life through for example, adult literacy programmes. Consequently this signals to the need to give health its due importance in development policy. In brief the Capability Approach does not present a set of ready-made solution, but rather a means to arrive at solutions.

REFERENCES

The Commission on Social Determinants of Health: Tackling the Social Roots of Health Inequities (2008) : Irwin A, Valentine N, Brown C, Loewenson R, Solar O, *etal. PLoS Medicine* Vol. 3, No. 6.

Balgobin, J. (2009) : Variations of dissolved nutrients lead and zinc in the coastal system at *Flic en Flac*, Mauritius. B.Sc. Thesis, University of Mauritius 89 pp.

Bissembur, Z.B. (2009) : A comparative study of lead and zinc in sediments and some nutrients at GRNW and Albion. B.Sc. Thesis, University of Mauritius 83 pp.

Burnett, W.C., Aggarwal, P.K., Aureli, A., Bokuniewicz, H., Cable J.E., Charette, M.A., Kontar, E., Krupa, S., Kulkarni, K.M., Loveless, A., Moore,W.S., Oberdorfer, J.A., Oliveira, J. , Ozyurt, N. Povinec, P., Privitera, A.M.G., Rajar R., Ramessur, R.T., Scholten, J., Stieglitz, T., Taniguchi and M.,Turner, J.V. (2006): Quantifying Submarine Groundwater Discharge in the Coastal Zone via Multiple Methods. Science of the Total Environment, 367, pp. 498–543.

Convey, G.R. and Pretty, J.N. (1988) : Fertilizer risks in the developing countries : a review. International Institute for Environment and Development, London.

Deneulin, S., Shahani, L. (2009) : An Introduction to the Human Development and Capability Approach: Freedom and Agency. London: Earthscan.

FAO (Food and Agriculture Organization) (1986a) : Irrigation in Africa south of the Sahara.

FAO (Food and Agriculture Organization) (1986b) : Atlas of African agriculture African agriculture : the next 25 years. FAO, Rome, Italy.

FAO (Food and Agriculture Organization) (1986c) : *The cost of soil erosion in Zimbabwe in terms of the loss of three major nutrients.* M. Stocking. Consultant's Working Paper No. 3, Soil Conservation Programme, Land and Water Development Division, FAO, Rome.

Global Environmental Monitoring System (GEMS) Water Programme (2004) : State of water quality assessment reporting at the global level (R. Robarts). Presentation at the UN International Work Session on Water Statistics.

http://www.fao.org/docrep/w2598e/w2598e04.htm. Accessed 4th October 2010.

Kamrin, M. (1995) : Environmental Hormones. Discussion paper taken from EXTOX -NET of the Internet. Access on WWW through: http://www.oes.orst.edu:70/1/ext, choose EXTOXNET.

Lawrence, A.R. and Kumppnarachi (1986): Impact of agriculture on groundwater quality in Kalpitiya, Sri Lanka. British Geological Survey, Report WD/03/86/20.

Nussbaum, M. (1997) : *Cultivating Humanity,* Harvard University Press, Cambridge, MA.

Rickert, D. 1993. Water quality assessment to determine the nature and extent of water pollution by agriculture and related activities. In : *Prevention of Water Pollution by Agriculture and Related Activities.*

Murugan, J. (2004) : Determination of dissolved nutrients in coastal waters at *Flic en Flac,* Mauritius. B.Sc. Thesis. University of Mauritius 73pp.

Parsons, T.R., Maita Y., Lalli C.M. (1984) : Chemical and Biological methods for seawater analysis. Pergamon Press.

Ramessur, R.T., Parry, S.J., Jarvis, K.E. (1998) : Characterization of some trace metals from the Export Processing Zone and a coastal tourist area in Mauritius. Environment International, 326 24 (7) pp 773-781.

Ramessur, R.T., Parry, S.J., Ramjeawon, T. (2001) : The relationship of dissolved Pb to some dissolved trace metals (Al, Cr, Mn and Zn) and to dissolved nitrate and phosphate in a freshwater aquatic system in Mauritius. Environment International, 26 (4) pp 223-230.

Ramessur, R.T., (2002): Anthropogenic driven changes with focus on the coastal zone of Mauritius, south-western Indian Ocean. Regional Environmental Change, 3, 99-106.

Ramessur, R.T., Ramjeawon, T. (2002) : Determination of Pb, Cr and Zn from an urbanized river in Mauritius. Environment International 28 (4) pp 315-324.

Ramessur, R.T. (2004) : Statistical comparison and correlation of Zn and Pb in estuarine sediments along the Western Coast of Mauritius. Environment International. 30 (8) pp : 1039-1044.

Ramessur, R.T., Boodhoo, K., Balgobin, J., Aurelio, A. (2010): Statistical comparison between consecutive winter and summer concentrations in zinc and lead from sediments in a tropical urban estuary in Mauritius. Environmental Monitoring and Assessment. 168 (1) pp: 345-351.

Ramessur, R.T. (2010) : Trace Metal Contamination in Urbanized Estuaries and Implications of Submarine Groundwater Discharge for Coastal Zone Management in Mauritius. In: Natural and Man Made Disasters Section-4: Chapter 16. Editors: Singh, K.K and Singh, A.K, 1, pp329-342 Times Press, New Delhi.

Singh, A.M. and Singh, M. (2006) : Lead Decline in the Indian Environment Resulting from the Petrol-Lead Phase-out Programme. Science of The Total Environment, 368, pp: 686-694.

Tyrrell T, Law CS (1997): Low [nitrate] : [phosphate] ratios in the global ocean. Letters to Nature 387, 793-796.

United Nations Environment Programme (UNEP) (1993) : *The Aral Sea: Diagnostic study for the development of an Action Plan for the conservation of the Aral Sea.* Nairobi.

United Nations Environment Programme (UNEP) (1996) : Groundwater: a threatened resource. UNEP Environment Library No. 15, UNEP, Nairobi, Kenya.

World Health Organization (1993) : *Guidelines for Drinking-Water Quality, Volume 1: Recommendations.* (Second Edition), WHO, Geneva.

World Health Organization and United Nations Children's Fund. (WHO and UNICEF) (2000): Global Water Supply and Sanitation Assessment 2000 Report. WHO and UNICEF Joint Monitoring Programme for Water Supply and Sanitation.

World Health Organization (WHO) (2002) : World Health Report: Reducing Risks, Promoting Healthy Life. France.

World Health Organization (WHO) (2008) : Closing the Gap in a Generation - Health Equity through Action on the Social Determinants of Health Commission on Social Determinants of Health. ISBN 978 92 4 156370 3.

18

Pollution Studies of The Water of River Ganga in Haridwar District at Uttarakhand

Namita Agrawal

ABSTRACT

A systematic study has been carried out to assess the water quality index and irrigation water quality of River Ganga in Haridwar District. 90 water samples from five sampling stations were collected and analysed for WQI, electrical conductivity (Ec), total dissolved salts (TDS), magnesium content (MC), sodium per cent (SP), sodium adsorption ratio (SAR), residual sodium carbonate (RSC) and permeability index (PI). The study area has been divided into three seasons : Winter (November-February), summer (March to June) and Rainy (July to October). Water quality variables were measured in the river over a period of two years (Nov. 2006 to Oct. 2008). Each parameter was compared with the standard desirable limit of that parameter in river water as prescribed by different agencies.

The analytical data of various physico-chemical parameters indicates that some parameters like pH, electrical conductivity, total dissolved solids, total suspended solids, turbidity and sodium are found to be in excess than the prescribed limit in some water samples of the study areas. The WQI value indicates that water samples of some sampling stations are quite unfit for drinking purpose because of high value of dissolved solids and sodium. Study of electrical conductivity (EC), total dissolved salts (TDS), and magnesium content (MC), sodium per cent

(SP), sodium adsorption ratio (SAR), residual sodium carbonate (RSC) and permeability index (PI) indicates that river water in rainy season is not suitable for irrigation purpose because of high values of total dissolved salts, EC and SP. It was also observed that the water in the year 2007 was of a better quality than in the year 2008.

Keywords

Water pollution, Ganga river water, Water quality index, potability, irrigation water quality, SAR, RSC.

INTRODUCTION

Pollution of river is a global problem. In India it is reported that about 70% of the available water is polluted. The chief source of pollution is identified as sewage constituting 84 to 92 per cent of the waste water. Industrial waste water comprised 8 to 16 per cent._Pollution of a river first affects its chemical quality and then systematically destroys the community disrupting the delicate food web. Diverse uses of the rivers are seriously impaired due to pollution and even the polluters like industry suffer due to increased pollution of the rivers. River pollution has several dimensions and effective monitoring and control of river pollution requires the expertise from various disciplines (Trivedy R.K., Khatawkar S.D., Kulkarni A.Y., and Shastri A.C., 1990). The indiscriminate and large scale deforestation and over grazing in the watershed areas of river basins have caused soil erosion resulting in considerable silting of dams and shrinkage of river flows. This leads to the flooding of the rivers at the time of excessive rains (Goel P.K., 2006). The disposal of waste leads to contamination of river and lakes chronically affecting the flora and fauna. According to surveys carried out on selected stretches of important rivers, it has been found that most of the rivers are grossly polluted. The domestic sewage gives rise to numerous water-borne diseases like typhoid, cholera, dysentery, poliomyelitis and cysticercosis, thereby affecting the human health and deterioration of the water quality (Sharma R.N., Baruah A.K., Bora G.C. and Choudhary P.K., 1996). Haridwar is a city in Northern India on the bank of the Ganga River north east of Delhi. It is a Hindu pilgrimage centre. Haridwar lies along the Ganga River at the boundary between the Indo-gangetic plain (South) and the Himalayan foothills (North). The water supply of the Ganga system is partly dependent on the rains brought by the monsoon winds from July to October as well as on the flow from melting Himalayan glaciers in the hot season from April to June. The religious importance of Ganga may exceed than that of any other river in the world.

The location of Haridwar on the Globe is on Latitude 29°58'N and longitude 78°10' E, while the height from sea level is 285.56 meters. Present study was carried out on the stretch of 10 Kms from Bhooma Niketan to Pul Jatwara, which is having a width of 15-40 meters and depth of 2-10 meters. Total surface area and volume of water for this area was recorded 220000 m^2 and 1760000 m^3. The climate for this region is temperate, dry winter and warm to hot summer, with an ambient temperature for winter, summer and monsoon 9.3 to 16.3°C, 26.8 to 40.4°C and 12.9 to 28.5°C respectively. Annual precipitation here is over 31 centimeters.

This study was performed at five stations starting from sampling station A (Bhooma Niketan-situated on the north of Haridwar on the right bank of Ganga) sampling station B (Jai Ram Ashram-the stream of Ganga is separated from Malviya Point and flows through Kharkhari via Jai Ram Ashram and confluenced with second stream of Ganga river on the left side of Pantdweep) Sampling station C (Har-ki-Pauri-sacred place, people come here to take holy bath and also immerse ashes in the river) Sampling station D (Prem Nagar Ashram-situated between Kankhal and Jwalapur connecting bridge. The water at this point is polluted by a newly constructed colony on the northeast bank of the canal) and Sampling station E (Pul Jatwara-located near the small town of Jwalapur. Entire fecal matter, decaying leaves, and flowers, wooden parts, waste clothes, food materials and ashes, charcoal etc. can be observed in the Ganga at this point. The sewer of B.H.E.L. Nullah also mixes in the water, which may affect the aquatic ecology of the river. The bottom of river at this point is stony and sandy). Fig. 1 : shows all the sampling stations.

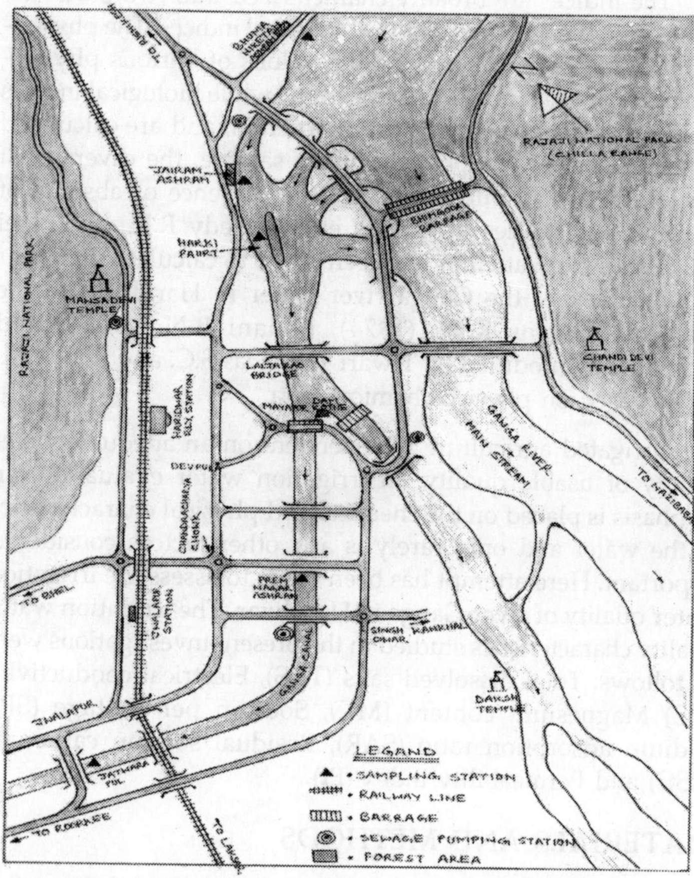

Fig. 1 : Map of River Ganga showing sampling sites.

Accurate and timely information on the quality of water is necessary to shape a sound public policy and to implement the water quality improvement programmes efficiently. One of the most effective ways to communicate information on water quality trends is with indices. Water quality index (WQI) is commonly used for the detection and evaluation of water pollution and may be defined as **"a rating reflecting the composite influence of different quality parameters on the overall quality of water"** (Mishra P. Patel R.K., 2005).

The indices are broadly characterized into two parts: the physico-chemical indices and the biological indices. The physico-chemical indices are based on the values of various physico-chemical parameters in a water sample, while biological indices are derived from the biological information and are calculated using the species composition of the sample, the diversity of species, their distribution pattern, the presence or absence of the indicator species or groups etc. (Trivedy R.K., and Goel P.K., 1984). Here attempt has been made to calculate the water quality index of the Ganga river water in Haridwar on the basis of Harkins R.D., (1974), Lohani B.N., (1981) and subsequently modified by Tiwari T.N., Das S.C. and Bose P.K., (1986) based on physico-chemical data.

Irrigated agriculture is dependent on an adequate water supply of usable quality. In irrigation water evaluation, an emphasis is placed on the chemical and physical characteristics of the water and only rarely is any other factors considered important. Here attempt has been made to assess the irrigation water quality of river Ganga in Haridwar. The irrigation water quality characteristics studied in the present investigations were as follows: Total dissolved salts (TDS), Electrical conductivity (EC) Magnesium content (MC), Sodium per centage (SP), Sodium adsorption ratio (SAR), Residual sodium carbonate (RSC) and Permeability index (PI).

MATERIALS AND METHODS

A total of 90 water samples were collected from five different spots during different seasons over a period of two years (November 2006 to October 2008). The samples were taken in BOD bottles and plastic jerry canes and brought to the laboratory with necessary precautions. All samples were labeled properly. Some parameters like temperature, velocity, pH and dissolved oxygen were measured on site. Grab sampling was generally applied during the sampling. Water samples were analysed by standard methods (Mathur R.P., 1982.; Toss, Fedrick C., 1983; Manivaskam N., 1986; Trivedi P.K. and Goel P.K., 1986; Khanna D.R., 1993; APHA standard methods for examination of water and waste water, 1998; Silvia Fernandez,

Unai Villanveva, Alberto de Diego Gorka Arana, Juan Manuel Madariaga, 2008). The samples were analyzed for following physico-chemical parameters :

Water temperature (ºC), velocity (m/s), pH, hardness (mg/l), turbidity (JTU), total dissolved solids (mg/l), total suspended solids (mg/l), electrical conductivity (μmho/cm), free CO_2(mg/l), dissolved oxygen (mg/l), B.O.D. (mg/l), C.O.D. (mg/l), alkalinity (mg/l), chloride (mg/l), calcium (mg/l), magnesium (mg/l), sodium (mg/l), potassium (mg/l), carbonate (mg/l), bicarbonate (mg/l) and sulphate (mg/l). Eleven parameters were taken for calculation (Kelly W., 1951) of water quality index : Ca, Mg, Na, K, NO_3^-, SO_4^{2-}, Cl^-, hardness, TDSD, B.O.D. and total alkalinity. MC, SP, SAR, RSC and PI were calculated as follows :

Water Quality Index (WQI)

It is an established fact that the more harmful a given pollutant is, the smaller is its standard permissible value recommended for drinking water. Therefore, the "Weights" for various water quality characteristics are assumed to be inversely proportional to the recommended standards for the corresponding parameters (Botton P.W., J.C. Currie and Ternet D.J. 1978), that is,

$$W_i = K/S_i$$

where W_i is the unit weight and S_i is the recommended standard for the *i*th parameter P_i. The constant of proportionality K in equation can be determined from the condition

$$\Sigma W_i = K\Sigma(1/S_i)$$

The quality rating q_i for the *i*th parameter P_i is calculated from the following equation :

$$q_i = 100(V_i/S_i)$$

Where V_i is the observed value. The subindex S_i for the parameter P_i is given by

$$(S_i) = (q_i w_i)$$

The overall WQI can be calculated by aggregating the quality rating (q_i) or subindices, linearly, and taking their weighted mean, *i.e.*

$$WQI = [(\Sigma q_i w_i / \Sigma w_i)]$$

Water samples were analyzed by standard methods.

Magnesium Content

Magnesium content of water is considered as one of the most important qualitative criteria in determining quality of water for irrigation. Magnesium content is calculated by the following formula (Pitchaiah, P.S. 1995).

$$\text{Mg content} = [Mg^{2+} / (Mg^{2+} + Ca^{2+})]\ 100$$

(Concentrations are in meq/l)

Sodium Per centage (SP)

Doneen method is used to calculate the sodium per centage (Pathak., S.S. : Upadhyaya K.K. and Joshi, S.K. 2002).

$$Na\% = [(Na^{+} + K^{+}) / (Ca^{2+} + Mg^{2+} + Na^{+} + K^{+})]\ 100$$

(Concentrations are in meq/l)

Sodium Absorption Ratio (SAR)

Sodium absorption ratio is an important parameter to determine the suitability of irrigation water and is calculated by the following formula (Raghunath, H.M. 1987).

$$SAR = Na^{+} / [(Ca^{2+} + Mg^{2+}) / 2]$$

(Concentrations are in meq/l)

Residual Sodium Carbonate (RSC)

The concept of residual sodium carbonate (RSC) is employed for evaluating high carbonate waters and is calculated by the formula given below (Domenico P.A. and Schwartz, F.W 1990).

$$RSC = (CO_3^{2-} + HCO_3^{-}) - (Ca^{2+} + Mg^{2+})$$

(Concentrations are in meq/l)

Permeability Index (PI)

Permeability index is calculated by the method suggested by Domenico and Schwartz (Domenico P.A., and Schwartz, F.W. 1990).

$$PI = [(Na^+ + HCO_3^-) / (Ca^{2+} + Mg^{2+} + Na^+)]\ 100$$

PI is used to evaluate the sodium hazards of irrigation water.

Results and Discussion

The results obtained from analysis of water samples of river Ganga are shown in table 1. The reported values refer to the mean value of water samples collected in different seasons at different areas along the stretch of Ganga River. The results indicate that the quality of water varies considerably from location to location. A summary of the findings is given below: The water temperature of the Ganga at Haridwar ranged between 10.18°C to 19.73°C. The maximum water temperature started decreasing due to the melting of snow at the peaks of the Himalaya. The water temperature showed an upward trend from winter season to summer season followed by a downward trend from rainy season onwards. The velocity was found to be directly proportional to the flood level and also with gradient of the river stretch. The water level and its velocity started increasing from winter season onwards due to melting of snow at the place of origin of the river. The maximum velocity 2.18 m/s of the Ganga at Haridwar was recorded in monsoon season and the minimum velocity 0.39m/s were observed in winter season.

The conductivity of water is affected by the suspended impurities and also depends upon the amount of ions in the water. The highest conductivity 415.66μmho/cm of the Ganga water was observed in monsoon season. From monsoon season onwards the conductivity decreased and minimum conductivity 95.89μmho/cm was observed in winter season. The turbidity in the river Ganga at Haridwar was lowest during winter season. From summer season onwards the water became turbid due to

Table 1 : Mean and Standard Deviation of Different Parameters at Different Sampling Stations

Sampling Station	*Bhooma Niketan*		*Jai Ram Ashram*		*Har ki Pauri*		*Prem Nagar Ashram*		*Pul Jatwara*	
Parameters	*Mean*	*S.D.*	*Mean*	*S.D.*	*Mean*	*S.D.*	*Mean*	*S.D.*	*Mean*	*S.D.*
Water Temp	16.70	4.2482	16.04	4.1811	16.45	4.1014	15.55	4.0425	15.82	4.0277
Velocity	1.18	0.7398	1.25	0.4964	1.52	0.5370	1.21	0.4834	1.32	0.5202
pH	8.16	0.1516	7.49	0.2675	7.47	0.3172	7.37	0.2777	7.42	0.3250
DO	9.28	1.4677	9.04	1.5955	9.53	1.8844	9.50	1.9003	9.38	1.7284
Free CO_2	3.00	1.5812	3.38	1.5684	3.05	1.5456	3.10	1.6316	3.13	1.6302
COD	3.89	1.3495	9.41	2.7837	8.77	3.0168	9.00	2.9982	9.38	2.9195
BOD	2.63	0.8041	2.95	0.7039	2.50	0.7390	2.60	0.7014	2.92	0.8602
CO_3^{2-}	2.46	1.2290	0.22	0.1241	0.20	0.0775	0.12	0.0628	0.15	0.0635
HCO_3^-	63.68	20.5654	52.46	13.7336	50.81	14.4990	50.26	14.0571	50.32	13.1820
Total Alkalinity	65.71	19.7929	52.69	13.6743	51.05	14.4619	50.51	14.2193	50.68	13.3678
Hardness	100.44	9.9355	109.97	7.5452	102.62	11.3952	117.09	26.6345	108.65	10.3600
Turbidity	235.63	281.4771	211.95	282.6695	208.57	278.0215	208.91	276.7748	203.11	267.0415
Ca	16.72	2.0371	19.18	2.4334	20.78	3.5819	15.57	2.3822	13.62	2.3524
Mg	4.18	0.5636	5.65	0.7868	6.78	0.5565	4.23	0.6947	3.48	0.7985
Na	12.1563	6.0644	14.5095	6.9261	15.8938	8.3367	15.5512	7.8066	15.4475	7.2250

Contd...

Sampling Station	*Bhooma Niketan*		*Jai Ram Ashram*		*Har ki Pauri*		*Prem Nagar Ashram*		*Pul Jatwara*	
Parameters	*Mean*	*S.D.*	*Mean*	*S.D.*	*Mean*	*S.D.*	*Mean*	*S.D.*	*Mean*	*S.D.*
K	2.21	0.5645	2.05	0.5812	2.07	0.8077	2.05	0.8115	2.07	0.8486
NO_3^-	0.0375	0.0354	0.0484	0.0372	0.0428	0.0315	0.0411	0.0233	0.0465	0.0335
PO_4^{3-}	0.0640	0.0565	0.1082	0.0749	0.0820	0.0508	0.1007	0.0676	0.0987	0.0662
Cl^-	6.11	5.1546	5.78	4.8231	5.71	4.8209	5.78	4.9050	5.84	4.9166
SO_4^{2-}	28.24	6.1971	24.52	6.3475	23.51	4.2921	22.76	3.4552	22.96	3.7560
EC	223.56	147.5192	200.47	108.0951	197.93	119.8590	199.83	120.4835	197.77	117.9629
TDSD	275.87	229.9665	336.20	294.7332	397.26	396.4792	409.91	401.9551	407.83	397.6374
TSSD	1151.76	1348.9397	1123.55	1369.5643	1169.96	1452.3666	1143.09	1406.9099	1155.23	1401.1927

melting of snow and rains. The maximum turbidity 608.15 JTU was observed in monsoon season and minimum 19.15 JTU was observed in winter season. Total solids may affect water quality. Water with high total solids generally is of inferior potability. Total dissolved solids were observed maximum 540.68 mg/l in rainy season and minimum 42.58 mg/l in winter season. Total suspended solids were recorded maximum 3125.76mg/l in monsoon season and minimum 110.28 mg/l.

The pH of the Ganga river at Haridwar was slightly alkaline. It ranged from 7.06 to 8.35. The Ganga water contained highest dissolved oxygen during winter season, followed by a gradual decrease to its lowest values during monsoon season. The higher concentrations of dissolved oxygen during winter season was probably due to low water temperature, no turbidity and increased photosynthetic activity of the green algae found on the submerged stones and pebbles. The maximum 11.71 mg/l oxygen content of water was recorded in winter season and minimum 7.08 mg/l in rainy season. From monsoon season the water of Ganga starts becoming turbid which reduces the photosynthetic activity of the algae and thus decreases oxygen concentration. Free carbon dioxide in the Ganga water was invariably present throughout the year. It fluctuated from 1.15mg/l in winter season to 5.39 mg/l in rainy season. The free carbon dioxide was found to be maximum in monsoon season and minimum in winter season. The C.O.D. ranged from 4.58mg/l to 13.72mg/l. The minimum C.O.D. was recorded in monsoon season and maximum in winter season.

The B.O.D. was maximum 3.90 mg/l in monsoon season and minimum 1.35 mg/l in winter season. Total alkalinity throughout the year ranges from 34.35 mg/l in winter season to 90.5 mg/l in summer season. The alkalinity due to carbonates was more or less nil. Maximum calcium 27.4 mg/l was found in rainy season. Minimum calcium 10.9 mg/l was found in winter season. Similarly, maximum magnesium 7.4 mg/l was found in rainy season and minimum magnesium 2.3 mg/l was found in winter season. Concentration of calcium was always greater than that of magnesium. The hardness was higher in the monsoon season (120.62mg/l) and lower in the winter season

(87.55mg/l). Calcium ions make major contribution to the hardness of river water.

Maximum concentration of sodium 28.320 mg/l was found in rainy season and minimum 6.720 mg/l in summer season. Similarly, maximum concentration of potassium 3.425 mg/lit was found in rainy season and minimum 1.216 mg/l in summer season. Sodium was found to have greater values than potassium throughout the study.

Maximum amount of nitrate (0.105mg/l) was found in rainy season and minimum amount (0.0115mg/l) was found in summer season. Nitrate concentration depends upon the activity of nitrifying bacteria. The total phosphate was highest in monsoon season (0.23mg/l) and lowest in winter season (0.037mg/l). Total sulphate was recorded maximum in monsoon season (37.15mg/l) and minimum in winter season (16.45 mg/l). The chloride was observed maximum (13.48 mg/l) in rainy season and minimum (1.97 mg/l) in winter season. Water quality index represents the integrated effects of the relevant water quality variables. Table 3 shows drinking water quality standards and unit weights for all the parameters used in calculating WQI. For Ganga river water, the rating of WQI of water samples was calculated and represented in table 4. Graphical representation of variation of WQI at all sampling stations in different seasons is shown in Fig. 2. It may be stated that the water quality requirements differ from one age to another and thus any polluted water may be considered suitable for some of the beneficial uses but may remain unsuitable for other purposes, gave the rating of water quality as shown below (Maruthi, Rao, 2004).

WQI Level	*Water Quality Rating*
0-25	Excellent
26-50	Good
51-75	Poor
76-100	Very poor
>100	Unfit for drinking purpose.

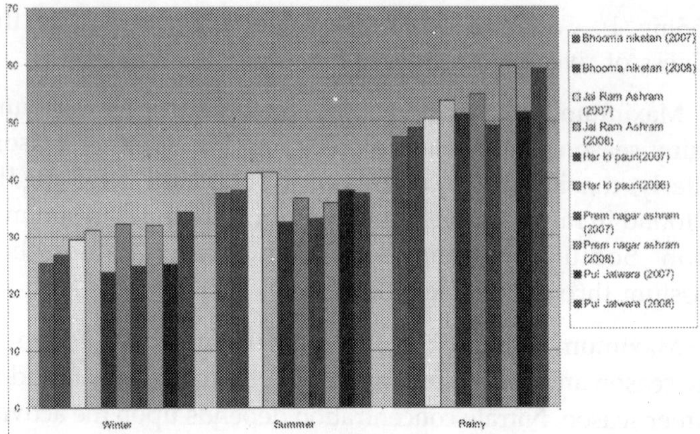

Fig. 2 : Seasonal trends of WQI at different sampling stations.

In the present study water of river Ganga was found to be in excellent quality in winter season at all the five sampling sites as the WQI ranged from 24.80 to 34.32 for both the years. Water of River Ganga was found to be of poor quality in rainy season as WQI at all sampling stations ranges from 47.30 to 59.81. The WQI starts increasing from winter to summer and it further increases from summer to rainy season. It was also observed that the water in the year 2007 was of better quality than in year 2008. It may be due to industrialization because of development of SIDCUL.

Table 2 : Mean of Different Parameters in Different Sampling Stations

S.N.	*Parameters*	*Values*	*Mean*
1.	Water Temp	10.18-19.73	16.11
2.	Velocity	0.39-2.18	1.30
3.	pH	7.06-8.35	7.58
4.	DO	7.08-11.71	9.34
5.	Free CO_2	1.15-5.39	3.13
6.	COD	4.58-13.72	8.09
7.	BOD	1.35-3.90	2.72
8.	CO_3^{2-}	0.02-3.50	0.63

Cont...

S.N.	Parameters	Values	Mean
9.	HCO_3^-	34.0-89.25	53.50
10.	Total Alkalinity	34.35-90.5	54.13
11.	Hardness	87.55-120.62	107.75
12.	Turbidity	19.15-608.15	213.63
13.	Ca	10.9-27.4	17.17
14.	Mg	2.3-7.4	4.87
15.	Na	6.72-28.32	14.71
16.	K	1.216-3.425	2.09
17.	NO_3^-	0.0115-0.105	0.04
18.	PO_4^{3-}	0.037-0.23	0.09
19.	Cl^-	1.97-13.48	5.84
20.	SO_4^{2-}	16.45-37.15	24.40
21.	EC	95.89-415.66	203.91
22.	TDSD	42.58-540.68	365.41
23.	TSSD	110.28-3125.76	1148.72

Table 3 : Drinking Water Quality Standards (Maximum Permissible Limit)

Parameters	*Standard Values (Si)*	*Wi = 1/Si*
Ca	75	0.0133
Mg	50	0.02
Na	20	0.05
K	10	0.1
NO_3^-	20	0.05
SO_4^{2-}	200	0.005
Cl^-	250	0.004
Hardness	300	0.0033
TDSD	500	0.002
BOD	5.0	0.2
Total Alkalinity	**120.00**	**0.0083**

Standards are taken according to WHO, ISI, ICMR.

Table 4 : WQI Result for the Year 2007–2008

Sampling Station	*Winter*	*Summer*	*Rainy*
Bhooma niketan (2007)	25.31	37.48	47.30
Bhooma niketan (2008)	26.84	38.03	48.79
Jai ram ashram (2007)	29.42	41.02	50.37
Jai ram ashram (2008)	31.17	41.14	53.57
Har ki pauri (2007)	23.81	32.41	51.25
Har ki pauri (2008)	32.17	36.67	54.64
Prem nagar ashram (2007)	24.8	33.11	49.29
Prem nagar ashram (2008)	32.08	35.81	59.81
Pul Jatwara (2007)	25.13	37.92	51.69
Pul Jatwara (2008)	34.32	37.54	59.23

Total Dissolved Salts (TDS)

Water used for irrigation can vary greatly in quality depending upon type and quantity of dissolved salts. Salts are present in irrigation water in relatively small but significant amounts. They originate from dissolution or weathering of the rocks and soil, including dissolution of lime, gypsum and other slowly dissolved soil minerals. These salts are carried with the water to wherever it is used. In the case of irrigation, the salts are applied with the water and remain behind in the soil as water evaporates or is used by the crop a salinity problem exists if salt accumulates in the crop root zone to a concentration that causes a loss in yield. Yield reductions occur when the salts accumulate in the root zone to such an extent that the crop is no longer able to extract sufficient water from the salty soil solution, resulting in a water stress for a significant period of time. If water uptake is appreciably reduced, the plant slows its rate of growth. Water with TDS less than 450mg/l is considered good and that with greater than 2000 mg/l is unsuitable for irrigation purpose. In the present study the minimum value of total dissolved solids was found in winter season with a range of 42.58 mg/lit to 70.12mg/lit. The maximum value was found in rainy season with a range of 545.68 mg/lit to 950.15mg/lit.

Electrical Conductivity (EC)

The most influential water quality guideline on crop productivity is the water salinity hazard as measured by electrical conductivity (EC) (Ahmed, S.S. Mazumder, Q.H. Jahan, C.S. Ahmed M and Islam, S. 2002). The primary effect of high EC water on crop productivity is the inability of the plant to compete with ions in the soil solution for water (physiological drought). The higher the EC, the less water is available to plants, even though the soil may appear wet. Because plants can only transpire "pure" water, usable plant water in the soil solution decreases dramatically as Ec increases. Water with EC less than 250μmhos/cm is considered good that with greater than 750 μmhos/cm is unsuitable for irrigation. In the year 2007 and 2008, the minimum EC was recorded in the winter season ranging from 95.89 μmho/cm to 130.68 μmho/cm. The maximum EC was recorded in rainy season ranges from 336.72 μmho/cm to 415.66 μmho/cm.

Magnesium Content

Magnesium content of water is considered as one of the most important qualitative criteria in determining the quality of water for irrigation. Generally, calcium and magnesium maintain a state of equilibrium in most waters. More magnesium in water will adversely affect crop yields as the soils become more alkaline. In the present study, the magnesium content of the water of river Ganga varies from 27.2 mg/l to 37.7mg/l. So, water is suitable for irrigation purpose in terms of magnesium content.

Sodium Per cent (SP)

Sodium per cent is another important factor to study sodium hazard. It is calculated as the per centage of sodium and potassium against all cationic concentration. It is also used for adjudging the quality of water for the use of agricultural purpose. The use of high per centage sodium water for irrigation purpose stunts the plant growth. Sodium reacts with soil to reduce its permeability (Todd D.K., 1980). Sodium per cent in water is a parameter computed to evaluate the suitability for irrigation (Mishra, 2005). Usually little or only minor problems occur when SP values are less than 15%. When SP>15%, reduced permeability will occur. The finer the soil texture and the greater the organic matter content, the greater the impact of sodium on water infiltration and aeration. The sodium per cent values of river Ganga were 23.56 to 52.35. These values are high in rainy seasons. Gypsum can be added to the soil to reduce the effect of high per centage of sodium in irrigation water.

Sodium Adsorption Ratio (SAR)

Excess sodium in waters produces the undesirable effects of changing soil properties and reducing soil permeability (Biswas, Hemlata Mohabey and Malik, 2002). Hence the assessment of sodium concentration is necessary while considering the suitability for irrigation. The degree to which irrigation water tends to enter into cation-exchange reactions in soil can be indicated by the sodium adsorption ratio (Singh and Chandel, 2004). Sodium replacing adsorbed calcium and magnesium is a hazard as it causes damage to the soil structure. It becomes compact and impervious. SAR is an important parameter for the determination of suitability of irrigation water because it is responsible for the sodium hazard. The waters were classified in relation to irrigation based in the ranges of SAR values. Water with SAR ranging from 0 to 3 is considered good and with greater than 9 is considered unsuitable for irrigation purpose. In the present study SAR was found to be 0.397 to 1.49. The low value of SAR for Ganga river water can be categorized as water of excellent category. It can be used for irrigation on almost all soil types with little danger of the development of harmful levels of exchangeable sodium.

Residual Sodium Content (RSC)

The concentration of bicarbonate and carbonate also influences the suitability of water for irrigation purpose. One of the empirical approaches is based on the assumption that all Ca^{2+} and Mg^{2+} precipitate as carbonate. Considering this hypothesis, Eaton proposed the concept of residual sodium carbonate (RSC) for the assessment of high carbonate waters. The water with high RSC has high pH and land irrigated with such water becomes infertile owing to deposition of sodium carbonate; as known from black colour of the soil. According to US Salinity Laboratory, an RSC value less than 1.25 meq/lit is safe for irrigation. A value between 1.25 and 2.5 meq/lit is of marginal quality and value more than 2.5meq/lit is unsuitable for irrigation. In the present study RSC values are below 1.25 meq/lit at all sampling stations. So water of Ganga river can be considered safe for irrigation purpose as mentioned according to above considerations.

Table 5 : Irrigation Water Characteristics in Winter Season, 2007

Parameter	*BN*	*JRA*	*HKP*	*PNA*	*PJ*
EC	128.52	110.28	97.86	95.89	100.06
DS	42.58	49.50	47.50	58.65	50.37
MC	29.412	31.800	35.976	30.246	26.018
PS	31.226	30.358	28.043	38.249	43.773
SAR	0.571	0.629	0.593	0.754	0.871
RSC	0.474	-0.601	-0.804	-0.304	-0.160
PI	115.513	74.361	66.607	91.017	101.644

Table 6 : Irrigation Water Characteristics in Summer Season, 2007

Parameter	*BN*	*JRA*	*HKP*	*PNA*	*PJ*
EC	125.58	147.67	139.25	142.27	138.25
TDS	217.95	246.20	226.18	234.46	246.15
MC	29.817	33.074	37.703	34.700	32.064
PS	23.566	24.036	23.654	28.409	33.163
SAR	0.396	0.458	0.486	0.534	0.588
RSC	–0.147	–0.168	–0.353	0.027	0.236
PI	89.838	86.130	78.555	98.859	116.566

Table 7 : Irrigation Water Characteristics in Rainy Season, 2007

Parameter	*BN*	*JRA*	*HKP*	*PNA*	*PJ*
EC	412.32	336.72	345.25	346.47	338.87
TDS	545.68	685.74	865.75	885.42	862.65
MC	29.817	33.074	37.703	34.700	32.064
PS	40.639	40.321	41.911	49.670	52.357
SAR	1.027	1.091	1.202	1.364	1.424
RSC	–0.559	–0.687	–0.801	–0.334	–0.189
PI	77.597	75.490	73.548	89.432	95.697

Table 8 : Irrigation Water Characteristics in Winter Season, 2008

Parameter	*BN*	*JRA*	*HKP*	*PNA*	*PJ*
EC	130.68	117.58	106.95	110.15	110.68
TDS	51.75	59.25	58.38	70.12	65.28
MC	28.744	32.258	35.032	27.206	27.304
PS	30.741	29.239	26.482	33.924	38.778
SAR	0.593	0.618	0.583	0.706	0.812
RSC	0.386	–0.684	–0.945	–0.491	–0.318
PI	107.001	71.160	62.578	80.067	89.304

Table 9 : Irrigation Water Characteristics in Winter Season, 2008

Parameter	*BN*	*JRA*	*HKP*	*PNA*	*PJ*
EC	128.60	150.28	141.55	143.80	140.58
TDS	235.12	265.85	258.55	262.15	272.38
MC	28.571	32.258	35.821	30.284	28.331
PS	22.902	22.719	21.580	25.689	29.724
SAR	0.402	0.460	0.463	0.525	0.581
RSC	-0.195	-0.343	-0.526	-0.215	0.021
PI	84.584	77.580	71.183	83.998	97.600

Table 10 : Irrigation Water Characteristics in Rainy Season, 2008

Parameter	*BN*	*JRA*	*HKP*	*PNA*	*PJ*
EC	415.66	340.28	356.72	360.39	358.15
TDS	562.15	710.65	927.18	948.68	950.15
MC	29.842	32.305	31.040	32.637	31.800
PS	41.281	39.994	39.787	50.125	49.882
SAR	1.070	1.158	1.235	1.499	1.450
RSC	–0.559	–0.915	–1.180	–0.514	–0.454
PI	77.727	70.495	66.115	83.634	85.164

Table 3-10

Permeability Index (PI)

The soil permeability is affected by long term use of irrigation water sodium, calcium, magnesium and bicarbonate content of the soil influence it. Doneen evolved a criterion for assessing the suitability of water for irrigation based on the permeability index. Accordingly, waters can be classified as Class I, Class II and Class III orders. Class I and Class II waters are categorized as good for irrigation with 75% or more maximum permeability. Class III water are unsuitable with 25% of maximum permeability. In the present study the minimum value of PI is 62.57. Hence water of River Ganga is of good irrigation quality. In the present study, TDS, MC, SP, RSC, SAR and PI were found to be within permissible range except in rainy season. In rainy season, TDS and PS were a little higher. Hence irrigation water quality of river Ganga was found to be good except in rainy season.

CONCLUSION

From present investigations it was concluded that the quality of the water samples under study was not suitable for drinking purpose in rainy season. In rainy season WQI increases due to increased concentration of sodium and dissolved solids. Because of high concentration of sodium, there is potential risk of getting cardiovascular diseases and in women toxemia associated with pregnancy. From WQI values, it is suggested that further improvement is required to treat the Ganga water at Haridwar. In the present study MC, SAR, RSC and PI were found in the permissible range for Ganga river water in Haridwar. But EC, TDS and ESP are higher in rainy season. Hence Ganga river water in haridwar is suitable for irrigation purpose except in rainy season.

REFERENCES

Ahmed, S.S. Mazumder, .Q.H. Jahan, C. S.Ahmed M and Islam, S. (2002): Hydrology and classification of ground water, Rajshahi city corporation area, Bangladesh, *Journal, Geologicalsociety of India*, Vol. 60, 411-418.

Biswas, S.N. Hemlata Mohabey and. Malik M.L. (2002) : Studies on Physico-chemical parameters to assess the water quality of Chandidongri Distt. Rajnandgaon for drinking and agriculture purpose, Asian journal of chemistry, Vol. 16, No. 2.

Botton P.W., J.C.Currie and Ternet D.J. (1978) : An index to improve water Pollution, Cent.20:653-659.

Domenico P.A and Schwartz, F.W (1990) : Physical and Chemical hydrology, John Wiley and sons, New York, pp 410-420.Goel P.K., (2006). Water Pollution: Cause, effects and control Second revised Edition, New Age International Publishers, ISBN : 8122418392.

Harkins R.D., (1974) : An objective water quality index, *J. Water Poll Cont. Fed*, 46 : 589.

Kelly W., (1951) : Alkali soils-their formation, properties and reclamation. Reinhold Publications, New York.

Khanna D.R., (1993) : Ecology and pollution of Ganga River, Ashish Publishing House, Delhi, 1-241.

Lohani B.N., (1981) : Water quality Indices, Water pollution and management reviews (Ed C.K.Varshney) South asian publications, New Delhi, pp 53-69.

Manivaskam N., (1986) : Physico-chemical examination, sewage and industrial, Pragati Prakashan Meerut.

Maruthi Y.A., Rao S.R., (2004) : Physico-chemical, biological and bacteriological study of Sarada River. *Asian J. of Chemistry*, 16(1) 122-130.

Mathur R.P., (1982) : Water and waste water testing, Nem Chand and Brothers, Publishers, Roorkee, 1-54.

Mishra. P.C., (2005) : Thesis- Some aspects of water quality of water in and around Rourkela. Department of Chemistry, National Institute of Technology, Rourkela.

Pathak, S.S. : Upadhyaya K.K. and. Joshi, S.K. (2002) : Some physico-chemical parameters and primary productivity of River Ramganga (Uttaranchal) *Him. J. Env. Zool*, 151-158.

Pitchaiah, P.S. (1995) : Ground water, Scientific Publ Jodhpur, Rajasthan pp. 304.

Raghunath, H.M. (1987) : Groundwater Wiley Eastern Ltd, Delhi India, p. 563.

Sharma R.N., Baruah A.K., Bora G.C. and Choudhary P.K., (1996): Assessment of physico-chemical parameters of the surface water around oil installation at Rydra Sagar Ashram India, *Pollu. Res.* 15 : 19-27.

Silvia Fernandez Unai Villanveva, Alberto de Diego Gorka Arana, Juan manuel Madariaga, Journal of Marine systems, Vol-72, Issue1-4, pages 332-341.

Tiwari T.N., Das S.C. and Bose P, K., (1986) : Water quality index for the river Jhelum in Kashmir and its seasonal variation *Poll. Res.* 5(1): 1-5.

Trivedy R.K., and Goel P.K., (1984) : Chemical and biological methods for water pollution studies Environmental Publications India, pp. 215.

Trivedy R.K., Khatawkar S.D., Kulkarni A.Y., and Shastri A.C., (1990): Ecology and pollution of the river Krishna in Maharashtra. Physico-chemical characteristics. River pollution in India, Ashish Publishing House, New Delhi, 26 : 99-193.

Todd D.K., (1980) : Groundwater Hydrology, 2nd Ed. John Wiley and Sons, New York, p. 535.

Toss, Fedrick C., (1983) : Introductory Microbiology, Charles E. Merill Publishing Company, A. Bell and Howell Company, Columbus, Ohio, pp. 1-615.

Vigendra Singh and Chandel C.P. Singh, (2004) : Assessing the quality of wastewater for farming purpose in and around various industrial areas of Jaipur city, Rajasthan In : *Asian Journal of Chemistry,* vol.18 no. 4, 2813-2817.

19

Physico-Chemical Characterization of Untreated and Treated Effluents of R.B.N.S. Sugar Mills at Laksar (Haridwar)

D.R. Khanna, R. Bhutiani, G. Matta, Prashant Tyagi and Bharti Tyagi

ABSTRACT

The present study was carried out to characterize the physico-chemical parameters from untreated and treated effluents from R.B.N.S. sugar mill. The samples were taken from two different sites of the industry *i.e.* before and after the treatment. The parameters studied were pH, Temperature, Odour, Colour, Total solids, Total dissolved solids, Total suspended solids, DO, BOD, COD and Oil & grease. The present study proved that after complete treatment the physico-chemical parameter were found within the permissible limit of BIS and Uttarakhand Pollution Control Board and the average per centage reduction in parameters like temperature, TS, TSS, TDS, BOD and COD were satisfactory. It can be concluded that the efficiency of ETP at R.B.N.S sugar mill is very good.

Keywords

BOD, DO, TDS, Effluent, physico-chemical, treatment.

INTRODUCTION

In recent years, much of the water bodies have become polluted by sewage, industrial wastes and wide range of synthetic chemicals (Hariharan and Varshya, 2002). The principal type of industries which contribute to water pollution of rivers in India are chemicals and pharmaceuticals, coal, soaps and detergents, pulp and paper, dyeing tanneries, steel mills, fertilizers, sugar industries etc. These industries produce large amount of waste water as effluent. All types of effluents and most of the byproducts from any kind of industry create most serious pollution to the water bodies and soil properties (Baskaran *etal.* 2009). The sugar industries in India are playing an important role in the economic development of the country because our country is the mainly agriculture based country. Generally the sugar mills are in rural areas where effluents pollute small rivers but these industries are of seasonal nature and operate for about 6 months in a year. Sugar industries generates large volumes of effluent and also emit harmful stock gases from its biogases fires, which contributes nuisance in the surroundings. The effluents are discharged during manufacture of sugar and they contain high polluted contents. Sugar industry is one of the major contributors towards pollution. In India molasses, a byproduct of sugar industry contains about 50% fermentable sugar which is a predominant raw material used in the production of alcohol. Use of molasses generates waste water containing highly biodegradable dissolved solids, high ash contents, high temperature, low pH, high concentration of calcium, chloride, sulphate ions. Thus the effluent of sugar industries contains higher amount of suspended solids, dissolved solids, BOD, COD, chloride sulphate, nitrates, calcium and magnesium. Sugar mill waste mainly consists of liquid containing organic and inorganic impurities which are harmful to human body, plant and animal life. Mill house, filter cloth washing, floor washing and leakage of juice from pipe are the main sources of water pollution from sugar industries. Sugar industries also generate air pollution though to a small extent. These give out large amount of suspended particulate matter in the form of fly ash in the atmosphere, this effect vegetation by inhibiting their photosynthetic activity. Various works have been done on a number of sugar mills so as to figure out the impact of their effluent on water bodies and land. David (1956) worked on studies on pollution of river Daha (N. Bihar) by sugar and distillery wastes. Marciniak and Lackowski, (1988) described the operation of biological waste water treatment plant (anaerobic fermentation followed by activated sludge process) at Melno Sugar Mill, Poland. Sinha (1993) observed the physico-chemical characteristic of effluent discharged from Lohat sugar factory in Bihar. Uttarakhand state has half dozen sugar mills in the 13 districts including R.B.N.S Sugar mill.

MATERIALS AND METHOD

The area under study was R.B.N.S Sugar mill, situated 27 km away from Haridwar. It is situated on 29.75° latitude and 78.05° longitude near Laksar Railway station at a distance 500 meters in the east. R.B.N.S Sugar Mill comes under the Green Category and does not affect human health directly. Samples were collected from two different sites; these were Site-1 (Before Treatment) and Site-2 (After Treatment). The physico-chemical parameters analysed were BOD, COD, Colour, DO, Odour, Oil & Grease, pH, TDS, Temperature, TS, and TSS using standard method outlined in Trivedi and Goel (1984), Khanna and Bhutiani (2004) and APHA (2005). The data were statistically analyzed and correlation coefficients were figured out.

RESULTS AND DISCUSSION

The average values of all the analyzed physico-chemical parameters are given in the table numbers 1 and 2 represents the ISI standards for discharge of sugar industry effluents. While the comparison of physico-chemical parameters before treatment and after treatment of effluent samples are graphically represented in Fig. 1-6. The mean pH of untreated and treated wastes was found 9.12±0.36 and 7.41±0.07 respectively. According to IS: 2490 (Part 2nd)-1990, the pH should ranges in between 5.50 to 9.00 for effluents of sugar industries. The parameter of temperature is basically important for its effects on the chemistry and biological reactions of the organism in the water. A rise in temperature in the water leads to the speeding up of chemical reactions in water and amplifies the taste and odours. The temperature of untreated and treated wastes was recorded 41.50±0.97 and 13.12±50.41. Chopra and

Arya, (1996), Patil (1998) and Bhutiani and Khanna (2003) also observed same pattern of temperature of industrial effluent. Total solids denote the various kinds of minerals present in the water. In the polluted waters, the concentration of other substances increases, depending upon the type of pollution. Total solids of untreated and treated effluents were found 1122.50 ± 43.81and 230.50 mg/ l ± 36.37 respectively. The weight of total suspended solids was found 595.00 ± 32.61 and 12.50 mg/l±1.41 for untreated as well as for treated effluents sequentially. According to (IS: 2490, Part-1990) standards the specific tolerance limit for industrial effluents is 100mg/l. Dissolved oxygen is one of the most important parameter in water quality assessment. Low oxygen concentration is generally associated with heavy contamination by organic matter. According to Khanna *et al.* (2002), DO is one of the most important parameters of water and it also reflects the physical and biological process which maintain higher form of biological life in the water and waste water discharged. The value of Dissolved oxygen for treated effluent was 3.22mg/l ± 0.88 while for untreated it was found nil. Similar trend have been recorded by Chopra and Arya (1996) in industrial effluents. Biochemical oxygen demand is the amount of oxygen utilized by microorganisms in degrading the organic wastes in a given sample of water. In this study values of Biochemical oxygen demand for untreated and treated effluents were observed 1512.50 ± 66.21and 16.00 mg/l ± 1.93. According to ISI (IS: 2490, Part 11) the specific tolerance for effluent is 30.00 mg/l. Similar trend was observed by Pathade and Kale (1999) and Patil (1998). Chemical oxygen Demand is the oxygen required by the strong oxidizing agent in degrading the organic as well as inorganic substances in given water sample. In general chemical oxygen demand is more than Biochemical demand for most of the industrial effluents. The values of chemical oxygen demand of untreated and treated effluents were 1127.50 ± 246.00 and 132.00 mg/l ± 9.79 respectively. This value was found under limit. Pathade and Kale, (1999) found same pattern of COD in industrial effluent. The value of oil and grease was found 0.07 mg/l for treated while in case of untreated effluent it was 6.02 ± 0.68.

Table 1 : Variation in Physico-Chemical Parameters

Parmeters	*Site – I (Before Treatment) Average Value*	*Site – II (After treatment) Average Value*
pH	9.12±0.36	7.41±0.07
Temperature (°C)	41.50±0.97	13.12±12.90
Odour	Decaying	Absent
Colour	Black	Absent
Total Solids (mg/l)	1122.50±43.81	230.60±36.37
Total Dissolved Solids (mg/l)	532.50±37.62	218.20±16.97
Total Suspended Solids (mg/l)	595.00±32.61	12.50±1.41
Dissolved Oxygen (mg/l)	NIL	3.22±0.88
Biochemical Oxygen Demand (mg/l)	1512.50±66.21	16.00±1.93
Chemical Oxygen Demand (mg/I)	1127.50±246.00	132.50±9.79
Oil & Grease (mg/l)	6.02±0.68	0.07±0.08

Table 2 : Standards for Sugar Mills Effluents (ISI : 2490 (Part-2) 1990)

S.N.	*Characteristics*	*Tolerance Limit*
1.	Colour and Odour	Absent
2.	pH	5.5 to 9.0
3.	BOD	30 mg/l
4.	Total Dissolved Solids	2199 mg/l
5.	Suspended Solids	100 mg/l
6.	COD	250 mg/l

Table 3 : Correlation Coefficient Among the Different Parameters at Site–I (Before Treatment)

Parameters	*pH*	*Temp*	*TS*	*TDS*	*TSS*	*BOD*	*COD*	*Oil and Grease*
pH								
Temp	0.73							
TS	–0.78	–0.53						
TDS	0.61	0.78	–0.08					
TSS	–0.96	–0.87	0.66	–0.79				
BOD	0.62	0.09	–0.87	–0.23	–0.39			
COD	0.15	–0.53	–0.32	–0.51	0.11	0.74		
Oil & Grease	–0.37	0.02	0.82	0.48	0.15	–0.93	–0.67	

Table 4 : Correlation Coefficient Among the Different Parameters at Site–II (After Treatment)

Parameters	*pH*	*Temp*	*TS*	*TDS*	*TSS*	*BOD*	*COD*	*Oil and Grease*
pH								
Temp	0.76							
TS	0.03	–0.07						
TDS	0.00	–0.10	0.99					
TSS	0.54	0.56	–0.81	–0.83				
BOD	0.40	0.14	0.92	0.91	–0.54			
COD	–0.26	–0.20	0.94	0.95	–0.92	0.74		
Oil & Grease	0.54	0.22	0.84	0.82	–0.40	0.98	0.62	

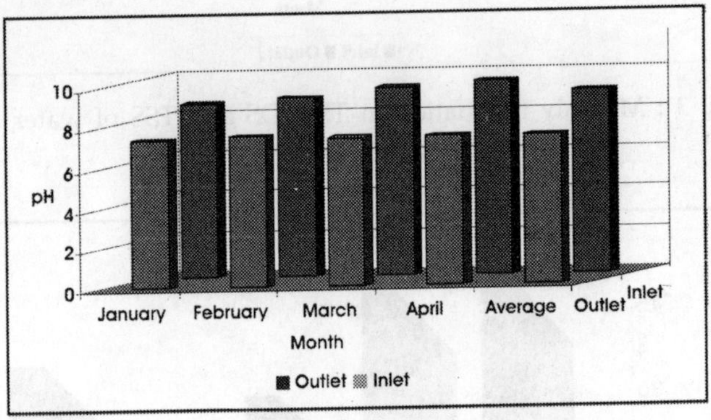

Fig. 1 : Monthly fluctuation in pH of water at inlet site and outlet site.

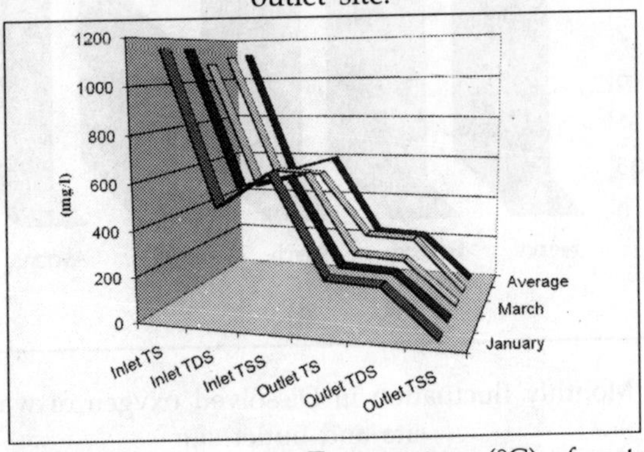

Fig. 2 : Monthly fluctuation in Temperature (°C) of water at inlet site and outlet site

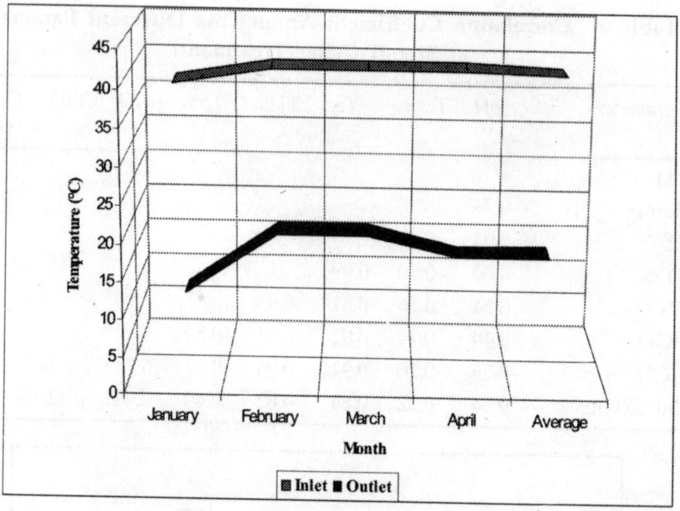

Fig. 3 : Monthly fluctuation in TS, TDS and TSS of water at inlet site and outlet site

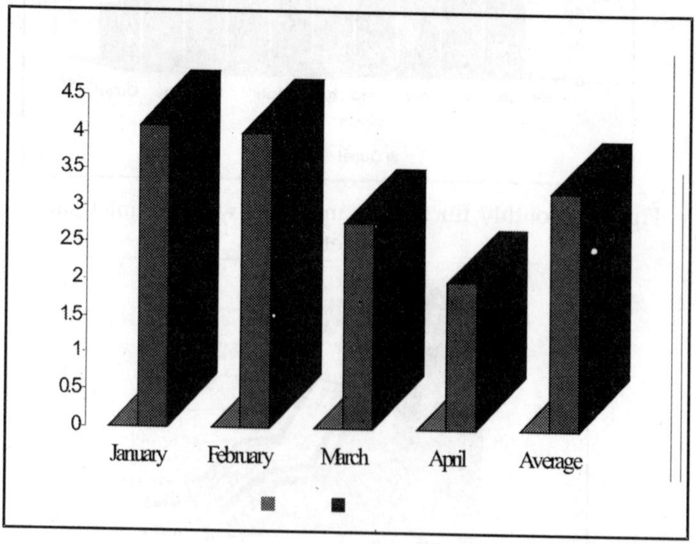

Fig. 4 : Monthly fluctuation in Dissolved oxygen of water at inlet site and outlet site

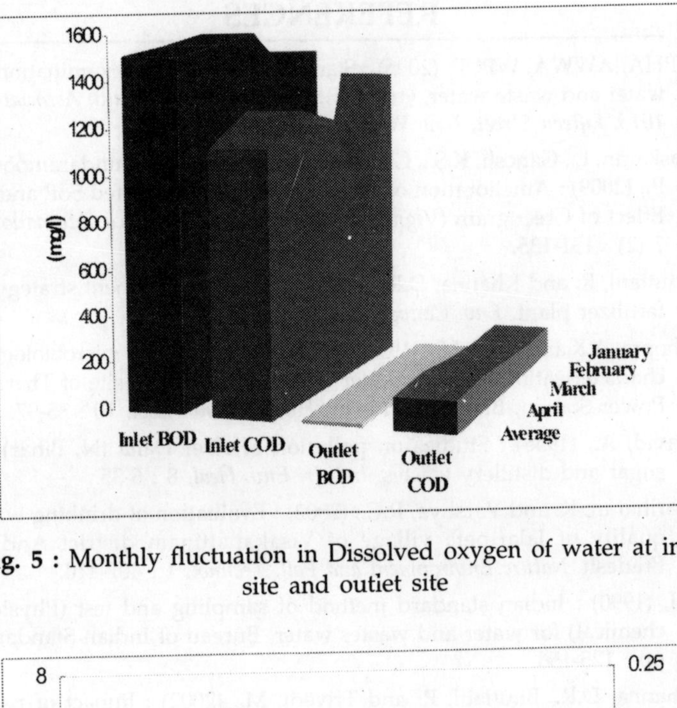

Fig. 5 : Monthly fluctuation in Dissolved oxygen of water at inlet site and outlet site

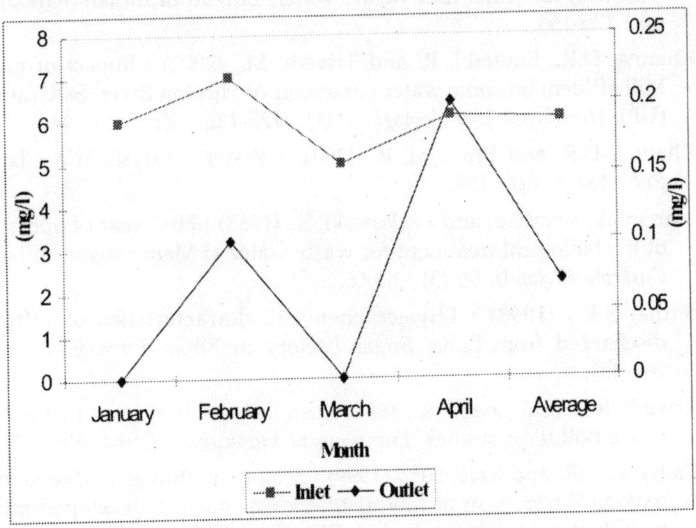

Fig. 6 : Monthly fluctuation in oil and grease of water at inlet site and outlet site

REFERENCES

APHA, AWWA, WPCF, (2005) : Standard methods for examination of water and waste water, (16th ed.), *American public Health Association; 1015, Fifteen Street, New Washington, 15 : 1-1134.*

Baskaran, L., Ganesh, K.S.,. Chidambaram, A.L.A. and Sundaramoorth, P., (2009) : Amelioration of Sugar Mill Effluent Polluted Soil and its Effect of Greengram (*Vigna radiata* L.). *Botany Research International,* 2 (2) : 131-135.

Bhutiani, R. and Khanna, D.R., (2003) : Waste management strategy of fertilizer plant. *Env. Conver. J.,* 4 (1-3) : 109-122.

Chopra, A.K.and Arya, J.S., (1996) : Physico-chemical and microbiological characterization of waste water effluent of different site of Thermal Power Station, BHEL, Haridwar. *Him. Jr. Env. Zool,* 10 : 35-37.

David, A., (1956) : Studies on pollution of river Daha (N, Bihar) by sugar and distillery wastes. *Ind. Jr. Env. Heal,* 8 : 6-35.

Hariharan, K. and Varshya, R.C., (2002) : Evaluation of drinking water quality of Jalaripeta village of Visakapattinam district Andhra Pradesh. *Nature Environment and Poll. Technol,* 1 : 407-410.

ISI, (1990) : Indian standard method of sampling and test (Physical-chemical) for water and wastes water, Bureau of Indian Standards. pp : 125-156.

Khanna, D.R., Bhutiani, R. and Trivedi, M., (2002) : Impact of paper Mill effluent on some water parameter of Hindon River, Saharanpur (UP). *Himalayan J. of Zoology,* 16(1) : 125-128.

Khanna, D.R. and Bhutiani, R., (2004) : Water Analysis at a Glance. *Pub. ASEA,* pp : 108.

Marciniak, Krystian, and Lackowski, S., (1988) : First year of operation of the biological treatment for waste water at Memo sugar mill. *Guz Curkrow (Poland),* 96 (3) : 64-66.

Sinha, S.K., (1993) : Physico-chemical characteristics of effluent discharged from Lohat Sugar Factory in Bihar. *Envi. Ecol,* 11 (2) 265-268.

Trivedi, R.K. and Goel, P.K., 1984. Chemical and Biological method for water pollution studies. *Environment publication, Karad,* New Delhi.

Pathade, G.R. and Kale, S.C., (1999) : Sclape up studies on the aerobic biological treatment of raw distillery waste using developed mixed microbial seed culture. *Jr. Ind. Poll. Cont,* 15 (1) : 21-34.

Patil, S.S., (1998) : Occurrence and tolerance of algae in distillery waste water. *Jr. Ind. Poll. Cont,* 14 (2) : 89-93.

20

Quality Assessment of Waste Water Being Utilized for Cultivation of Crops and Vegetables

Subrata Pani, Amit Dubey and Mukesh Dixit

INTRODUCTION

The large scale urbanization, a consequence of economic development is leading to production of huge quantities of effluents in India and posing serious environmental problems for their disposal. Sewage sludge (bio solids) generation in India is also increasing at a faster rate as more and more waste water treatment facilities with enhanced efficiencies are being developed. Sewage sludge and effluents are frequently disposed off on agricultural lands for irrigation/manures purposes that create both opportunities and problems.

The use of sewage sludge and other organic materials as fertilizers for plant growth improvement is one of the opportunities (Bahri, 2000). Many countries have used human and animal excreta for soil fertility enhancement throughout their history. These materials have been a source of nitrogen, phosphorous, potassium and sulphur for plants, and have contributed to the maintenance of desirable physical properties of soils, thus allowing for more abundant food and fiber production than would have been possible without them. Yet, in recent times, man has come to recognize that application of excessive quantities of these materials to land poses potential pollution problems, and a rising concern has become evident in many localities (Asano *etal.* 1985). Therefore keeping the above facts in mind the present study was conducted to evaluate the characteristics of wastewater of Patra Drain which is mainly used for agriculture purposes and also to identify major impacts of wastewater irrigation on soil and selective vegetables.

STUDY AREA

Photo-1 : Irrigation of agriculture land with sewage water.

The proposed study has been done in one of the major drains of Bhopal commonly known as Patra Nalla to assess the impact of use of sewage water on agricultural activity. Patra Nalla is of the major channels of Bhopal, carrying huge quantity of waste water of the densely habitat city area.

Material and Methods

Samples of waste water were collected at monthly intervals from three stations of the drain. Standard methods (APHA, 1995), Adoni (1985), OLSEN *et al.* 1982 have been followed in analyzing the identified parameters.

Similarly, vegetables (spinach, onion and beetroot) were also collected from the areas where they are being grown with the irrigation of sewage water pumped from the drain (Photo-1). Soil samples were also collected from the three these areas.

Sample Collection

For collection of water samples, containers were thoroughly washed with detergent, rinsed with water and then with distilled water before soaking in 5% HNO_3 for about 24hrs.

Containers were finally rinsed with distilled water before being used for sampling. Water and sludge samples were collected following the procedure recommended by American Public Health Association (1995). Sewage water samples were kept cooled in the laboratory. Whereas soil samples, were collected and placed in a polyethylene bags, labeled and taken to the laboratory. Vegetable samples were placed in polyethylene bags, labeled and taken to the laboratory. Plant samples were thoroughly rinsed with water and with distilled water to remove any attached soil particles, cut into smaller pieces and placed in a large crucible where they were oven dried at 60°C over night. The dried vegetables were grinded into fine particles using a clean acid–washed mortar and pestle. A portion of the homogeneous soil samples were also air dried by circulating air in the oven over night and thereafter grinded and sieved mechanically using a 2mm sieve.

Results

Variations in different physico-chemical parameters in water samples collected from Patra Nalla are described below:

1. pH : pH concentration at different stations along Patra Nallah ranges from 6.5 to 6.7. The minimum value of pH was observed in Bhanpur *i.e.,* station no. 3 whereas the maximum value was rexcorded in Islamnagar (behind quila) *i.e.;* station no. 2.

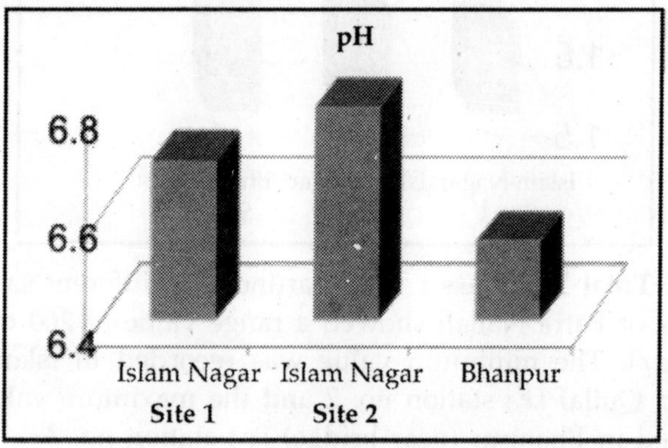

2. Dissolved Oxygen : Dissolved oxygen during the period of investigation showed a range value of 2 mg/l to 6 mg/l. The minimum value of dissolved oxygen was observed in Bhanpur *i.e.*, station no. 3 while the maximum value was noticed in Islamnagar (near bridge) *i.e.*; station no. 1.

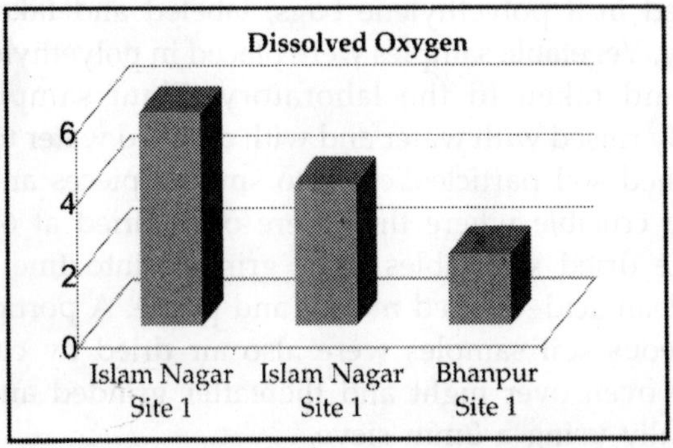

3. Conductivity : Conductivity in different stations of Patra Nallah was found within the range of 1.61 mS/cm to 1.69 mS/cm. The minimum value was observed in Islamnagar (near bridge) *i.e.*; station no.1 and the maximum value was observed in Bhanpur (near bridge) *i.e.*; station no. 3.

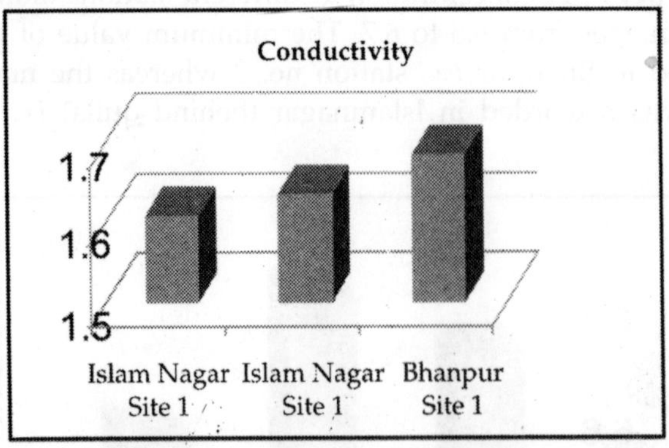

4. Total Hardness : Total hardness in different sampling stations of Patra Nallah showed a range value of 260 mg/l to 300 mg/l. The minimum value was recorded in Islamnagar (behind Quila) *i.e.*; station no. 2 and the maximum value was observed in Bhanpur (near bridge) *i.e.*; station no. 3.

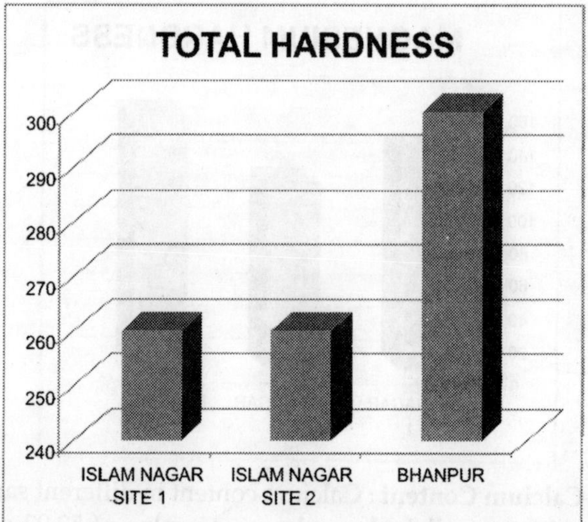

5. Calcium Hardness : Calcium hardness in different sampling stations of Patra Nallah showed a range value of 126 mg/l to 147mg/l. The minimum value was observed in Islamnagar (near bridge) *i.e.* station no.1 and the maximum value was recorded in Bhanpur (near bridge) *i.e.* station no. 3.

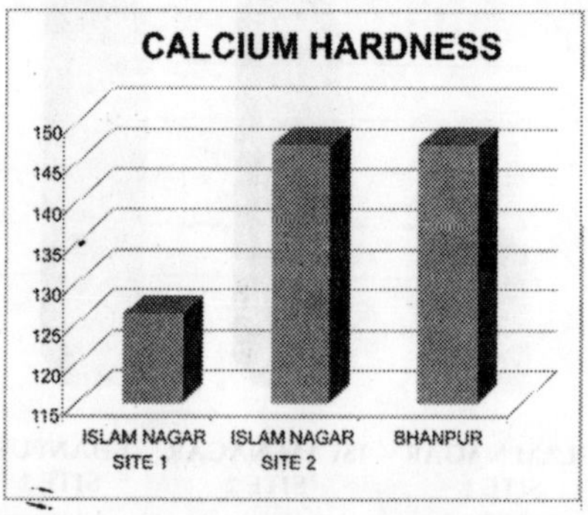

6. Magnesium Hardness : Magnesium hardness in different sampling stations of Patra Nallah showed a range value of 113 mg/l to 153 mg/l. The minimum value was found in Islamnagar (behind Quila) *i.e.;* station no. 2. The maximum value during the period of investigation was observed in Bhanpur (near bridge) *i.e.;* station no. 3.

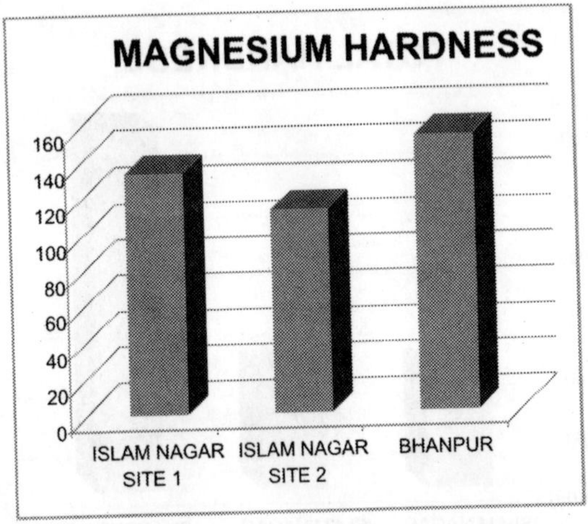

7. Calcium Content : Calcium content in different sampling stations of patra nallah showed a range value of 52.92 mg/l to 61.74 mg/l. The minimum value was observed in Islamnagar (near bridge) *i.e.;* station no.1 and the maximum value were observed in Bhanpur (near bridge) *i.e.;* station no. 3.

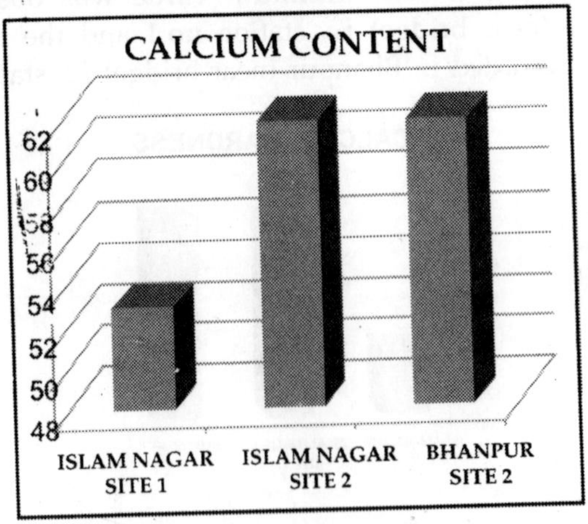

8. Magnesium Content : Magnesium content in different sampling stations of Patra Nallah showed a range value of 27.45 mg/l to 37.17mg/l. The minimum value was observed in Islamnagar (behind Quila) *i.e.;* station no. 2 while the maximum value was observed in Bhanpur (near bridge) *i.e.;* station no. 3.

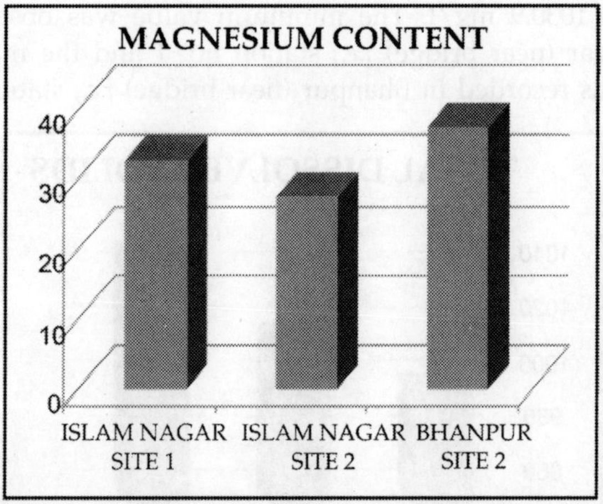

9. Chloride : The value of Chloride in different sampling stations of Patra nallah showed a range value of 54.94 mg/l to 139.86 mg/l. The minimum value was observed in Islamnagar (near bridge) *i.e.* station no.1 and the maximum value was observed in Bhanpur (near bridge) *i.e.;* station no. 3

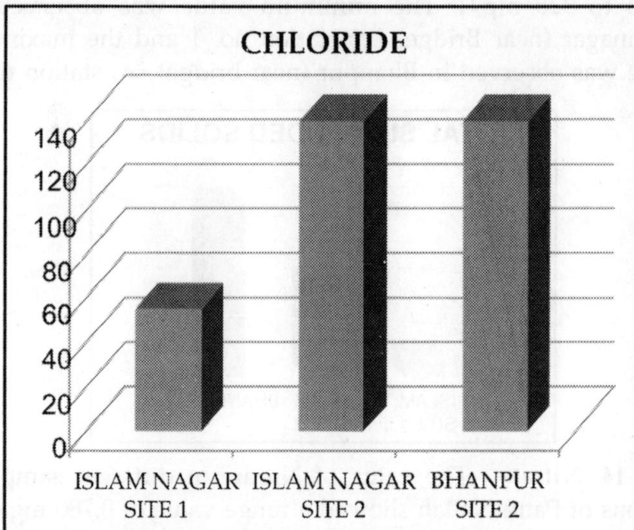

10. Total Dissolved Solids (TDS) : TDS in different sampling stations of Patra Nallah showed a range value of 982.1

mg/l to 1030.9 mg/l. The minimum value was observed in Islamnagar (near bridge) *i.e.;* station no. 1 and the maximum value was recorded in Bhanpur (near bridge) *i.e.;* station no. 3.

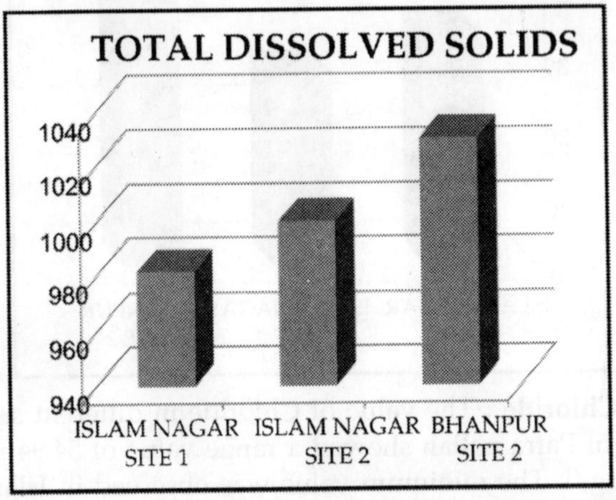

11. Total Suspended Solids (TSS) : The value of TSS in different sampling stations of Patra nallah showed a range value of 43 to 320 mg/l. The minimum value was observed in Islamnagar (near Bridge) *i.e.;* station no. 1 and the maximum value was observed in Bhanpur (near bridge) *i.e.;* station no. 3

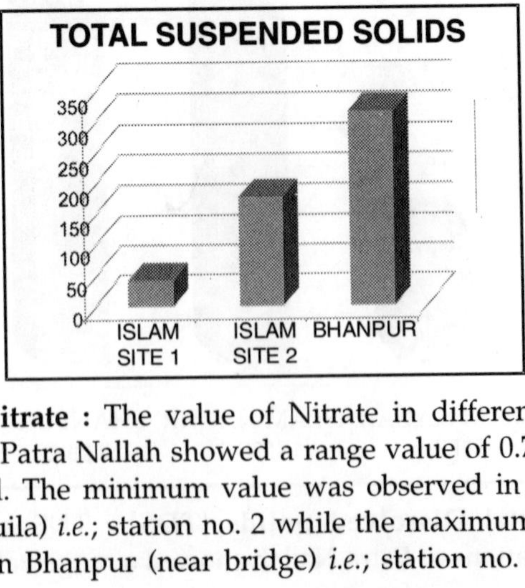

14. Nitrate : The value of Nitrate in different sampling stations of Patra Nallah showed a range value of 0.780 mg/l to 4.557 mg/l. The minimum value was observed in Islamnagar (behind Quila) *i.e.;* station no. 2 while the maximum value was observed in Bhanpur (near bridge) *i.e.;* station no. 3.

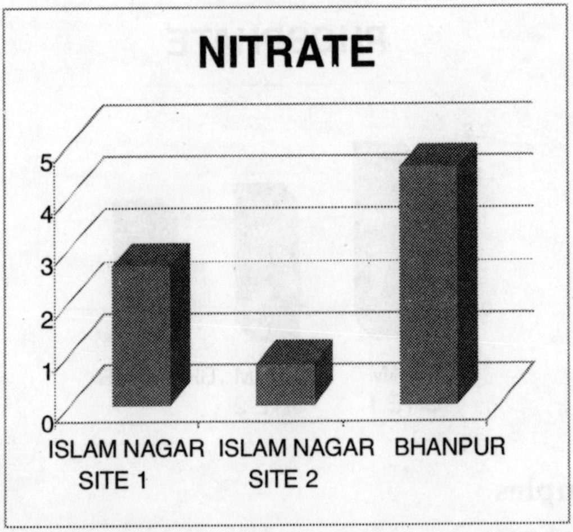

15. Total Phosphorous : The value of total phosphorous in different sampling stations of Patra Nallah showed a range value of 14.19 mg/l to 14.90 mg/l. The minimum value was observed in Islamnagar (near Bridge) *i.e.;* station no.1 and the maximum value was observed in Islamnagar (behind Quila) *i.e.;* station no. 2.

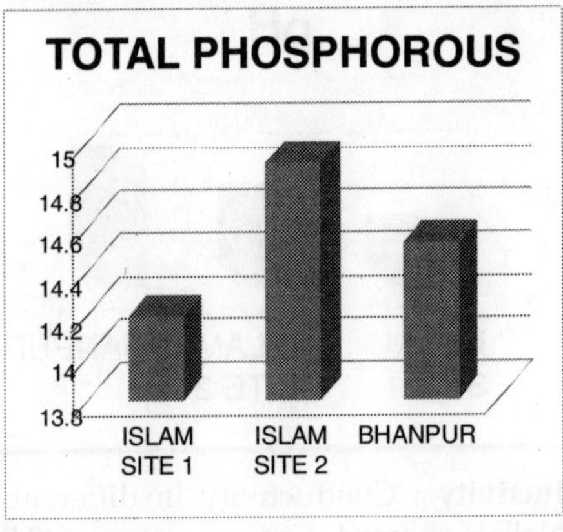

16. Orthophosphate : The value of orthophosphate in different sampling stations of Patra Nallah showed a range value of 1.413 to 2.017mg/l. The minimum value was observed in Bhanpur (near bridge) *i.e.;* station no. 3 and the maximum value was observed in Islamnagar (near Bridge) *i.e.;* station no.1.

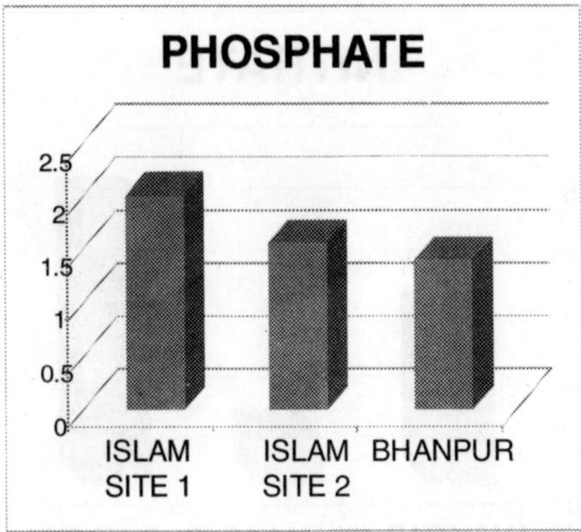

Soil Samples

1. pH : pH concentration in soil sample of different stations along Patra Nallah ranges from 7.17 mg/Kg to 7.44 mg/Kg. The minimum value was observed in Islamnagar (Prabhu lal Kushwaha field) *i.e.;* station no. 1 and the maximum value was observed in Bhanpur (near bridge) *i.e.;* station no. 3.

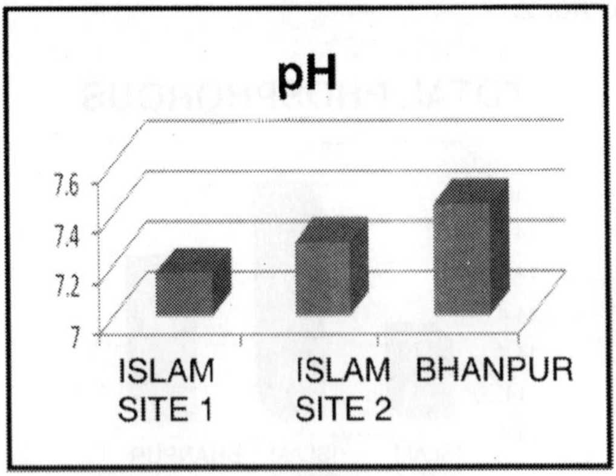

2. Conductivity : Conductivity in different soil stations along Patra Nallah showed a range value of 0.51 mS/cm to 0.66 mS/cm. The minimum value was observed in Islamnagar (Prabhu lal Kushwaha field) *i.e.;* station no. 1 and the maximum value of dissolved oxygen was observed in Bhanpur (near bridge) *i.e.;* station no. 3.

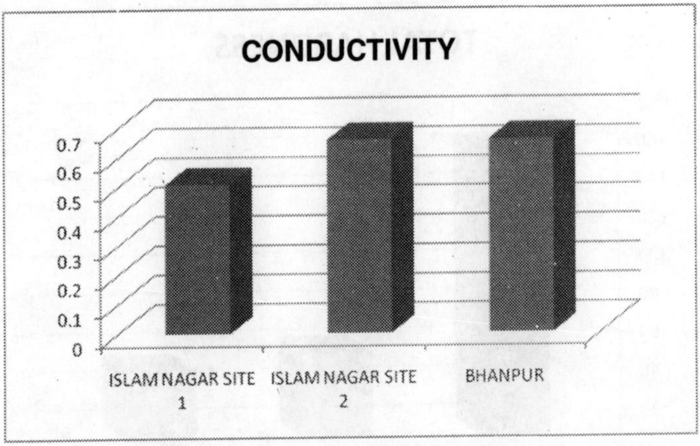

3. Moisture Content : Moisture content in different soil stations along Patra Nallah showed a range value of 3.29 to 4.26%. The minimum value was observed in Islamnagar (Ganesh Ram Kushwaha Field) *i.e.;* station no. 2 and the maximum value were observed in Islamnagar (Prabhu lal Kushwaha field) *i.e.;* station no.1 .

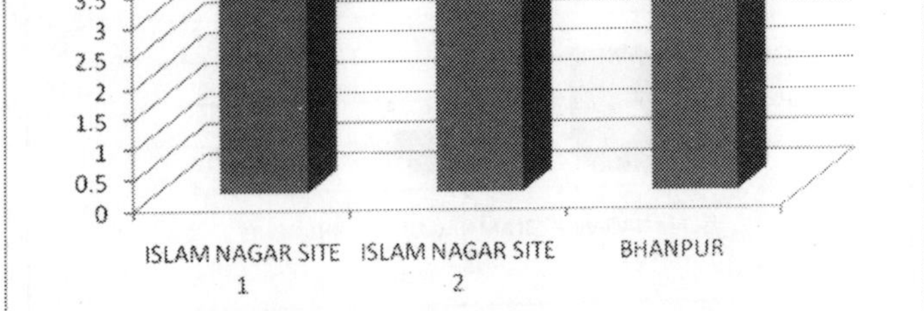

4. Total Hardness : Total hardness in different soil stations along Patra Nallah showed a range value of 36 mg/Kg to 148 mg/Kg. The minimum value was observed in Bhanpur (near bridge) *i.e.;* station no. 3 and the maximum value was observed in Islamnagar (Prabhu lal Kushwaha field) *i.e.;* station no. 1.

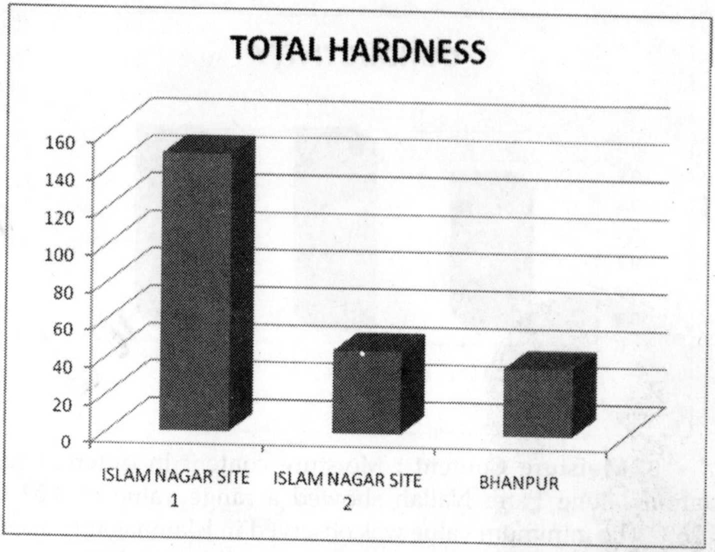

5. Calcium Hardness : Calcium hardness in different sampling soil stations along Patra Nallah showed a range value of 14.7 mg/Kg to 126 mg/Kg. The minimum value was observed in Bhanpur (near bridge) *i.e.;* station no. 3 and the maximum value was observed in Islamnagar (Prabhu lal Kushwaha field) *i.e.;* station no. 1.

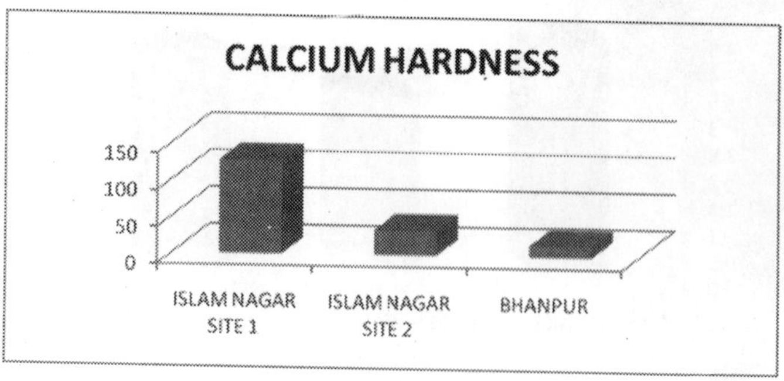

6. Magnesium Hardness : Magnesium hardness in different soil sampling stations along Patra Nallah showed a range value of 1.1 mg/Kg to 22 mg/Kg. The minimum value was observed in Bhanpur (near bridge) *i.e.;* station no. 3 and the maximum value was observed in Islamnagar (Prabhu lal Kushwaha field) *i.e.;* station no. 1.

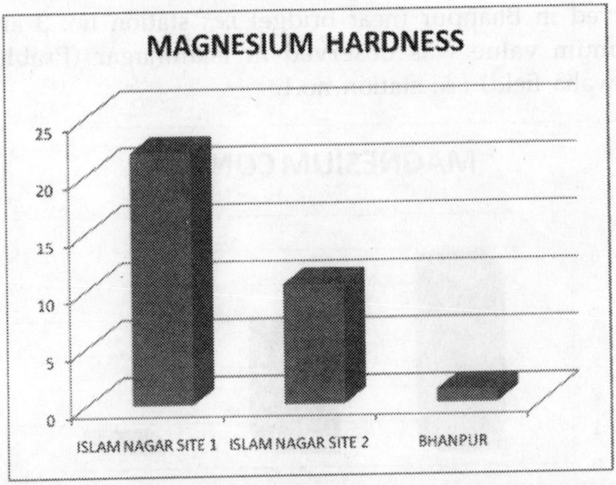

7. Calcium Content : Calcium content in different soil sampling stations along Patra Nallah showed a range value of 6.17 mg/Kg to 52.92 mg/Kg. The minimum value was observed in Bhanpur (near bridge) *i.e.;* station no. 3 and the maximum value was observed in Islamnagar (Prabhu lal Kushwaha field) *i.e.;* station no. 1.

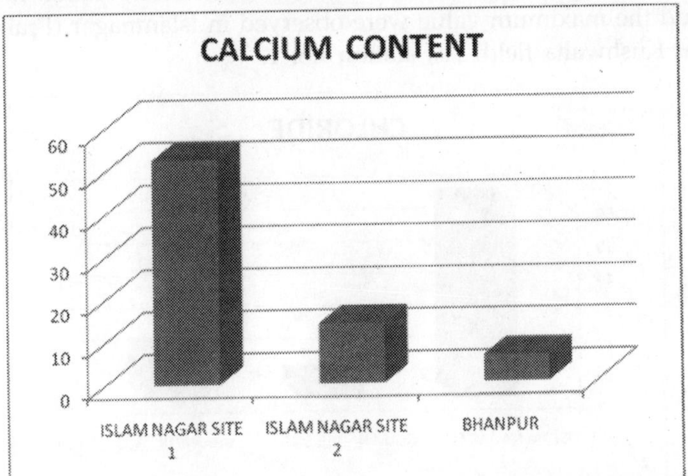

8. Magnesium Content : Magnesium content in different soil sampling stations along Patra Nallah showed a range value of 0.26 mg/Kg to 5.34 mg/Kg. The minimum value was observed in Bhanpur (near bridge) *i.e.;* station no. 3 and the maximum value was observed in Islamnagar (Prabhu Lal Kushwaha field) *i.e.;* station no.1.

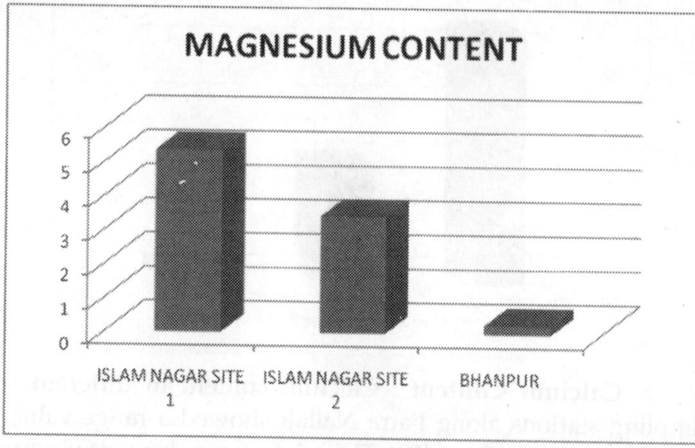

9. Chloride : The value of Chloride in different soil sampling stations along Patra Nallah showed a range value of 9.99 mg/Kg to 59.94 mg/Kg. The minimum value was observed in Islamnagar (Ganesh Ram Kushwaha Field) *i.e.;* station no. 2 and the maximum value were observed in Islamnagar (Prabhu lal Kushwaha field) *i.e.;* station no. 1.

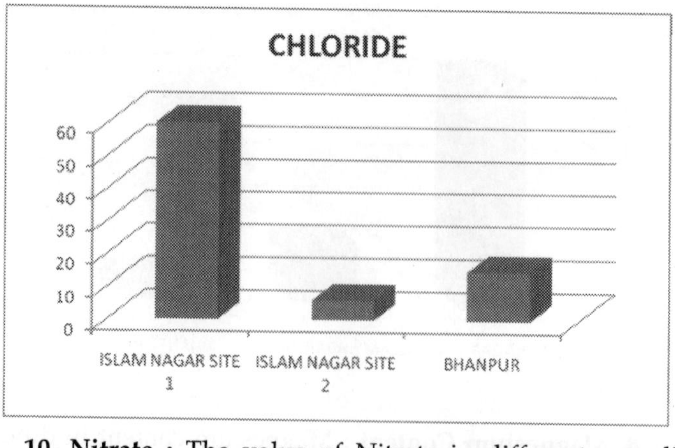

10. Nitrate : The value of Nitrate in different sampling soil stations along Patra Nallah showed a range value of 1.180

to 4.923 mg/l. The minimum value was observed in Islamnagar (Prabhu lal Kushwaha field) *i.e.*; station no. 1 and the maximum value was observed in Islamnagar (Ganesh Ram Kushwaha Field) *i.e.*; station no. 2.

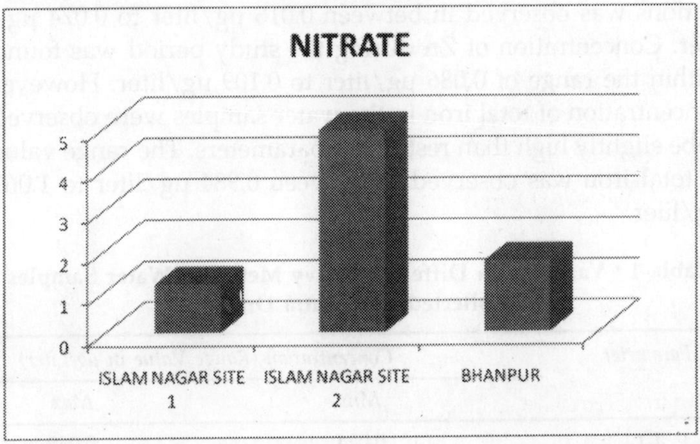

11. Total Phosphorous : The value of total phosphorous in different soil sampling stations of Patra Nallah showed a range value of 2.293 to 4.413 mg/l. The minimum value was observed in Bhanpur (near bridge) *i.e.*; station no. 3 and the maximum value was observed in Islamnagar (Prabhu Lal Kushwaha field) *i.e.*; station no. 1.

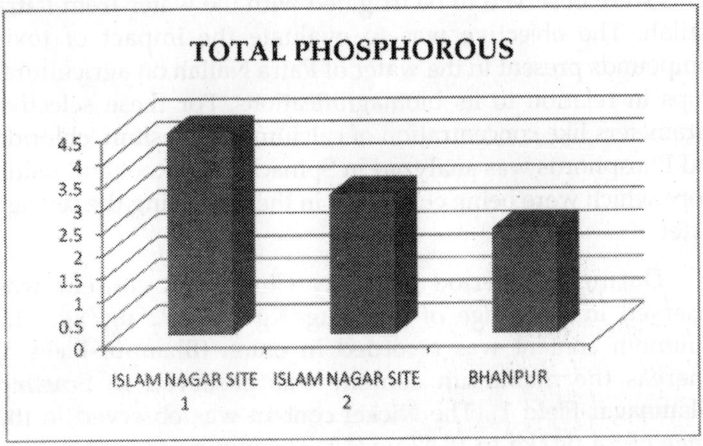

Toxicological Parameters

The concentrations of toxic metals in water observed to be low (Table-1). The range value of chromium was within 0.024 µg/liter to 0.068 µg/liter. Nickel during this period in different stations was observed in between 0.016 µg/liter to 0.074 µg/liter. Concentration of Zn during the study period was found within the range of 0.086 µg/liter to 0.109 µg/liter. However concentration of total iron in the water samples were observed to be slightly high than rest of the parameters. The range value of total iron was observed in between 0.984 µg/liter to 1.069 µg/liter.

Table 1 : Variation in Different Heavy Metals in Water Samples Collected from Patra Drain

Parameter	*Concentration (Range Value in µg/Liter)*	
	Min	*Max*
Cr	0.024	0.068
Ni	0.016	0.074
Zn	0.086	0.109
Fe	0.984	1.069

During the period of investigation vegetables samples were collected from agricultural fields adjoining Patra nallah. These fields were observed to be irrigated with the water from Patra Nallah. The objective was to evaluate the impact of toxic compounds present in the water of Patra Nallah on agricultural crops in relation to its biomagnifications. For these selective parameters like concentration of calcium, magnesium, chloride and Phosphorus was analyzed in Spinach, Beet Root and onion crops which were being cultivated in the area using the sewage water.

During the period of study Chromium content was observed in the range of 26.46 µg/Kg to 52.92 µg/kg. The minimum content was recorded in onion (Bhanpur-Field 3) whereas the maximum content was observed in Spinach (Islamnagar-Field 1). The Nickel content was observed in the range of 3.4 µg/kg to 18.47 µg/kg. The minimum content was recorded in Spinach (Islamnagar-Field 1) whereas the maximum content was observed in Beet root (Islamnagar-Field 2).

The Zinc content was observed in the range of 1.14 µg/kg to 1.42 µg/kg. The minimum content was recorded in Onion (Bhanpur-Field 3) whereas the maximum content was observed in Spinach (Islamnagar-Field 1). The Iron was observed in the range of 4.103 µg/kg to 7.198 µg/kg. The minimum content was recorded in Onion (Islamnagar-Field 1) whereas the maximum content was observed in Spinach (Islamnagar-Field 1).

Vegetable	*Parameter*			
	Concentration (Range Value in µg/kg)			
Cr	Ni	Zn	Fe	
Spinach	0.068	0.016	1.42	7.198
Beet root	0.034	0.074	1.17	5.664
Onion	0.024	0.048	1.14	4.103

DISCUSSION

It is the inevitable truth that agriculture is the main thriving lifeline for sustaining our country. Certainly, in relation with this, farmers occupy the most dependable job of our country. With the population of our country reaching 1 billion marks on previous year (2005), the emerging question lies in front of us *i.e.*, how long we can continue to fulfill the demand and supply of the food for this ever increasing population load.

Several technologies have been introduced for producing higher yield of food as per requirement of the demand. In the process of attaining higher levels of food production for matching the demands of the growing population during the past four to five decades, emphasis had been laid on intensive agricultural practices. But to add the woe the technologies generated over the past years has proved, having detrimental effect on the environment not only in our country but also in every parts of the world. So, during this very process of increasing the crop yield from modern farming techniques reaching a landmark in most countries, and the environmental problems due to excessive use of chemical fertilizers and pesticides becoming a matter of concern, the need of sustainable agriculture is increasingly felt throughout the world.

The excessive fertilization (eutrophication) of natural waters is one of the most significant causes of water quality deterioration. This increasing eutrophication, resulting principally from the cultural activities of man, is occurring because of the excessive input of aquatic plant nutrients into water bodies. Some water bodies are naturally eutrophic in that they receive sufficient supplies of aquatic plant nutrients, mainly phosphorus and nitrogen, from natural sources to produce excessive growths of algae and macrophytes. The weightage of environmental degradation has drawn the attention of scientist and planners towards finding out ecologically sound, more viable and sustainable farming system for different agro climatic condition, keeping in view the needs of the future and present generation. The farming practices that are govern by the principles of ecology has proved to be highly productive and often sound to meet the demands of today. Thus in present study an attempt has been made to assess the irrigation impact of sewage fed water on agricultural crops being grown along Patra Nalla. The study of transport of nutrients in the soil is important for a good irrigation–fertilization management, particularly when using sewage water. Therefore the present study was focused on effect of few macro and micro nutrients on agricultural crops in relation to their productivity and toxicity. Effects of some of the macronutrients are described below.

NITROGEN

Nitrogen is an essential nutrient element for living organisms, it is usually added in large quantities to agricultural lands to obtain high crop yields. Irrigation with sewage results in the addition of considerable amounts of N to the soil, depending on the content of N in the sewage. Consequently, the actual quantity added to crops varies widely and is greatly affected by climatic conditions, soil and effluent properties, and crop and irrigation management (Dahama, and Sinha, 1985).

Since sewage N eventually enters the N cycle in soils, irrigation with sewage affects nitrogen uptake by the crop, the level of available N in the soil, gaseous N losses and N leaching below the rooting depth. Irrigation management determines the distribution of N application over time, since N is added in irrigation with sewage (Feigen *et al.* 1978). In present investigation the identified crops were observed to have nitrogen affinity, higher concentration of nitrogen was recorded in some vegetables.

PHOSPHOROUS

Phosphorous is an essential nutrient element for all organisms and is added in large quantities to obtain enhanced crop yields. Sewage effluent often has high Phosphorous content, making it an important source of P for irrigated soil. Excessive levels of available P may result in nutrient imbalances, such as Cu, Fe, and Zn deficiencies (Riden and Pratt, 1980) necessitating soil or foliar application of micro nutrients. Application of P to the soil, as a fertilizer, manure or in effluent, results in an immediate rise in the level of water soluble P in the soil. In present investigation phosphorous concentration was found in between. The existing phosphorous concentration was observed to be enough for growth of the identified vegetables as these crops did not require supplementation of any chemical fertilizers.

POTASSIUM

Potassium is taken up by crops in large quantities, similar to, or even greater than those for Nitrogen. Potassium is considered an effective means for obtaining a high quality product, and is extremely important in high productive agriculture. The level of K in sewage effluent varies greatly, typical concentrations being 7 to 20 mg K/l. Irrigation with sewage effluent often does not satisfied plant needs in K, especially when the level of available K in the soil is low. During present investigation the potassium per centage in plant tissue was found to be relatively high, which indicates high crop up take.

Table 3 : Conclusion Based on Comparison of Values with Bureau of Indian Standard

Parameter	*Range*	*BIS Limit (Class A) : Drinking*	*BIS Limit (Class E) : Irrigation*	*Conclusion Based on Present Observation*
Chlorides	109.89 to 147 mg/l	250	600	Values are well within limits
pH value	5.93 to 6.8	6.5- 8.0	–	Moderate values
Dissolved Oxygen	2 to 10 mg/l	6.0	–	Highly fluctuating
Total Dissolved Solids	329.4 to 1030.9	500	–	Values were observed slightly on higher side
Total Hardness	180 to 300	300	–	Although within the limit but are alarming
Calcium Hardness	126 to 210	200	–	Within the limit but are alarming
Magnesium Hardness	22.5 to 153	100	–	Highly fluctuating
Nitrate	0.780 to 7.181	20	1000	Well below the limit when considered with the BIS limit

CONCLUSION

On the basis of present investigation, it can be concluded that the water irrigated from Patra Nalla did not have any toxic effect on the crops that are being cultivated in the region. This was also confirmed from local farmers/cultivator. However, the water quality of Patra Nalla is worst for drinking purpose when compared with the values prescribed by BIS under Class A. As the water is used for irrigation activities, Class E of BIS standard has been considered for the present purpose and most of the values of parameters investigated fall well below the limit as per the standard (Table 3). Therefore under present condition the water from Patra Nalla is safe for irrigation uses.

ACKNOWLEDGEMENT

The authors are grateful to Chief Executive Officer, Lake Conservation Authority of Madhya Pradesh for the support and help in conducting this study.

REFERENCES

APHA (1976) : Standard. Method for the examination of water and waste water. 14th edition, American Public health association, Washington D.C.

A.D. Adoni, (1985) : Work book of Limnology, Pratibha publisher.

Asano, T. Smith R.G. and Techobanoglons, G. (1985) : Municipal waste water characteristic. Irrigation with Reclaimed Municipal waste water A guidance Manual, G.S. Petlygrove and T. Asano (eds.) Lewis publishers inc, Chelsea, Mississippi.

Bahri, A. (2000) : The experience and challenges of reuse of wastewater and sludge in Tunisia. Water Week 2000, April 3-4, 2000. Washington DC, USA : World Bank. 15 p.

Bureau of Indian Standard (BIS, 10500), (1995) : Dahama, A, J., and Sinha, M.N-1985.A study on NP concentration and their uptake. *Indian J.Agron.* 30 (4) : 422-428.

Feigen, A., Bidomi, H.Dag, Y., Kipnis, T, Giskin. M. (1978) : The nitrogen factor in the management of effluent-irrigated soils. *Soil Sci.* 125 : 248-254.

OLSEN, S.R., Sommers, L.E (1982) : Phosphorous. In : Page, A.L., Miller, R.H., Keeney, D.R.(eds.). Methods of soil anaylsis, to : Chemical and micro biological properties. Am. Soc. Agron., Madison. *Agronomy,* 9 : 403-430.

Riden, J.C., Pratt, P.F. (1980) : Phosphorous removal from waste water applied to land. Hilgardia 48(1) : 1–36.

21

Watershed Management Through Bioengineering Measures of Torrents at Doon Valley, Shiwalik Himalaya

Nitin Kamboj

INTRODUCTION

The Himalaya consists of ranges which extend parallel and are interspersed by valleys or platforms. The Himalayan eco-system is one of the most important and at the same time the most threatened life support system on the earth. The population pressure and its mounting demand for fuel and fodder have already denuded large tracts of Himalayan ranges. The Himalayas, which are the youngest mountain ranges are geologically very complex and need more in depth study to restore the ecosystem. The Himalayan system is broadly divided into three parallel or longitudinal zones *viz.*; Great Himalayas, lesser Himalayas and Shiwalik Himalayas, which distinctly differ from one another in geomorphologic, physiological and vegetation characteristics. Shiwalik are the outer Himalayas, which join plains with foothills of Himalayas and lies between 300 meter to 1500 meter above mean sea level and many rivers of the region pass through this tract. The activities, which take place in the Shiwalik foot-hills also directly effect the habitants living in the plains and lead to other environmental hazards. Hence, the Shiwalik forest region plays an important role in the lives of the people down below. Shiwalik range extends from Jammu to Uttarakhand and onwards and in five States of Punjab,

Haryana, J&K, Himachal Pradesh and Uttarakhand covers about 30 lakhs hectares. Geologically Shiwalik are unconsolidated and extremely erodible in nature and are considered as most fragile ecosystem, which has a tendency towards degradation. Due to increase in human & cattle population, the process of degradation has been further accelerated. Uttarakhand a state located in the northern part of India known for its natural beauty, was carved out of Himalayan adjoining districts of Uttar Pradesh on 9 November 2000, becoming the 27th state of the Republic of India. It borders China (Tibet) on the north, Nepal on the east and the Indian states of Uttar Pradesh to the south, Haryana to the west and Himachal Pradesh to the north west. The region is traditionally referred to as Uttarakhand in Hindu scriptures and old literature, a term which derives from Sanskrit *uttara* meaning north, and *khand* meaning country or part of a country. It has an area of 53,566 km^2. In January 2007, the name of the state was officially changed from Uttaranchal, its interim name, to Uttarakhand. The provisional capital of Uttarakhand is Dehradun, is also a rail-head and the largest city in the region. Uttarakhand is characterized by two types of climate, sharply differentiated in the plains and the mountainous regions. Rainfall is quite favorable in this area and is more than 1100 mm per annum. However its main concentration is in monsoon season (about 80% of the total precipitation), which is the major cause of erosion in the rainy season. Frost conditions also occur in the winter months. The soils have poor structure and their water retention capacity is quite low. As a result of poor water holding capacity, it supports less biological activities and that is why drought conditions prevail in the summer months. Due to constant erosion over a period of time, top soils have been washed and left with very poor nutrients status.

Torrents are ephemeral mountainous streams emanating from the outer Himalayas and Shiwalik empty into major rivers and causes extensive damage to life and property in the valley areas by scouring the beds, undermining and eroding their banks. Torrent have mostly taken as small mountainous streams rushing down to the slopes with flashy floods and often loaded with sediments. On rushing the milder slopes such as foothills, they develop into channels with large volumes of debris and low velocity of flow. Flash floods with high sediment load fill up the channel beds and flow keeps on shifting its course and damaging fertile lands. Decreasing land man ratio, increase in the animal population, shift in the consumption pattern due to change in the forest ownership at the time of independence, lack of environmental concerns, non-participation of the community for micro level planning and inadequate enforcement of the land, forest and wild life preservation acts have all contributed to the problems of flash floods. Ultimately lack of adequate vegetative cover is the main cause of flash floods, erosion transport, deposition of sand and debris. The problem has been a serious one and is becoming acute with the increasing pressure of population decreasing the land man ratio. It may not be feasible to avoid utilization of such risk prone areas for producing biomass that the society needs. Consequently a degree of development of infrastructure and settlements in these areas can not be avoided. Therefore, proper investigations and surveys need be carried out to assess extent of the problems and the areas likely to be damaged through this problem. The investigations should explore the means to provide better protection and those which will at least moderate the hazards.

In India, torrents have been mostly taken as small mountainous streams rushing down the slopes with flashy floods and often loaded with sediments. On reaching the milder slopes such as foot hills they develop the channel problem of large volume and low velocity of flow. This is aggravated by high sediment load and depositions obstructing the channel flow. The torrents in Indian foot hills have, therefore, been linked with the problem of channel as well as riverine land and the area affected by depositions, shifting of course and inundation of banks. In fact, the history of deforestation is the history of the torrent problem. During the later half of the 19th century, heavy deforestation took place to meet the demands of the defence personnel, railways and cities etc. The mountainous, sub-mountainous and foot hill areas of the entire Himalayas have the torrents as a major land degradation problem. Its impact, whether natural or induced, has been felt increasingly as human societies extended their settlements and activities to areas where torrents have been having their unabated sway. The fast growing rivers and their tributaries are primarily glacier based snow fed or rain fed stream. These torrential streams exhibited physical properties like perennial or seasonal based on total discharge fluctuations, diurnal fluctuations in the water level and high current velocity along with the topographical situation of the riverbeds, making the study difficult and interesting to evaluate their irrigation harvest potential.

River or fluvial systems of all sizes compose a large part the earth's surface and have been the gravitating factor for the great civilizations of the world. With our increasing understanding about the role of water as an erosional and depositional agent and the disturbance that occurs to its process equilibrium, more vigorously by artificial stimuli, invariably leads to consequences like destruction of forest and agricultural lands, decimation of urban property by floods, accelerated removal of top soil, pollution and sedimentation of water bodies. Serious torrent erosion problem is being faced in dissected mountains of northern and north-eastern Himalayas and in Shiwalik hills. The National Commission on Agriculture observed that the devastation by torrents in the Punjab state alone is over 2.5m ha. Torrent affected erosion is due to the high run-off down the hill slopes with heavy gravel, sand and silt loads that fill up the channel beds thereby reducing their capacity to carry the run-off and sediment and as a consequence, the flow shifts from the original course. Torrents also known as Choes, thus spoiled vast stretches of fertile land swamped with coarse sediments and involved in gully erosion of the fresh areas. The erodibility of soil is one of the important factors responsible for prediction of torrent risk. Fine textured soils contribute most in the formation of torrents. Therefore, evaluation of soil erodibility factor necessitates the information on soil types of a torrent watershed catchment area. The primary concern of watershed management is to continuously increase the productive capacity of land, water, vegetation and other resources by selecting and developing the best way of managing the renewable and non-renewable resources in the watershed to meet the present and future needs of the community. It requires the collection and analysis of a great deal of information of the physical relationship of vegetation-soil water to land management which would ensure sustainable economic and social progress of a region.

The population pressure on the existing resources have led to severe environmental degradation in different watersheds of Doon valley (Garhwal Himalaya) to have more land for agricultural and horticultural purposes. The pressure on forests for the fuel wood (5.4-38.5 kg/ha/year) timber (4.6-33.0 kg/ha/year) and grazing (22-77%) have also led to the indiscriminate felling of the precious tree cover leaving the steep slopes bare. The excessive human interferences like over-lopping, deforestation, overgrazing, over ploughing, respectively have stimulated local geo-catastrophic events of land sliding, slumping, gulling, soil creeping excessively damaging landform and torrent problems. The human interferences like over grazing, over lopping, over fallowing influence watershed in many ways especially modifying the other factors as and when land is cleared of forest for pasture and cultivation. Landforms are encroached by terraces or drainage ditches and soil patches are changed by the effects of cultivation, construction of roads and buildings. In the absence of local and other alterative sources of energy, the people of foot hills region have to depend upon scarce dung and fuel wood obtained from the nearby forests for their domestic requirements. However, dry stems, barks and leaves are seldom enough to meet the requirements of an average village person, thus villagers are forced to exploit more than what is permitted which results in the over pressure on the existed forest. The immense pressure of population on the forest cover in recent years has resulted in deforestation, which has made bare patches of land due to the soil erosion in rainy season is very high. The land use of the Sabhawala watershed has been changed rapidly within the period of past thirty years.

The pressure of the population to meet the requirement of fuel wood, timber and other products per hectare of forest cover is very high. The steep slope and high relative relief in this catchment has led to the low availability of agricultural land. The agricultural land is only available in some flat alluvial terraces in river valleys and along the rivers on moderate slopes. Due to this pressure on cultivated land, there has been large scale encroachment on the forest land for extension of agriculture and thus deforestation had denuded the forest cover. Due to an excess of human interference severe soil degradation is experienced in the catchment basin of watershed. The top soil of study region has been severely depleted due to active erosional processes. The loss of valuable top soil is very high in most of the region. Due to deforestation the vegetative cover is being depleted leaving the soil cover bare and prone to erosion. The integrated watershed management approach, which takes into account all components of a watershed have been found to be very effective, sustainable, socially acceptable, economically viable and technically feasible and environment friendly. This approach aims at regulation of excessive biotic interference through people's participation, catchment treatment and improvement in the productivity of arable lands, non-arable lands and other allied activities. Characteristics of sediment particles, gravels and boulders for size, bed profile, and its variation along the stream are some of the related basic features that need to be determined. Identification and quantification of the factors influencing the formation and behaviour of different types of torrents relate catchment and morphological characteristics (such as channel dimensions, bed material composition, bed configurations and vegetation on banks) with channel behaviour; study the relative performance and parameter influences of selected torrent control measures (mechanical and vegetative) in channelizing the flow and reducing the bank and bed erosion and evaluate their structural and functional performances. Technologies comprising construction of gabion spurs for prevention of bank cutting and toe wall for bank stabilization should be developed. Vegetative measures for similar purposes may be adopted either simultaneously or independently depending upon flood volume and sediment characteristics. Seasonal river treatments works are basically aimed to control meandering tendency of the hill river by restricting it in a defined and narrow course. If this is achieved, problems of overtopping and undermining of banks and debris deposition in the flow channel can be eliminated. A study was taken up to evaluate the relative efficacies of different mechanical and vegetative control measures in terms of their effects on the flow, sediment transportation and scour or deposition patterns and also in terms of structural stability, longevity, cost of construction and periodic maintenance.

MATERIALS AND METHODS

Study Area

Doon valley is bounded on the North by Mussooriee ridge in the Northeast by lesser Himalayan ranges; on the Southeast by Shiwalik ranges; river Ganga in the Southeast; and river Yamuna in the Northeast. The Valley has a total area of 1,853 km^2, with an average length of 80 km and width of 30 km. Dehradun is situated around 235 km from Delhi at the centre of the Doon valley. Dehradun is the part of Garhwal Himalayan region, which was earlier known as Kedar Khand. Dehradun at present is the capital of Uttarakhand and poised for major developments in the coming future. The climate in the northern part of Uttarakhand exerts an appreciable extent of influence on monsoon and rainfall patterns. Within the Himalayas, climate differs depending on altitude and latitude position Climate ranges from subtropical in the southern foothills with average summer temperatures of about 30°C and winter temperatures of about 18°C. Warm temperate conditions prevail in the middle Himalayan valleys, with summer temperatures usually hovering about the mark of 25°C and cooler winters. Cool temperate conditions dominate the higher areas of the middle Himalayas, where the summer temperature are usually around 15 to 18°C and winters drop below the freezing point. Cold alpine climate prevail at higher hill reaches, where summers are cool and winters are harsh. At altitudes over 4880 meters (16,000 feet), the climate is bitterly cold with temperatures consistently below the freezing point and the area perennially shrouded in snow and ice. The eastern flanks of the Himalayan ranges are subject to heavy rainfall while the western section is relatively dry.

Many torrential watersheds originating from Shiwalik Himalaya foothill regions have been identified for the purpose of find out various morphological characteristics. All the watersheds drain caused natural hazards in forest as well as agricultural plain basin and lastly drained into Asan river, a tributary of Yamuna river. In analytical observations of watersheds, various morphological characteristics of the watersheds analysis and the frequency of sampling were set as pre existing conditions during the rainy seasons or flood periods.

RESULT AND DISCUSSION

The Uttarakhand state is particularly sensitive to land disturbing activities. Steep slopes, high rainfall and weak geology of the Uttarakhand state accentuate the land degradation and soil erosion process at much faster rate than in the plains. In the head water reaches the sediment is flushed with the high velocity currents. But when it reaches the relatively flatter terrain foothills, its debris carrying capacity diminishes and the debris starts accumulating on the river bed causing change of river course and flooding its banks. These rivers with flash flows and high sediment loads are known as torrents. Thus, the real problem of torrent formation lies in geomorphology of lesser Himalaya further aggravated due to misuse and mismanagement of upstream catchments. The problem of torrent menace has been increasing in Uttarakhand with the rise in population pressure and related mismanagement of upstream watersheds.

In the foothills of Shiwalik Himalaya, torrents are the prominent land features and causing vast area submergence and damage to life, property and infrastructure almost every year. This is the most common problem in foothills of Shiwalik, spread over the northern states of India. The torrents are seasonal in nature and characterized by high sediment ladder flash flow during monsoon period. These torrents have low banks and thus the flow frequently over tops the banks and cause floods. Moreover, the torrents have meandering nature *i.e.* they often change. Physico-hydrology of watershed and torrent behaviour is closely related. Rates and volumes of runoff are directly dependent on watershed characteristics Juyal (2005). It is in this context that characterization of watersheds in relation to torrent behaviour is essential. Some of the watershed characteristics that directly influence the rates and volumes of runoff are compactness coefficient, drainage density, watershed and channel slopes. These characteristics influence width of torrent beds and cross sectional areas of flow at different places for the same drainage areas (**Table-1**). A protective vegetative cover was established in the torrent of Sabhawala watershed area for the identification of effective protective vegetative cover with the help of mechanical measures and appropriate soil and water conservation techniques. Grasses used for the establishment of the initial vegetative cover by the species as *Crysopogon fulvus, Eulaliopsis binata (Bhabhar), Pennisetum purpureum (Hybrid Napier), Saccharum munja, Pueraria hirsute (kudzu), P.antidotale* and *V.zizanioides*. The plants used as *Leucaena leucocehapla (Subabool), Salix tetrasperma, Acacia catechu* (Khair) *Cedrela toona, Baunhinia* spp. *Erythrina suberosa, Lannea grandis, Pithecelobium dulce (Jungle Jalebi) etc.* **(Table-2).** In the present study, various plant species effectively served as vegetative barrier and had better soil binding capacity in the layer, which helped in soil aggregation, increased infiltration rates, improve soil moisture level and conserved maximum runoff. Napier *(P.purpureum)* could be suitable vegetative barrier, where fodder production is a priority while *S. munja* as a barrier is suitable where conservation and crop production is the main objective. Burdak (1982): reported that *S. munja* is recent advances in desert afforestation. Various research works in the rooting behaviour of some grasses in the Shiwalik were carried out by Saha *etal.*, (1994) on *Saccharum munja, Eulaliuopsis binata* and *Vetivaria zizanionides* and described at the roots of S. *munja* (Munj) reached the maximum depth followed by Khus and Bhabbar, Highest root densities were confined to the plough layer in all the three species, the trend being *E. binata* (89%) > *Vetivaria* (77%) > *S.*

Table 1 : Different Watershed Characteristics of the Himalayan Foothill Torrent and Shiwalik Torrent

S.No.		*Himalayan Foothill Torrent Watershed*				*Shiwalik Torrent Watershed*			
A.	*Watershed Characteristics*	*Bansiwala*	*Banikhala*	*Sudhowala*	*Average*	*Sabhawala*	*Ramgarh*	*Kaluwala*	*Average*
1.	Area ha	633.30	1220.80	153.0	919.03	1100.0	731.10	405.10	745.40
2.	Perimeter, a	20.40	22.66	21.00	21.35	21.4	13.05	10.30	14.92
3.	Max. length of flow	10.98	12.14	9.55	11.5	7.85	6.96	4.54	5.75
4.	km	0.66	1.02	0.84	0.84	0.48	1.31	0.92	0.90
5.	Av width, km	533-841	518-811	518-826	525-	53-904	564-858	600-875	556-
6.	Elevation range, m	208	293	250	826	285	294	275	879
7.	Fall in elevation, m Not of stream order	13	16	39		43	23	17	
	I	3	4	11	23	11	8	4	27.67
	Ii	1	2	03	6	03	3	2	7.67
	Iii	1	1	01	2	01	1	1	2.67
	Iv	18	23	54	1	58	35	24	1
8.	Total				32				39
	Length of streams in each	7.6	16.26	26.25		25.5	13.15	8.71	
	I	5.91	5.28	8.5	16.70	9.00	5.54	1.58	15.79
	II	2.75	6.6	8.9	6.56	4.25	3.64	1.03	5.37
	III	4.86	—	2.5	6.17	6.25	—	3.17	2.97
	IV	21.12	28.4	46.15	2.45	45	22.33	14.49	3.14
	Total				31.89				27.27

Cont...

S.No.		Himalayan Foothill Torrent Watershed				Shiwalik Torrent Watershed			
B.	Watershed Geomorphic								
1.	Parameter	9.00	6.70	7.85	7.85	13.09	14.8	13.00	13.63
2.	Average slope (%)	0.06	0.08	0.17	.28	0.18	0.18	0.20	0.19
3.	From factor (A/lm^2)	2.57	2.52	2.94	2.68	4.09	3.49	3.44	3.67
4.	Drainage density (km/km^2)								
	Bifurcation Ratio	—	—		—	—	—	—	—
	I	4.33	5.00	4.6	4.64	3.5	2.87	4.25	3.56
	II	3.00	2.00	2.5	2.5	2.33	2.66	2.00	2.33
	III	—	2.00	2.0	1.3	2.5	3.00	2.0	2.5
5.	IV	3.66	2.67	3.03	3.12	2.79	2.84	2.75	2.79
	Average								

Table 2 : Species for Protective Vegetative Cover

S.No.	Vegetation Type	Names
1.	Grasses	*Crysopogon fulvus, Eulaliopsis binata* (Bhabbar), *Pennisetum purpureum* (Hybrid Napier), *Saccharum* (munja), *Pueraria hirsute* (kudzu), *P.antidotale, V.zizanioides,*
2.	Shrubs	*Vitex negundo, Agave americana, Ipomoea carnea*
3.	Trees	*Leucaena leucocehapla* (Subabool), *Salix tetrasperma, Acacia catechu,* (Khair) *Cedrela toona, Baunhinia* spp. *Erythrina suberosa, Lannea grandis, Pithecelobium dulce* (Jungle Jalebi) *etc.*

munja (48%). Samra and Sharma (1995) made experiment the soil binding factor for three perennial grasses *(S.munja, S.spontaneum, Arundo don ax)* in sandy choes at Rei Majra and observed that the *S.spontaneum* had the highest above ground biomass (398 kg) followed by *S. munja* (28.3 kg) and least *A. donax* (3.7 kg). Soil binding capacity in 0-10 cm layer was recorded highest for *S. spontaneum* (1590) followed by *S.munja* (788) and *A.donax* (31).

CONCLUSION

The present study concludes that the vegetative methods are nature's free tool to rehabilitate for controlling soil erosion leading to floods. Grasses have their own role to play in torrent treatment technology. The three tier system, if properly utilized by keeping trees, shrubs and grasses as an efficient land use can be of great help. The scare inflicted on the mother earth by man's greed can easily be healed and covered to reduce run off and help control torrents. Planting along staggered trenches reduces the run off as they work as mini-reservoirs, thus helping plant growth, recharging ground water and controlling flood. The solutions aim at moderating or regulating the runoff, debris and sediments coming down in periodic flashes of varying durations. Catchment area of torrent treatment depends on runoff peaks and rainwater volume, erosion and sediment load calculation are most important data for management practices. Treatments to modify the channel grade, flow velocity, erosion, transport and deposition of sediments are important for reclamation of already affected areas by various remedial strategies.

REFERENCES

Juyal, G.P., Bankey Bihari, Ghosh B.N. and Rathor A.C., (2005) : *Cost effective technology for torrents treatment in Uttaranchal Shiwaliks*. In : Proceeding of *research conservation technology*. pp. 125.

Juyal, G.P., Vittal, N and Sastry, G. (2005) : Studies on performance and score around spur-A review, In Torrent Menace–Challenges and Opportunities, Eds. G.Sastry, V.N. Sharda, G.P. Juyal and J.S. Sharma, CSWCRTI, Dehradun, pp 283-319.

Samra, J.S. and Sharma, N.K. (1995) : Soil binding factor above and below ground phytomass of three perennial grasses for stream course stabilization. *Ind. Jr. Soil Cons. 23(1) : 39-43.*

Burdak K.J.M (1982) : Selection of desert tolerance grass and shrub species for protection of soil upper layer in desert region. Soil conservation and management Vol-II pp. 352-360.

Saha, B., Samra, J.S. and Mittal, S.P. (1994) : Studies on the effect of roots of different grasses on water transmission characteristics of soil. *Annual Report*, CSWCRTI, Dehradun. pp. 47.

22

The Uttar Pradesh Ground water Conservation, Protection and Development (Management, Contol and Regulation) : A Bill, 2010*

Director, Department of Hydrogeology,
Ashok Marg, Lucknow (U.P.)

INTRODUCTION

A bill to provide for the management, control and regulation of the conservation, protection and development of ground water in the State and for matters connected therewith or incidental thereto. Whereas uncontrolled and rapid extraction of ground water has resulted in alarming situation of declining ground water levels and depletion of ground water reservoirs in many parts of the State, both in rural and urban areas. Whereas development of ground water is the need of the State, its management, control and regulation specially in overexploited and critical areas is also the need of the hour for protection and preservation of this precious resource; and where as it is also expedient to provide for conservation, protection and development of ground water resources for the purpose of proper recharge of ground water and to prevent the pollution thereof in the State; and whereas the State Government has, after careful examination of all aspects, decided that it is expedient and necessary in the public interest to manage, control and regulate the extraction and use of ground water in any

*Draft recieved from the courtesy of Janhit foundation, Meerut.

form and to conserve and recharge ground water in the State. It is hereby enacted in the Sixty-one Year of the Republic of India as follows :

CHAPTER – I

PRELIMINARY

1. Short Title, Extent & Commencement

1. This Act may be called the Uttar Pradesh Ground water Conservation, Protection and Development (Management, Control and Regulation) Act, 2010.
2. It extends to the whole of the State of Uttar Pradesh.
3. It shall come into force on such date as the State Government may, by notification in the Gazette, appoint, and different dates for different areas may be appointed.

2. Definitions

In this Act, unless the context otherwise requires :

(*a*) **Artificial Recharge to Ground Water** means the process based on geo-scientific methods by which ground water reservoir is augmented at a rate exceeding that under natural condition of replenishment.

(*b*) **"Authority"** means the Uttar Pradesh Ground Water Authority, established under section-3.

(*c*) **Drinking Water"** means water for consumption or use by human population for drinking and for other domestic purposes, which shall include consumption or use of water for cooking, bathing, washing, cleansing and other day to day activities. It shall also include water meant for consumption by the livestock.

(*d*) **"Ground water"** means the water, which exists below the ground surface in the zone of saturation and can be extracted through wells or any other means or emerges as springs and base flows in streams and rivers.

(*e*) **"Industry"** means any business, trade, undertaking, manufacture or calling of employers, carried out with a motive to make any gain or profit, and includes any calling service, employment, handicraft, or industrial occupation or avocation of workman or any systematic activity carried on by co-operation between an employer and his workman (whether such workman are employed by such employer directly or by or through any agency including a contractor) for the production of goods.

(*f*) **"Notified area"** means the area notified as such under section-5.

(*g*) **"Over Exploited, Critical and Semi Critical area in Rural and Urban areas"** shall be such as may be notified by the authority.

(*h*) **"Owner"** includes any mortgagee in possession, lessee, common manager, receiver appointed by a court and any person holding any property in trust and also include a Court of Wards in respect of property under the superintendence or charge of such Court.

(*i*) **Rain Water Harvesting** means the technique of collection and storage of rain water at surface or in sub-surface aquifer by adopting site specific scientific guidelines.

(*j*) **Resident Welfare Association"** means an association of residents or a group of people of a ward/locality/ area residing in urban areas recognized by the authority.

(*k*) **Self Regulation** a type of regulation to implement the provisions and standards of ground water management, control and regulation as may be prescribed.

(*l*) **Service Provider** an organization having agricultural engineer, civil/mechanical engineer, Master's degree holder in Geology/Geophysics or individual having qualification of B.Tech. / B.E. (Civil / Mechanical / Agriculture) or M.Sc. (Geology) or M.Sc. (Geophysics) with at least 20 years of working experience in relevant field.

(*m*) **"Sink"** with all its grammatical variations and cognate expression in relation to a well includes digging, drilling or boring of new well or deepening and modification (of radials and galleries) of the existing wells.

(*n*) **"Technical Officer"** means a subject specialist so designated by the authority to give advice on technical matter.

(*o*) **"User of ground water"** means the person or a class of persons or an institution who or which own or use or sell ground water for any purpose including domestic use made either on personal or community basis and includes an industry, a commercial user, a bulk user, a company or an establishment whether government or not but does not include a person or a class of persons or an institution who or which use ground water drawn from well by manual or animal devices such as hand pump, rope and bucket, Persian wheel etc.

Explanation

For the purpose of this clause, the expression :

(*p*) "Commercial user of ground water" means any person, trade, undertaking, business institution which withdraws, use or sell ground water for commercial purposes.

(*q*) "bulk user of ground water" means a ground water user engaged in use of ground water in excess of quantity as notified by the authority for drinking, domestic & related use for a particular area.

(*r*) **"Water User Association "** means an association of water users in rural areas in a unit/area, utility, industrial user's association, other user's associations or any other group or individual, formed under section-6.

(*s*) **"Well"** means a structure sunk for the search or extraction of ground water and shall include open well, dug well, bore well, dug-cum bore well, tube well, filter point, collector well, infiltration gallery, recharge well, disposal well or any of their combinations or variations, which when installed, could be utilized for extraction of ground water.

CHAPTER – II

3. Establishment of Ground Water Authority

1. The State Government shall, by notification in the Gazette, establish an Authority to be known as the Uttar Pradesh Ground Water Authority.

2. The Authority shall consist of :

(*i*)	The Agriculture Production Commissioner, Government of Uttar Pradesh.	Chairman
(*ii*)	The Principal Secretary /Secretary, Minor irrigation & Ground Water, Government of Uttar Pradesh.	Vice-Chairman
(*iii*)	The Principal Secretary/Secretary, Irrigation, Government of Uttar Pradesh.	Member
(*iv*)	The Principal Secretary/Secretary, Housing and Urban Planning, Government of Uttar Pradesh.	Member
(*v*)	The Principal Secretary/Secretary, Nagar Vikas, Government of Uttar Pradesh.	Member
(*vi*)	The Principal Secretary/Secretary, Finance, Government of Uttar Pradesh.	Member
(*vii*)	The Principal Secretary/Secretary, Agriculture, Government of Uttar Pradesh.	Member
(*viii*)	The Managing Director, Uttar Pradesh Jal Nigam.	Member
(*ix*)	The Member Secretary, Uttar Pradesh Pollution Control Board.	Member
(*x*)	Nominated Representative of National Institute of Hydrology, Roorkee.	Member
(*xi*)	A representative of the Central Ground Water Board (CGWB), New Delhi, to be nominated by the Chairman, CGWB	Member
(*xii*)	Nominated representative of U.P. Water Management & Regulatory Commission	Member

Cont...

(*xiii*)	Nominated representative of State Water Resources Agency, U.P.	Member
(*xiv*)	An eminent NGO working in the field of Ground Water Management and Conservation.	Member
(*xv*)	The Director, Ground Water Department, Uttar Pradesh.	Member-secretary

3. The term of office and the manner of filling the vacancies and other conditions of service of the Chairperson and other members shall be such as may be prescribed.

4. The Chairperson shall be the principal executive officer of the Authority and the Vice-Chairperson shall exercise the powers, perform the function and discharge the duties of the Chairperson in the absence or under the direction and control of the Chairperson.

4. 1. In order to enable the Authority to perform its function properly or exercise the powers under this Act, the State Government may appoint such number of technical personnel and other staff as it may consider necessary.

2. The functions and the terms and conditions of service of such employees shall be such as may be prescribed.

5. 1. The Authority shall function under the overall control and supervision of the State Government.

2. If the Authority, after consultation with various expert bodies, including Central Ground Water Authority is of the opinion that it is necessary or expedient in the public interest to control and/or regulate the extraction or the use or both of ground water in any form in any area including over exploited, critical and semi-critical areas, it will advise the State Government to declare by notification, such area to be a notified area for the purposes of this Act with effect from such date as may be specified in such notification; Provided that—

(*a*) The date specified in the notification under this

sub-section shall not be earlier than three months from the date of publication of the notification,

(*b*) Geo-hydrologically areas specific suitable & feasible rain water harvesting and recharge schemes based on scientific guidelines *viz.* construction of ponds and check dams, rejuvenation of old ponds, *in-situ* (on farm) water harvesting in rural areas and water conservation through ponds or lakes, Roof Top Rain Water Harvesting Schemes in urban areas shall be prepared for the notified area by the concerned departments in consultation with local bodies, institutions, industry, company, government departments, individuals or any other concerned under the supervision and guidance of the Authority before its declaration as a notified area under this section.

(*c*) Crops requiring less water with suitable changes in cropping pattern as per the availability of water shall also be promoted in such areas.

3. Every notification in Hindi as well as English or vernacular language under this section shall, in addition to its publication in the Gazette, be published in not less than two daily regional newspapers having wide circulation in that region and also be served in such other manner as may be prescribed.

4. The Authority shall also take steps through registered service provider to ensure that exploitation of ground water resources does not exceed the natural replenishment to the aquifers. Wherever, there is mismatch, steps will be taken to ensure augmentation of ground water resources in addition to regulatory measures.

5. The Authority shall maintain and upkeep the data-base on ground water related information.

6. There shall be formed Water User Association in rural area in such manner as may be prescribed.

7. The functions of the Water User Association shall be

(*a*) To inform the concerned authorities in case of any violation of the provisions of the Act.

(*b*) To carry out Mass awareness and sensitization programme,

(*c*) Such other functions as may be prescribed.

8. There shall be formed Resident Welfare Association in urban area in such manner as may be prescribed.

9. The functions of the Resident Welfare Association shall be

(*a*) To inform the registered Service provider appointed by the authority if any user does not comply with the provisions of the Act.

(*b*) To carry out Mass awareness and sensitization programme.

(*c*) Such other functions as may be prescribed.

10. 1. Any organization or individual having requisite qualification specified in clause-l of section-2 may apply to the Authority for the grant of a certificate of registration. The application and the certificate of registration shall be issued in such form and in such manner as may be prescribed.

2. On receipt of an application, if the Authority is satisfied that the applicant fulfils the requisite qualification, it may issue a certificate of registration of Service Provider to the applicant on payment of such fees as may be prescribed. If the Authority decides not to grant the certificate of registration, it will inform the applicant about the same with reason with in thirty days.

3. The certificate of registration shall be in such form as may be prescribed.

4. The registration shall be valid for a period of two years and it will require the renewal after evaluation of his performance by the Authority.

5. If it is found that any registered Service Provider is not performing his duties sincerely and slackness in duties is observed or any wrong reporting is being communicated to the authority the registered Service Provider will be liable for punishment as prescribed in the Act.

11. 1. Registered Service Provider will be responsible for service *e.g.* inspection of limits of ground water withdrawal by different type of ground water users, size of the pump set, suction & delivery pipe construction of well/Rain Water Harvesting and Recharge structures, efficacy of Rain Water Harvesting and Recharge structures, quality of water being recharged to ground water and any other works as may be prescribed.

2. If it is found that a registered Service Provider is not performing his duties sincerely or a wrong reporting is being communicated by him to the authority the registered Service Provider will be liable for such punishment as may be prescribed.

3. A registered Service provider may charge such fee as may be determined from time to time by the authority from the users of ground water for providing the services.

4. A registered Service provider may select any drilling agency registered at district level according to the provisions of this Act.

12. 1. Every existing bulk user of ground water of different areas, within a period of one hundred and twenty days from the date of commencement of this Act in the area thereof and new user before the start of construction of well shall apply to the Authority on payment of such fee as may be prescribed for the grant of a certificate of registration, in such form and in such manner as may be prescribed, provided that the Authority may entertain any such application after the expiry of the said period of one hundred twenty days, if it is satisfied that the user was prevented by sufficient cause for making the application in time.

2. On receipt of an application under sub-section (1), the Authority shall issue a certificate of registration subject to such conditions and restrictions as may be specified, authorizing the continued bulk use of ground water.

3. The certificate of registration shall be in such form as may be prescribed.

4. Pending the communication by the Authority of the decision on an application under sub-section (1), every existing bulk user of ground water shall be entitled to the continued bulk use of the ground water in the same manner and to the same quantity as he was entitled prior to the date of his application.

5. If a registered well becomes defunct, this fact shall immediately be brought to the notice of the Authority by the bulk user of ground water.

6. If any user of ground water, having certificate of registration wants to carryout any modification or alteration in a registered well, he shall have to obtain clearance for the same.

7. If any existing/new bulk user is found extracting ground water without registration, he shall be liable to be punished under section-34.

13. (*a*) All drilling agencies of the State shall apply to the Authority for grant of registration at district level, in such form and in such manner as may be prescribed.

   1. On receipt of an application, the Authority shall issue a certificate of registration within sixty days after due enquiries and satisfaction about the veracity of applicant's claim to the drilling agency on payment of such fee as may be prescribed.

2. The certificate of registration shall be in such form as may be prescribed.

3. The registration will be valid for a maximum period of three years, which may be renewed after due enquiries and satisfaction about the veracity of applicant's claim maximum for a period of three years on payment of such fee as may be prescribed.

(*b*) *i*th regard to construction of bore well/tube well and abandoned bore well, the Authority or the State Government may issue directions specifying measures for prevention of cases of children being trapped or falling into bore well and closure of abandoned well. The District Magistrate or any person authorized by the Authority will have powers to verify that guide lines issued are being followed and proper monitoring & check about the status of bore holes/tube wells are being taken care through the concerned drilling agencies.

(*c*) The owner of well/drilling agencies shall abide by the directions/duties to follow the such safety norms/ directions issued by the Government.

(*d*) In case of 'Abandoned' bore well/tube well, all the agencies/individuals will obtain a certificate from the Authority or the department/person authorized by it stating that 'Abandoned' bore well/tube well is filled up to the ground level.

(*e*) The registered drilling agency shall deposit security money in the form of fixed deposit etc. as prescribed by the authority. If it is found that the agency does not follow the directions/duties issued by the Government, the certificate of registration will be cancelled and the security money will be forfeited.

14. 1. Every existing user of ground water of over-exploited/ critical areas, within a period of sixty days from the date of commencement of this Act in the area thereof and new user before the start of construction of well shall apply to the Authority on payment of such fee as may be prescribed by the authority for the grant of a certificate of registration, in such form and in such manner as may be prescribed. Provided that the Authority may entertain any such application after the expiry of the said period of sixty days, if it is satisfied that the user was prevented by sufficient cause for making an application in time.

2. On receipt of an application under sub-section (1), the Authority shall issue a certificate of registration subject to such conditions and restrictions as may be specified, authorizing the continued use of ground water.
3. The certificate of registration shall be in such form as may be prescribed.
4. Pending the communication by the Authority of the decision on an application under sub-section (1), every existing user of ground water shall be entitled to the continued use of the ground water in the same manner and to the same quantity as he was entitled prior to the date of his application.
5. If a registered well becomes defunct, this fact shall immediately be brought to the notice of the Authority by the user of ground water.
6. If any user of ground water, having certificate of registration wants to carryout any modification or alteration in a registered well, he shall have to obtain clearance for the same.
7. If any existing/new bulk user is found extracting ground water without registration, he will be liable to be punished under section-34.

15 1. Every existing commercial & industrial user of ground water in non-notified/semi-critical areas shall, within a period of one hundred and twenty days & new commercial & industrial users before the construction of well and every existing commercial & industrial user of ground water in over-exploited/critical areas shall, within a period of sixty days from the date of commencement of this Act in the area thereof, apply to the Authority on payment of such fee as prescribed by the authority for the grant of a certificate of registration, in such form and in such manner as may be prescribed. Provided that the Authority may entertain any such application after the expiry of the said period of one hundred twenty days in non-notified/semi-critical areas and sixty days in over-exploited/critical areas, if it is satisfied that the user was prevented by sufficient cause for making an application in time.

2. On receipt of an application under sub-section (1), the Authority shall issue a certificate of registration subject to such conditions and restrictions as may be specified, authorizing the continued use of ground water.

3. The certificate of registration shall be in such form as may be prescribed.

4. Pending the communication by the Authority of the decision on an application under sub-section (1), every existing user of ground water shall be entitled to the continued bulk use of the ground water in the same manner and to the same quantity as he was entitled prior to the date of his application.

5. If a registered well becomes defunct, this fact shall immediately be brought to the notice of the Authority by the user of ground water.

6. If any user of ground water, having certificate of registration wants to carryout any modification or alteration in a registered well, he shall have to obtain clearance for the same.

7. If any existing/new commercial and industrial user is found extracting ground water without registration, he will be liable to be punished under section-34.

16.1. Urban Areas

(*a*) **Ground water users using pump set up to 0.5 H.P. :** There will be no regulation for the time being for any user of ground water for installing pump set of power not exceeding 0.5 H.P. But in future it may be as prescribed by the Authority from time to time.

(*b*) **Ground water users using pump set of more than 0.5 H.P.**

(*i*) All users of ground water will self regulate themselves through Resident Welfare Association to install submersible pumpset of power exceeding 0.5 H.P. or as may be prescribed by the Authority, however, withdrawal of ground water will be up to the limit notified for bulk users for each urban area.

(*ii*) These users of ground water will adopt Rain Water Harvesting/Recharging activities as per prevailing Government order/as prescribed by the Authority. For these users, mass awareness and sensitization programme will be carried out through Resident Welfare Association.

(*iii*) It will be mandatory provisions for sellers to inform the authority about selling of submersible pumpsets of power exceeding 0.5 H.P. with verification of identity and address of the person who purchases the pumpset.

(*c*) **Bulk Users**

(*i*) Existing and new bulk users to mandatory adopt area specific Rain Water Harvesting/Recharging technique. Minimum amount of Rain Water Harvesting/Recharging to be fixed by the Authority.

(*ii*) New bulk users will be allowed to construct well/ tube well on their own, however, limit of ground water extraction to be fixed by the Authority.

(*iii*) Bulk users may be charged fee for ground water withdrawal annually by the Authority/ Government.

(*iv*) To ensure and check efficacy of rain water harvesting and recharge structure, limit of ground water withdrawal, size of pump set, suction & delivery pipe, quality of water being recharged to ground water and any other works etc. ground water user will get inspected the recharge structure twice a year by the concerned registered service provider and registered service provider will charge the fee from ground water user as prescribed by the Authority. Period of inspection in a year will be as prescribed by the Authority. If Rain Water Harvesting/Recharging structure is not found working properly, and amount of ground water withdrawal & quality of water are not as per limit/norms, the registered service provider will advise the ground water user to make it functional and ensure other provisions as per sanction order and directions. In case ground water user does not act on the advice of registered service provider, he will inform the Authority to take further action.

(*v*) Any modification/alteration in a registered well will require clearance. No person shall cause any such activity in urban area, which is against the provisions of this act or which is prejudicial or in contravention of the act.

2. Rural Areas

(*a*) **Ground water users using pump set up to 7.5 H.P. :** There will be no regulation for the time being for any user of ground water for installing pump set of power not exceeding 7.5 H.P. But in future it may be as prescribed by the Authority from time to time.

(*b*) **Ground water users using pump set of more than 7.5 H.P.**

(*i*) All users of ground water will self regulate themselves through Water User Association to install pump set of power exceeding 7.5 H.P. or as may be prescribed by the Authority, however, withdrawal of ground water will be upto the limit notified for bulk users for each rural area.

(*ii*) These users of ground water will adopt Rain Water Harvesting/Recharging activities as per prevailing Government order/as prescribed by the Authority. For these users, mass awareness and sensitization programme will be carried out through Water User Association.

(*iii*) It will be mandatory provisions for sellers to inform the authority about selling of submersible pump sets of power exceeding 7.5 H.P. with verification of identity and address of the person who purchases the pump set.

(*c*) **Bulk users**

(*i*) Existing and new bulk users to mandatory adopt area specific Rain Water Harvesting/Recharging technique. Minimum amount of Rain Water Harvesting/Recharging to be fixed by the Authority.

(*ii*) New bulk users will be allowed to construct well/ tube well on their own. However, limit of ground water extraction to be fixed by the Authority.

(*iii*) Bulk users may be charged fee for ground water withdrawal annually by the Authority/ Government.

(*iv*) To ensure and check efficacy of rain water harvesting and recharge structure, limit of ground water withdrawal, size of pump set, suction & delivery pipe, quality of water being recharged to ground water and any other works etc. ground water user will be required to get an inspection done in April every year by the concerned registered service provider. Service provider will charge an inspection fee from ground water user as prescribed by the Authority. If Rain Water

Harvesting/Recharging structure is not found working properly, and amount of ground water withdrawal & quality of water are not as per limit/ norms, the registered service provider will advise the ground water user to make it functional and ensure that other provisions as per sanction order and directions are being complied with. In case ground water user does not act on the advice of registered service provider, he will inform the Authority to take further action as per the Act.

(*v*) Any modification/alteration in a registered well will require clearance. No person shall cause any such activity in urban area, which is against the provisions of this act or which is prejudicial or in contravention of the act.

17. 1. For Urban Areas

(*a*) **Ground water users using pump set up to 0.5 H.P.**

(*i*) All users of ground water will self regulate themselves through Resident Welfare Association by installing pump set of power not exceeding 0.5 H.P. or as may be prescribed by the Authority.

(*ii*) These users of ground water will adopt RWH/R activities as per prevailing Government order/as prescribed by the Authority. For these users, mass awareness and sensitization programme will be carried out through Resident Welfare Association.

(*b*) Ground water users using pump set of more than 0.5 H.P.

(*i*) If any user of ground water wants to install pump set exceeding 0.5 H.P., he will do so through the registered service provider as per the procedure prescribed by the Authority.

(*ii*) The Authority may specify/ fix the limit / rate of ground water withdrawal from one well with submersible pumpset exceeding 0.5H.P. in specific sub-unit/areas.

(*iii*) It will be mandatory for all sellers to inform the Authority about selling of submersible pumpset of power exceeding 0.5 14 H.P. along with verification of identity and address of the persons who purchase these pumpset.

(*iv*) All users of ground water will adopt Rain Water Harvesting/Recharging activities as per prevailing Government order/as prescribed by the Authority.

(*v*) The Authority will promote mass awareness programme on Rain water harvesting and recharge through Resident Welfare Associations. No person shall cause any such activity in area, which is against the provisions of this act or which is prejudicial or in contravention of the act.

(*c*) **Bulk Users**

(*i*) Construction of new well for human consumption/drinking water will be allowed in consultation & under guidance of registered service provider.

(*ii*) Existing and new bulk users to mandatory adopt area specific RWH/R technique through registered service provider. Minimum amount of rain water harvesting / recharging to be fixed by the authority.

(*iii*) Limit of ground water extraction will be fixed by the Authority.

(*iv*) Bulk users may be charged fee for ground water withdrawal annually by the Authority/Government.

(*v*) To ensure and check efficacy of rain water harvesting and recharge structure, limit of ground water withdrawal, size of pump set, suction & delivery pipe, quality of water being recharged to ground water and any other works etc. ground water user will be required to get an inspection done twice a year in April & October by the concerned registered service provider. Service provider will charge an inspection fee from ground water user as prescribed by the Authority. If Rain Water Harvesting/Recharging structure is not found working properly, and amount of ground water withdrawal & quality of water are not as per limit/norms, the registered service provider will advise the ground water user to make it functional and ensure that other provisions as per sanction order and directions are being complied with. In case ground water user does not act on the advice of registered service provider, he will inform the Authority to take further action as per the Act.

(*vi*) Any modification /alteration in a registered well will require clearance. No person shall cause any such activity in urban area, which is against the provisions of this act or which is prejudicial or in contravention of the act.

2. For Rural Areas

(*a*) **Ground water users using pump set up to 7.5 H.P.**

(*i*) All users of ground water will self regulate themselves through Water User Association by installing pump set of power not exceeding 7.5 H.P. or as may be prescribed by the Authority.

(*ii*) These users of ground water will adopt Rain Water Harvesting/Recharging activities as per prevailing Government orders/as prescribed by the Authority. For these users, mass awareness and sensitization programme will be carried out through Water User Association.

(*b*) **Ground water users using pump set of more than 7.5 H.P.**

(*i*) If any user of ground water wants to install pump set exceeding 7.5 H.P., he will do so through the registered service provider as per the procedure prescribed by the Authority.

(*ii*) The limit/rate of ground water withdrawal from one well with pump set exceeding 7.5 H.P. can be restricted in specific sub-unit/areas of the block as prescribed by the authority.

(*iii*) It will be mandatory for all sellers to inform the Authority about selling of pumpset of power exceeding 7.5 H.P. along with verification of identity and address of the persons who purchase these pumpset.

(*iv*) All users of ground water will adopt Rain Water Harvesting/Recharging activities as per prevailing Government order/as prescribed by the Authority.

(*v*) The Authority will promote mass awareness programme on Rain water harvesting and recharge through Water User Associations. No person shall cause any such activity in area, which is against the provisions of this act or which is prejudicial or in contravention of the act.

(c) **Bulk Users**

(*i*) Construction of new well for human consumption/drinking water will be allowed in consultation & under guidance of registered service provider.

(*ii*) Existing and new bulk users to mandatory adopt area specific Rain Water Harvesting/Recharging technique through registered service provider. Minimum amount of Rain Water Harvesting/Recharging to be fixed by the Authority.

(*iii*) Limit of ground water extraction will be fixed by the Authority.

(*iv*) Bulk users may be charged fees for ground water withdrawal annually by the Authority/Government.

(*v*) To ensure and check efficacy of rain water harvesting & recharge structure, limit of ground water withdrawal, size of pump set, suction & delivery pipe, quality of water being recharged to ground water and any other works etc. ground water user will be required to get an inspection done twice a year in April & October by the concerned registered service provider. Service provider will charge an inspection fee from ground water user as prescribed by the Authority. If Rain Water Harvesting/Recharging structure is not found working properly, and amount of ground water withdrawal & quality of water are not as per limit/norms, the registered service provider will advise the ground water user to make it functional and ensure that other provisions as per sanction order and directions are being complied with. In case ground water user does not act on the advice of registered service provider, he will inform the Authority to take further action as per the Act.

(*vi*) Any modification /alteration in a registered well will require clearance. No person shall cause any such activity in urban area, which is against the provisions of this act or which is prejudicial or in contravention of the act.

18. (*i*) In areas notified as over-exploited/critical there shall be complete ban on the construction of new wells (both private & govt.). The ban shall continue till such areas are de-notified by the Authority.

(*ii*) However, except commercial and industrial users of ground water, construction of new wells in these areas for water supply schemes for human consumption/ drinking water will be allowed to users of ground water in consultation and under the guidance of registered service provider. The procedure to avail the services of registered service provider will be as prescribed by the Authority.

(*iii*) In such areas, the existing and new users of ground water (both private and government) shall have to mandatorily adopt the area specific rain water harvesting techniques as prescribed by the Authority under the supervision & guidance of the registered service provider. The procedure to avail the services of registered service provider will be as prescribed by the Authority. The minimum amount of Rain Water Harvesting/Recharging will be fixed by the authority.

(*iv*) For bulk user limit of ground water extraction will be fixed by the Authority.

(*v*) To ensure and check efficacy of rain water harvesting & recharge structure, limit of ground water withdrawal, size of pump set, suction & delivery pipe, quality of water being recharged to ground water and any other works etc. ground water user will be required to get an inspection done twice a year in April & October by the concerned registered service provider. Service provider will charge an inspection fee from ground water user as prescribed by the Authority. If Rain Water Harvesting/Recharging structure is not found working properly, and amount of ground water withdrawal & quality of water are not as per limit/norms, the registered service provider will advise the ground water user to make it functional and ensure that other provisions as per sanction order and directions are being complied with. In case ground water user does not act on the advice of registered service provider, he will inform the Authority to take further action as per the Act. Authority, if on verification finds that the information about the working of Rain Water Harvesting/Recharging structure given by the registered service provider is not correct, he will be made accountable and penalized as per provisions of the act.

(*vi*) Bulk users may be charged fees for ground water withdrawal annually by the Authority/ Government.

(*vii*) No person shall cause any such activity in urban and rural area, which is against the provisions of this act or which is prejudicial or in contravention of the act.

19. 1. Any commercial user of ground water as well as any industry desiring to sink a well for any purpose or existing units in Semi-critical, level-1 and in non-notified area shall apply to the registered Service Provider for sinking the well and construct the well/tube well under the supervision & guidance of registered service provider. The limit of ground water will be fixed by the Authority. It will be mandatory for these ground water users in such areas to harvest rain water and execution of RWH/R works will be under the supervision and guidance of registered service provider with in time specified by the Authority. The quantity/limit of harvesting rain water will be as prescribed by the authority. To ensure and check efficacy of rain water harvesting & recharge structure, limit of ground water withdrawal, size of pump set, suction & delivery pipe, quality of water being recharged to ground water and any other works etc. ground water user will be required to get an inspection done twice a year in April & October by the concerned registered service provider. Service provider will charge an inspection fee from ground water user as prescribed by the Authority. If Rain Water Harvesting/Recharging structure is not found working properly, and amount of ground water withdrawal & quality of water are not as per limit/norms, the registered service provider will advise the ground water user to make it functional and ensure that other provisions as per sanction order and directions are being complied with. In case ground water user does not act on the advice of registered service provider, he will inform the Authority to take further action as per the Act. Authority, if on verification finds that the information about the working of Rain Water Harvesting/

Recharging structure given by the registered service provider is not correct, he will be made accountable and penalized as per provisions of the act. Commercial and Industrial Users may be charged fees for ground water withdrawal annually by the Authority/Government. Any modification/alteration in a registered well will require clearance of the Authority in such areas.

2. No commercial and industrial user of ground water shall install/sink any well in areas notified as Over-exploited/Critical area, level-2 and 3 and ban will continue till area is de-notified. It will be mandatory to establish Rain Water Harvesting/Recharging activities for existing users through registered Service provider. Minimum amount of Rain Water Harvesting/Recharging will be fixed by the Authority. To ensure and check efficacy of rain water harvesting & recharge structure, limit of ground water withdrawal, size of pump set, suction & delivery pipe, quality of water being recharged to ground water and any other works etc. ground water user will be required to get an inspection done twice a year in April & October by the concerned registered service provider. Service provider will charge an inspection fee from ground water user as prescribed by the Authority. If Rain Water Harvesting/Recharging structure is not found working properly, and amount of ground water withdrawal & quality of water are not as per limit/norms, the registered service provider will advise the ground water user to make it functional and ensure that other provisions as per sanction order and directions are being complied with. In case ground water user does not act on the advice of registered service provider, he will inform the Authority to take further action as per the Act. Authority, if on verification finds that the information about the working of Rain Water Harvesting/Recharging structure given by the registered service provider is not correct, he will be made accountable and penalized as per provisions of the act. Commercial and Industrial Users may be

charged fees for ground water withdrawal annually by the Authority/Government. Any modification/alteration in a registered well will require clearance of the Authority in such areas.

3. Every application to the registered service provider shall be made in such form, contain such particulars, and in such manner and accompanied by such fee as may be prescribed.

4. On receipt of an application if the registered service provider is satisfied that it shall not be against public interest to do so, it may advise and guide the applicant for extraction and use of ground water, which will include mandatory provisions of artificial recharge structures of appropriate size to be constructed by the applicant. Provided further that the commercial and industrial users shall not be refused if they wants to sink a new well of the same specification in the same plot or in the adjoining plot, as a replacement of an old well which has gone out of order on account of completion of life or for any other reason.

5. The registered Service Provider while advising the applicant shall have regard to :

   (*a*) The purpose or purposes for which water is to be used.

   (*b*) The existence of other competitive users.

   (*c*) Number of existing wells already in use.

   (*d*) Number of wells to be sunk.

   (*e*) Depth and design of well (including diameter, pipe etc), to be decided as per requirement of water.

   (*f*) The availability of ground water.

   (g) Quantity of ground water to be drawn.

   (*h*) Quality of ground water with reference to use.

(*i*) Spacing of ground water structures keeping in consideration the purpose for which water is to be used.

(*j*) Long term ground water level behavior.

(*k*) Its likelihood of adversely affecting water availability of any drinking water source in its vicinity.

(*l*) The effect on other sources of water supply.

(*m*) The compatibility with the existing water supply system.

(*n*) The availability of factors controlling or preventing pollution.

(*o*) Rain water harvesting.

(*p*) Any other factor relevant thereto.

20. 1. The authority shall by notification in the gazette fix the water rates for selling the ground water from private tube wells.

2. No owner of private well shall sell the ground water on rates more than the rates fixed under sub-section-1.

21. 1. The Authority shall take steps for promotion of Mass Awareness Programme for protecting ground water from pollution.

2. The Authority shall ensure that no person/institution/ industry pollutes ground water. It shall make establishment of effluent treatment plant mandatory wherever necessary. In case the polluter fails to set up effluent treatment plant within the time period prescribed, the Authority shall have the right to get the Effluent Treatment Plant constructed at polluter's cost and proceed against the polluter as per provisions of section 31(2).

22. Users of ground water in rural and urban areas especially in clusters shall have to maintain the geo scientific standards of minimum spacing between two

wells, as and when specified by the authority for such areas.

23. At any time after certificate of registration, as the case registration may be, has been granted or issued, the Authority may, for technical reasons, by order alter, amend or vary the terms of the certificate of registration, as the case may be, provided the user of ground water/ registered service provider/drilling agencies has been given an opportunity of being heard, provided that before taking such action, the Authority shall ensure that standing crops are not damaged.

24. If the Authority is satisfied either on a reference made to it in this behalf or otherwise, that :

(*a*) The certificate of registration granted, under section-10, 12, 13, 14 or 15, is not based on facts.

(*b*) The holder of the certificate of registration has without reasonable cause failed to comply with the conditions and responsibilities subject to which the certificate of registration has been granted or issued or has contravened any of the provisions of this Act or the rules made there under or fails to deliver his responsibilities or,

(*c*) A situation has arisen which warrants limiting of the use or extraction of ground water, then without prejudice to any other penalty to which the holder of the certificate of registration may be liable under this Act, the Authority may after giving the holder of the certificate of registration, an opportunity to show cause, cancel the certificate of registration.

25. 1. The Authority or any person authorized by the Authority in writing in this behalf, shall have power -

(*a*) To enter on any property (private or Government owned) having the right to investigate and make any measurements concerning the land or the water located on the surface or the under ground.

(*b*) To inspect the well which has been or is being sunk and the soils and other materials excavated there from.

(*c*) To take specimen of such soils or other materials or of water extracted from such wells.

(*d*) To require in writing the persons sinking a well to keep and preserve in the prescribed manner specimens of soils or any material excavated there from for such period not exceeding three months from the date of completion or abandonment of the work as may be specified by the Authority and thereupon such person shall comply with such requisition.

(*e*) To inspect and to take copies of the relevant record or documents and ask any question necessary for obtaining any information (including diameter or depth of the well which is being or has been sunk; the level at which the water is or was struck and subsequently restored/ rested, the types of strata encountered in the sinking of the well and the quality of the water struck) required for carrying out the purposes of this Act.

(*f*) To direct the user of ground water to install water measuring device on any ground water abstraction structures, Provided that where the user of ground water does not comply with the directions issued to him under this clause within a period of sixty *days*, the Authority itself may install such water measuring device and recover the cost from the defaulting user of ground water.

(*g*) To seize any equipment/device utilized for illegal sinking and destroy the work executed fully or partly.

(*h*) To inspect the premises of commercial, bulk users and industrial user of ground water, as and when required.

(*i*) To direct the ground water user in cluster of multi-storied buildings to maintain the geo-scientific standards of minimum spacing as specified between two tube wells.

(*j*) To require any user of ground water, who does not comply with the provisions of this Act and the rules framed hereunder, to close down any water supply or destroy any hydraulic work found to be illegal according to the provisions of this Act and the rules framed there under, Provided that where the user of ground water does not comply with the requisition issued to him within a period of sixty days, the Authority may carryout the necessary work and recover the cost from illegal user of ground water.

(*k*) To enter and search with such assistance, if any, as it considers necessary, any place in which it has reason to believe that offence under this Act has been or is being committed and order in writing the person who has been or is committing the offence not to extract or use the ground water for a specified period not exceeding thirty days.

(*l*) To issue directions to local bodies, institutions, industry, company, government departments, individuals or any other concerned, for undertaking specific measures for ensuring rain water harvesting and water conservation and such local bodies, institutions, industry, company, government departments, individuals or any other concerned shall be bound to comply with such directions within the time specified by the Authority in such directions.

(*m*) To issue directions to the District, Block, Village Panchayat and Urban Local Bodies to allocate certain percentage of funds available with them for the construction of Ground Water Recharge Structures.

(*n*) To exercise such other powers as may be necessary for carrying out the purposes of this Act or any rules made there under.

(*o*) To give directions to Service provider.

2. The power conferred by this section includes the power to break open the door of any premises where sinking, extraction and use of ground water may be going on, provided that the power to break open the door shall be exercised only after the owner or any other person in occupation of the premises, if he is present therein, refuse to open the door on being called to do so.

3. The provisions of the Code of Criminal Procedure, 1973 (Act no. 2 of 1974), shall so far as may be, apply to any search or seizure under this section as they apply to any search or seizure made under the authority of a warrant issued under section 93 of the said Code.

4. Where the Authority seizes any mechanical equipment or device under clause (g) of sub-section (1), it shall as soon as may be, inform the magistrate and obtain his orders as to the custody thereof.

26. Every order or direction issued under section-25 shall be served in such manner as may be prescribed.

27. No person shall be entitled to claim any damages or compensation from the Government for any loss sustained by him by virtue of any action taken under this Act.

28. The Authority may, by general or special order in writing, direct that all or any of the powers or duties which may be exercised or discharged by it shall, in such circumstances and under such conditions, if any, as may be specified in its order be exercised or discharged also by any person, body or other authority, specified in this behalf in the order.

29. All members and employees of the Authority shall when acting or purporting to act in pursuance of the provisions of this Act or of any rules made there under be deemed to be public servants within the meaning of section-28 of the Indian Penal Code.

30. No prosecution, suit or other legal proceeding shall be instituted against the Government, the Authority or any other officer of the Government or any member or other employees of the Authority for anything done or intended to be done in good faith under this Act, or the rules made there under.

31. 1. No prosecution for an offence under this Act shall be instituted except with the written consent of the Authority or a person authorized in this behalf.

2. No court inferior to that of a Metropolitan Magistrate or a Magistrate of first class shall try any offence under this Act.

CHAPTER – III

RECHARGE OF GROUND WATER

32. The over-exploitation of ground water due to ever increasing population and other development activities have let to fall in ground water table, drying up of wells, reduced sustainability of wells, environmental degradation etc. in many parts of the State. In order to improve the ground water situation in critical areas, it is, therefore, essential to adopt rain water harvesting for ground water recharge. In urban areas, rain water available from roof tops of buildings and other open areas can be utilized gainfully for ground water recharge. Rain water harvesting structures feasible in urban areas include recharge pits, trench, existing wells etc. In rural areas, ground water recharge be taken up considering watershed as a unit. The feasible recharge structures in rural areas include construction of gully plugs, contour bunding, Gabion structure, check dam/ weir, percolation tank, recharge shaft etc.

1. To improve the ground water situation, the Authority may identify the recharge worthy areas in the State and issue necessary guidelines for adoption of rain water harvesting for ground water recharge in these areas. In rural areas, watershed management to facilitate ground water recharge may be encouraged through community participation. The Authority may give appropriate directions to the concerned departments of the State Government to include Rain Water Harvesting in all developmental schemes falling under notified areas. In urban areas, falling in notified areas, the Authority may issue directives for constructing appropriate rain water harvesting structures in all residential, commercial and other premises having an area of 100 Sq.mt. or more in manner prescribed within the stipulated period, failing which the Authority may get such rain water harvesting structure constructed and recover the cost incurred along with a penalty as may be prescribed.

2. Not withstanding any thing contained in the relevant laws, the Municipal Corporation or any other local Authority as the case may be, may impose stipulated conditions for providing roof top rain water harvesting structures in the building plan in an area of 100 Sq. mt. or more, while according approval for construction, and permanent water and electricity connections shall be extended only after compliance of the directions given in this regard.

3. Provision of rain water harvesting for ground water recharge shall be made mandatory for the townships. For this the authority may give appropriate directions along with mandatory provisions for maintenance of such recharge structures.

4. The authority may issue directions to the Housing department for providing combined rain water harvesting system for existing urban colonies along with provisions of maintenance.

5. Existing ponds, lakes and other water bodies shall be maintained and governed by concerned District Magistrate or any other body appointed by District Magistrate.
6. Water bodies existing in rural areas at the time of enforcement of the Uttar Pradesh Zamidari Abolition and Land Reform Act and in other areas on 1st January, 1950 shall be maintained properly and any encroachment etc. made there upon shall be removed as per provisions of the respectable laws applicable on that particular area for that purpose which has been made out after the time mentioned in this act. However, any encroachment on the water bodies will be removed after providing an opportunity of hearing to the person who has made encroachment on water bodies.
7. No encroachment will be allowed after the commencement of this act.
8. Removal of encroachment shall be made effective by the Sub divisional Magistrates/City Magistrates under their respective jurisdiction.
9. Land records at the time of enforcement of Uttar Pradesh Zamidari Abolition and Land Reform Act-1950 in rural areas and on 1st January 1950 in other areas will be the conclusive proof for defining the water bodies.
10. The State Government may, by notification, exempt any building constructed before the enforcement of this Act by a detailed reasoned order on the grounds of public utility purpose.
11. The Authority shall take steps for promotion of Mass Awareness and Training Programmes on Rain Water Harvesting and Artificial Recharge to Ground Water through Government Agencies / Non Government Organisations /Volunteer Organisation / Educational Institutions/ Industries / Individuals.

12. Resident Welfare Associations in urban areas shall be motivated to adopt rain water harvesting for ground water recharge as a process of self regulation.

CHAPTER - IV

---

33. If any difficulty arises in giving effect to the provisions of this Act, by reasons of anything contained in this Act, or any other enactment for the time being in force, the State Government may, as the occasion requires, by order direct that this Act shall, during a period not exceeding twelve months after the date of such order have effect subject to such adaptations, whether by way of modifications, addition or omission, as it may deem to be necessary and expedient, provided that no such order shall be made after the expiry of the period of two years from the date of commencement of this Act.

34. 1. If any user of ground water, registered service provider or any drilling agency :

(*a*) Contravenes or fails to comply with any of the provisions of this Act or any rule made there under,

(*b*) Obstructs the Authority or any other person authorized by it to exercise any of the powers under this Act, shall be punishable :

(*i*) For the first offence, with fine which may extend to rupees Five Thousand and

(*ii*) For the second and subsequent offence, with imprisonments for a term which may extend to six months or with fine which may extend to rupees ten thousand or with both.

2. Any person, user of ground water, institution, industry, registered service provider, drilling agency contaminates ground water in any manner or helps in promoting ground water contamination by industrial and aquaculture waste disposal or directly disposes waste water including sewage into the aquifers shall be

punishable with serious penal action as such act shall be treated as an act of criminal offence. The person responsible for such offence shall be liable for punishment with imprisonment for a term which shall not be less than one month but which may extend to one year and/or with fine as may be prescribed. In addition to that the cost of its repairs or remedying the same shall be recovered as an arrears of land revenue, provided that the person responsible for abetment of such offence or connivance shall also be liable for punishment in the same manner.

35. Any offence under this Act may be compounded (leaving matters related to pollution of ground water) by the Authority, either before or after the institution of the proceeding subject to such conditions as may be prescribed.

36. 1. Whenever an offence under this Act has been committed by a company, every person who at the time of the commission of offence was in charge of, or was responsible to the company for the conduct of the business of the company, shall be deemed to be guilty of the offence and shall be liable to be proceeded against and punished accordingly.

2. Not withstanding anything contained in sub-section (1), where an offence under this Act has been committed with the consent or connivance of, or is attributable to any neglect on the part of, any director, manager, secretary or other officer of the company, such director, manager, secretary or other officer shall also be deemed to be guilty of that offence and shall be liable to be proceeded against and punished accordingly. Provided that nothing contained in this sub-section shall render any such person liable to any punishment under this Act if he proves that offence was committed without his knowledge or that he exercised all due diligence to prevent the commission of such offence.

Explanation : For the purpose of this section :

1. "Company" means any body corporate and includes a firm or other association or individuals, and
2. " Director ", in relation to a firm, means a partner in the firm.

37. 1. Any person aggrieved by the decision or action of the authority under this Act may, within a period of sixty days from the date on which the action is taken or the decision is communicated to him and on payment of such fees as may be prescribed, prefer an appeal to such an authority as may be specified by the State Government in this behalf, provided that appellate authority may entertain an appeal after the expiry of the said period of sixty days, if it is satisfied that the applicant was prevented by sufficient cause from filing the appeal in time.

2. On receipt of an appeal under sub-section (1), the appellate authority shall, after giving the appellant an opportunity of being heard, dispose of the appeal as expeditiously as possible.

38. 1. The State Government may, by notification in the Gazette, make rules for carrying out the purposes of this Act.

2. In particular and without prejudice to the generality of the foregoing power, such rules may provide for :

   (*a*) The term of office and manner of filling vacancies and other conditions of service of the Members and Chairman of the Authority.

   (*b*) The function and the terms and conditions of service of the employees of the Authority.

   (*c*) Any other manner of affecting service of the notification under sub-section (3) of section-5 of chapter- II.

(*d*) The formation of Water Users Association under the section-6.

(*e*) Any other power of Water Users Association under sub-section-3 of section-7.

(*f*) The formation of Resident Welfare Association under section-8.

(*g*) Any other power of Resident Welfare Association under section-9.

(*h*) The form of application under section-10(1) and the particulars that may be furnished with the application.

(*i*) The fees for issue of registration certificate of service provider under section-10(3).

(*j*) The responsibilities of registered service provider for ensuring the implementation of works under section-11.

(*k*) The fee charged by registered service provider for providing services under section-11(3).

(*l*) The punishment for existing/new user extracting ground water without registration under section-12(8), 14(8) & 15(viii).

(*m*) The severe punishment for registered service provider under section-11(2) & 19(i).

(*n*) The form of application under section-12(1), 13(a), 14(a) & 15(a) (i) and the particulars that may be furnished with the application.

(*o*) The fees for issue of registration certificate of service provider & other applicants under section-12(1), 13(a) (2), 14(a) (i) & 15(a) (i).

(*p*) The procedure for application to install the pump-set of power exceeding 7.5 H.P. and fixing limit/rate of ground water extraction under section-16 (2) (b) & 17(2) (b).

(*q*) The procedure to avail the services of registered service provider under section-18(iii).

(*r*) The quantity /limit of harvesting rain water under section-19(1) for commercial and industrial users.

(*s*) The manner and form of application to be submitted by commercial and industrial users to service provider for sinking well under section-19(3) & 19(4).

(*t*) The rates of selling ground water under section-20(ii).

(*u*) The cancellation of certificate of registration under section-24.

(*v*) The manner to serve every order or direction under section-26.

(*w*) The manner in which the specimens of soils or other material shall be kept and preserved under clause (d) of sub- section (1) of section-25 of chapter-II.

(*x*) Specifying the appellate authority under sub-section (1) of section-37 of chapter-IV and the fee to be accompanied with the application for appeal.

(*y*) Any other matter which is to be or may be prescribed.

39. Save as otherwise provided in this Act, the provision of this affected Act shall be in addition to and not interogation of the provisions of any other law for the time being in force.

23

Ground Water Policy of Uttar Pradash State and Working Draft

Anita Rana

INTRODUCTION

Preamble

Ground water is the major source of water on the planet. Its conservation is also emerging as a critical issue for cities and towns around the world. Sustainable and equitable use of ground water over millennia has been ensured by cultural adaptation to water availability through water conservation technologies, agricultural systems and cropping patterns adapted to different climatic zones, and conservation-based life styles. Ground water irrigation worldwide has been found difficult to curb and regulate. And often, resource depletion has been countered by augmenting water supply. Search for ways to augment water supply in closing basins will remain an enduring element in the strategy of coping with ground water depletion. However, high capital costs and paucity of resources will keep many countries from trying out these grandiose ideas for long time to come. In many situations, there are indications that water infrastructure in river basins is getting reconfigured by autonomous local action in private or community domain. Where ground water stress has deepened enough to threaten life, there are some signs of people and local institutions shaking off their generally passive attitude of dependency on the government, and have taken charge of the resource. Such signs are most visible in western Indian states of Rajasthan and Gujarat which are most critically dependent on overexploitation of ground water for sustaining their agriculture. However, even here, there is little accent on rule-making and demand management, but great mobilization has occurred around rain water harvesting and ground water recharge in a decentralized format.

Water is a vitally important substance and its role in the evolution of life now is widely recognized. Uttar Pradesh is endowed with bountiful water resources but because of increasing demand for various purposes namely irrigation, drinking and domestic, power (thermal and hydro), industrial and other uses, its scarcity is becoming apparent and more pronounced with increasing population. Water is a prime natural resource, its planning and development needs to be governed by the development perceptions in the state. For the proper management of ground water resource which should ensure optimal utilisation and avoid over exploitation, the following aspects should be duly addressed :

1. Demand side management and conservation through the spread of efficient irrigation technologies, such as piping drip and sprinkler irrigation is essential.
2. Opportunities for and constraints on the spread of low water intensity cropping patterns should be investigated as a component of demand side management.

Understanding Ground Water

Ground water is water that is found underground in the cracks and spaces in soil, sand and rock. The ground beneath the soil can be divided into two main parts. Firstly, unsaturated zone or zone of aeration and secondly, saturated zone or aquifer. The water in the zone of aeration does not flow in to the well. The upper surface of the zone of saturation is water table. In other words, the water table is the first occurrence of ground water. However, ground water is mainly stored in an aquifer. Aquifers are geological formations- layers of sand, gravel and rock-where significant amounts of water can be stored, transported or supplied to a well or spring.

Water as a Matter of Right

Water has not been guaranteed as a matter of fundamental rights in our constitution. However, courts by several judicial pronouncements have inferred that right to clean and safe water is an important aspect of Article 21 of the constitution of India.

In Narmada Bachao Andolan vs. Union of India the Supreme Court said that: Water is the basic need for the survival of human beings and is part of right to life and human rights enshrined in Article 21 of the Constitution of India.

In FK Hussain vs. Union of India the court was of opinion that an administrative agency could not be permitted to function in such a manner as to make inroads into the Fundamental Right guaranteed under Article 21 of the Constitution. The right to sweet water and the right to free air are the attributes of the right to life. They are the basic elements that sustain life. Right to water can be further inferred from Article 15(2) of the Constitution which states that no citizen shall be subjected to any limitation with regard to the use of wells, tanks, bathing ghats.

Right to Ground Water

It is customarily accepted across India that a well on a piece of land belongs to the owner of that land, and others have no right to extract water from the well or restrict the landowner's right to use water. Interpretation of the Transfer of Property Act of 1882 and the Land Acquisition Act of 1894 support the position that a landowner has proprietary rights to ground water; it is connected with dominant heritage and cannot be transferred apart from the land. *However, government can restrict these rights in the larger public interest as right to property is not fundamental right in India.*

The Doctrine of Public Trust

The recent judicial pronouncements have clearly stated

that deep underground water is the property of the State under the Doctrine of Public Trust. The holder of land has only a user right towards the drawing of water in tube wells. In M.C.Mehta vs. Kamal Nath the Court declared that :

The State is the trustee of all natural resources, as a trustee the State has a legal duty to protect the natural resources and these resources meant for public use cannot be converted into private ownership.

In M.C.Mehta vs. Union of India, the Court observed that: Ground water is a social asset, and the right of the citizen to the use of water, air and earth are protected under Article 21 of the Constitution. It further states that the environmental balance should be maintained and wherever ground water is required for domestic and agriculture needs, priority is to be given to these.

Policy Context

The National Water Policy (2002) and the Uttar Pradesh State Water Policy have both identified ground water protection as an important component of water resources management in general. The National Water Policy states that "the development and overexploitation of ground water resources in certain parts of the country have raised the concern and need for judicious and scientific resource management and conservation". The Policy recommends several measures on ground water management - periodical reassessment of the ground water potential on a scientific basis, regulation of ground water so as to not exceed the recharging possibilities as also to ensure social equity; artificial recharge for the augmentation of ground water resources, for integrated planning of surface and ground water resources incorporating quantity and quality aspects as well as environmental considerations. *For judicious ground water use and management, the Uttar Pradesh Water Policy prescribes demand side management and conservation through the spread of efficient irrigation technologies and low water intensity cropping patterns, conjunctive surface and ground water management, ground water recharge and regulation of ground water abstraction. The current policy framework recognizes the importance of ground water resources and recommends several measures for the protection of ground water quantity and quality.*

Legislative Framework

As a result of Government of India's initiatives on the conservation and protection of the environmental resources of the country, an elaborate legal framework exists for the protection of ground water in tandem with other water resources and the environment in general. These laws are applicable in Uttar Pradesh as well and shall be taken full recourse to in conjunction with this policy for the protection of the ground water resources of the state. In addition there are several statutes of Uttar Pradesh that are of relevance to ground water management and that need to be modified to include the objectives and strategic measures listed in this policy.

These Central and State laws are the following :

1. The Water (Prevention and Control of Pollution) Act, 1974 as amended in 1988.
2. The Water (Prevention and Control of Pollution) Rules, 1975
3. The Water (Prevention and Control of Pollution) (Procedure for Transaction of Business) Rules, 1975.
4. The Water (Prevention and Control of Pollution) Second Amendment Rules, 1976.
5. The Water (Prevention and Control of Pollution) Cess Act, 1977 as amended by Amendment Act, 1991.
6. The Water (Prevention and Control of Pollution) Cess Rules, 1978.
7. The Water (Prevention and Control of Pollution) Amended Rules, 1989.
8. The Environment (Protection) Act, 1986.
9. The Environment (Protection) Rules, 1986.

10. The Environment (Protection) Amendment Rules, 1987.
11. The Environment (Protection) Third Amendment Rules, 1987.
12. The National Environmental Tribunal Act, 1995.
13. The National Environment Tribunal Rules, 1995.
14. Notification on Emission Standards and Guidelines for Location of Industries, mining operation etc. for various areas.
15. The Public Liability Insurance Act, 1991.
16. The Public Liability Insurance Rules, 1991.
17. Hazardous Waste (Management and Handling) Rules, 1989 (amended in 2000) Batteries (Management and Handling) Rules, 2001.
18. Manufacture, Storage and Import of Hazardous Chemical Rules, 1989.
19. Manufacture, Use, Import, Export and Storage of Hazardous Micro-organisms, Genetically Modified Engineered Organisms or Cells Rules, 1993.
20. Hazardous Micro-organisms and Genetically Modified Organisms (Manufacture, Use, Import Expert and Storage) Rules, 1999.
21. Bio-Medical Waste (Management and Handling) Rules 1998.
22. Municipal Solid Waste (Management and Handling) Rules, 1999.
23. Recycled Plastics Manufacture and Usage Rules, 1999.
24. Guidelines for seeking Environmental Clearance.
25. National Conservation Strategy and Policy Statement on Environment and Development, 1992.
26. Ecomark Criteria for 16 Product Categories.

27. Temperature Limit for Discharge of Condenser Cooling water from Thermal Power Plant.
28. Environmental Standards for Gas/Naphtha Based Power Plants.
29. The Indian Forest Act, 1927.
30. Forest Conservation Act, 1980.
31. Uttar Pradesh Soil and Water Conservation Act, 1963.
32. UP Land Revenue Act 1901.
33. U.P. Village Panchayat Act, 1947.
34. U.P. Kshetra Panchayat & Zilla Panchayats Act, 1961.
35. U.P. Minor Minerals Concession Rules, 1963.
36. U.P. Water Supply and Sewerage Act, 1975.
37. U.P. Water Supply and Sewerage Act, 1975.
38. U.P. (Regulation of Building Operations) Act, 1958.
39. U.P. Housing and Development Board Act, 1965.
40. U.P. Urban Planning and Development Act. 1973.
41. U.P. Special Areas Development Authorities Act, 1986.
42. U.P. Municipal Corporation Act, 1959.
43. U.P. Municipalities Act, 1916.

WHY A NEED FOR POLICY?

Over the years, State Irrigation Departments have aimed to store and distribute water to farmers in the dry season, so that at least two crops can be grown in a year. It is generally assumed that the monsoon brings enough water to go round and that farmers don't need help. This is not always true. Monsoons are erratic and sometimes do not bring the promised rainfall. Another fact is that the torrential monsoon rains bring vast amounts of water that are not needed for agriculture at that very moment. Most of the monsoon rains flood fields, filling rivers and streams that rush unused to the sea. This is the very water that could help the farmer year-round, if it could be diverted and its flows slowed down enough to percolate through to underground aquifers. Fast-flowing torrents wash away topsoil, and waterlogged fields are a farmer's nightmare for such crops as pigeon pea that cannot tolerate excessively wet conditions. From the farmers' perspective, having better 'crop security' for monsoon crops by making irrigation water available to supplement an erratic supply of rain and still having enough water stored in aquifers to grow another crop later is an attractive proposition. But policy makers and water managers will have to clearly articulate the advantages and positive impact of the change if farmers are to be convinced. Farmers live with risks the year-round, so removing one of their main worries the possibility of drought caused by monsoon failure will give them the confidence to face their many other challenges. Farmers will need to be convinced that this new approach will work for them. If they can be sure that their hard-earned pumps will continue to find water in the aquifers after the monsoon, that the pumps will not need to be lowered into ever-deepening borewells, and that the stable water table means lower power costs, then they will support changing cropping patterns as part of a new ground water recharge strategy.

There is still a need, however, to enact a special, consolidated law for ground water conservation, protection and development in the state. Some of the basic policy measures can be adopted for Ground water Policy in Uttar Pradesh.

1. Policy on Database, Technology, Research and Development

The first step that is of fundamental importance in ground water regulation and management is the availability of a comprehensive and reliable database and information system on the basis of which appropriate decision making may be made. Data and information on ground water needs to cover all land areas-rural residential areas, cultivated areas, waste lands, urban residential areas, urban waste lands, dumping grounds, government sites, institutional areas [government and private sector], industrial areas, forest areas, identification of vulnerable areas etc. Under this policy, a comprehensive data base and information system shall be established. All land areas of the state shall be mapped for ground water, surface water and land use. Where such data base and information systems exist through the aegis of other policy initiatives such as watershed maps/soil survey maps, etc. ground water information shall be factored into such data bases. An integrated approach of remote sensing, geophysics and GIS (Geographical Information System) shall be applied in demarcation of the different ground water zones to delineate or map the land area that could contribute water and pollutants to the ground water aquifers. Community groups such as environmental watch groups, non-government organizations, watershed organizations, resident welfare organizations, rural community organizations such as women's groups, youth groups, farmers' organizations, and farmer's cooperative societies can enhance the state's assessment by conducting site-specific inventories of potential pollutant sources that might not be on state databases or maps. Remote Sensing and GIS can provide useful information for large-area water and salt balances and identification of parameters such as evapotranspiration, rainfall distribution, interception losses and crop types and intensities that can be used as indirect measures of salinity and water logging and as evidence for direct estimates. Interdisciplinary diagnostic studies involving field studies of the plant environment, farm management practices, water supply and removal, and institutional linkages are necessary to diagnose site related problems and potential solutions.

2. Policy on Quantity

The maintenance of quantity of ground water requires a twin pronged approach one, the protection from pollution, water below the ground surface in geological structures or formations known as 'aquifers', and secondly, the ecosystems from which these waters are recharged or into which they discharge. As such a coordinated approach to land-surface water–ground water protection is required and shall be ensured. In the identification and prioritization of aquifers/districts/blocks for ground water development, ground water quantity status shall be taken fully into account, and reflected in the ground water Management Plans. The concept of 'Critical Blocks' that is being adopted at present shall be modified to include levels of contamination/pollution in addition to degree of ground water extraction. It shall be ensured that all developmental departments of government take into account ground water quantity objectives in their policy and developmental planning.

3. Policy on Quality

Ground water Information and Ground water Quality Assessment and Monitoring Centres shall be set up in every block, with full access to the public for contributing to as well as for receiving information. These centres shall be functionally linked to similar centres at district level and to the State Ground Water Board to enable decision making at appropriate levels. Based on information provided by district level, Ground water Information and Ground water Quality Assessment and Monitoring Centres, prioritization norms shall be developed by the State Ground Water Board for selection of districts/blocks for development taking into account current levels of abstraction. Priority shall be given to blocks and districts that have low percentage of development to ensure equity in ground water access. These norms shall be given wide publicity. For the purpose of ground water quality protection, entry of the highest risk categories of pollutants shall be banned, while the pollution of ground water by other categories shall prevented until they are sufficiently treated as per norms laid down for this purpose. Or addressing pollution alleviation measures, the 'polluter pays' principle shall be adopted. In the identification and prioritization of aquifers/districts/blocks for ground water development, ground water quality status shall be taken fully into account, and reflected in the Ground Water Management Plans. Agricultural practices pose a risk to ground water in many ways – such as improper storage of chemicals like diesel, seepage from livestock shelters, overflow from effluent tanks related to agro-based industries, fertilizer and pesticides leaching chemical compounds into surface and ground water, the disposal of waste in farm landfills, inappropriate drainage facilities, and seepage of farm run-off into ground water via poorly protected, constructed and sited wells, etc.

4. Policy on Pollution and Polluters

Pollution assessment shall be undertaken on a state-wide basis comprehensively and rigourously. This activity will involve identifying potential sources of pollutants that could contaminate the surface or ground water supply, resulting in a list and a map of facilities and activities within the delineated area that might release contaminants into the ground water supply (for wells) or into the watershed of the river or lake (for surface water sources). Potential pollutant sources are garbage or waste disposal sites, underground or aboveground fuel storage tanks, residential or commercial septic systems, open and underground sewage drains, storm water runoff from streets and lawns, farms that apply pesticides and fertilizers, sludge disposal sites, industries and sewage treatment plants that discharge wastewater, hazardous waste sites, mining operations, particular land use categories (such as industrial, agricultural and urban areas), and various facilities that have environmental permits. The information so gathered on surface and ground water sources shall be used to prepare an index of contaminated or potential contamination sites in terms of the risk they pose to human health and environment. Risk areas shall be prioritized on the basis of identified potential contamination sources or specific chemicals that could pollute the water. Water sources can be assigned ranking as high, medium or low depending on the threats to ground water. This method of prioritization of water sources for ground water protection is critical for decision making and shall constitute an input into the formulation of Ground Water Management Plans. In addition to identifying and prioritizing sites, the specific polluting substances shall also be listed based on toxicity, persistence and potential for bioaccumulation and then classified into specific categories, with some categories having higher polluting potential and therefore ranked higher. Waste is a significant threat to ground water quality. In order to protect ground water, it is important to ensure that waste is recovered or disposed of without harming human health or causing pollution of the environment, including ground water. Waste management activities include the storage and handling of waste at the place of production and thereafter; the transfer/transport of wastes from one place to another; the treatment of wastes to reduce their quantity, hazardousness, and/or facilitate their handling or disposal; the disposal of waste (*e.g.* in landfills and incinerators). The risk to ground water from waste varies in relation to toxicity, biodegradability and the mobility of pollutants in soil and ground water. Waste management activities not only impact on ground water quality, but also on ground water quantity, *e.g.* by affecting ground water flow direction or recharge rates. Policy initiatives shall be taken through the appropriate agencies to promote solar irrigation pumps on a wide scale to replace diesel operated pump sets in order to reduce pollution to ground water from the transport and storage of this fuel. Irrigation commands are a source of threat to ground water quality as a result of the phenomena of water logging and soil salinization, which is a problem of significant proportion in Uttar Pradesh, needs restoration and rehabilitation. There is increasingly serious water logging, soil salinization and water quality deterioration in these commands. The water that is drained from irrigation commands undergoes a material degradation compared to the water that is applied. Salinization occurs as a result of evapo-transpiration, as well as due to the leaching of natural salts deposited in the soil and below ground. Through water draining from irrigated lands, the concentrated salts and other pollutants are conveyed to a receiving stream or ground water reservoir. Remote Sensing and GIS can provide useful information for large-area water and salt balances and identification of parameters such as evapo-transpiration, rainfall distribution, interception losses and crop types and intensities that can be used as indirect measures of salinity and waterlogging and as evidence for direct estimates. The State Ground Water Board shall institute specific Task Teams for different policy issues connected with ground water such as waste management, pollution from agriculture, conjunctive surface and ground water use and management, industrial ground water use and pollution, education, training and capacity building, inter-sectoral coordination, eco-system protection, etc.

5. Policy on Right to Ownership

Since the right to ground water is linked to land rights, landless communities whether rural or urban - are at a disadvantage in terms of access to ground water. The State has a responsibility to provide water supply to such communities. All state programs for supply of drinking water or irrigation shall be prioritized in favour of these communities. Legal rights to ground water in the state vest in the owner of the land by virtue of the law of property. Indian Easements Act, 1882 describes ground water right as a right attached to land. However, the nature of ground water is such that it cannot hydrologically be confined to a particular piece of land and therefore to a particular owner. The abstraction of ground water by any individual or entity impacts on the rights of other right holders at all times. Therefore logically, the concept of absolute private rights to ground water is without basis in the natural world. There is a need to view ground water as a public, common resource. Its use by any person needs to be regulated in common interest. There is no law at present in Uttar Pradesh for the regulation of this private right in the interest of sustainability and equity. Policy measures taken so far to regulate ground water rights indirectly have not had the desired effect. The Government of Uttar Pradesh shall therefore enact a law for the purpose of regulating ground water rights in public interest. The enactment of this law shall be on the basis of public consultations. Attempts will be made to bring about a public consensus to transfer ground water resources to the public domain, taking example from many countries around the world which have done so in order to preserve this precious resource for future generations. Through such a law, policy measures outlined in the sections above on protection and development of ground water resources shall be supported by legal provisions to ensure their implementation.

6. Policy on Water Right Market

Comparisons of efficiency and performance between the privatized companies and those that remained public are revealing: private companies showed an improvement, while public companies' performance actually deteriorated. However, efficiency in water use is clearly superior in private companies than in public ones, and improvements were made at an astounding pace. As a result of adjusted water rates, consumers reduced would reduce their water use. Markets for water rights hold the potential to unleash significant entrepreneurial activity in water production and consumption. We need many additional entrepreneurs, whether small or large, to focus on water at all levels because many people lack access to clean and safe water– a number which "has held constant for decades". In this environment, water rights have the potential to evolve into a more complex bundle than has so far been recognized with limited efforts to create water markets. Marketable water rights will acquire quality, temporal, and other characteristics, all of which would allow a wide range of transactions not currently possible. We cannot fully anticipate what these transactions will look like, because we lack all of the information necessary to do so. We do know that decentralized free markets can more effectively utilize information than even the most sophisticated administrative arrangement. Unless there are significant problems with markets, these advantages should create a presumption in favor of relying on markets. Without prices, administrative allocations often rely on crude proxies for use-value. Thus residential users often pay the same price for the water they drink and the water used for their lawns, even though watering a residential lawn would likely be a lower valued use. One of the most important aspects of markets is their dynamic nature. Market prices quickly respond to constantly changing information. The change in prices sends signals to participants about the impact of events on the goods and services sold in markets.

Markets allow the uses of water to vary with changes in knowledge and demand. Because they provide a dynamic rather than a static valuation, markets adapt to constantly changing circumstances. Markets encourage the production of new knowledge about water: new uses and new ways to think about these uses, new conservation methods, new delivery methods, etc. Thus, markets encourage investment in meeting human needs. Markets do not require large scale agreement among their participants on overall ends, allowing a diversity of individual ends to coexist peacefully. In large part, the argument about markets for water is about governance of resources more generally. Markets excel in generating positive-sum transactions that benefit all parties precisely the situation in which we should all aspire to be.

In the *absence of a market for water, one of the problems which hinder mutually beneficial trades between users (for instance, two figurative parties called Smith and Jones) is that they lack the information needed to know that the transaction is possible, and that the transaction will produce an increase in their own welfare.*

7. Policy on Users

The UP State Water Policy recognizes the situation of inequity with regard to all water resources and has declared as one of its objectives, the promotion of equity and social justice among individuals and groups of users in water resource allocation and management. A common but incorrect perspective among ground water users is that ground water is unlimited; users therefore do not include sustainability concerns in their water abstractions practices. Unless this perspective is changed fundamentally, user practice will continue to remain the same. In order to achieve public participation in ground water conservation, a participatory approach to indicator development on sustainability of ground water use shall be adopted. Public participation on this important task however may be mobilized only on the basis of appropriate information. Water users with respect to specific ground water aquifers shall be consulted for this purpose for decision making on consensus on extent of ground water abstraction and equitable use, based on scientific data on quantity and quality of ground water provided to them. Statutory force shall be given to the guidelines and codes of practice developed for the abstraction of ground water for domestic, agricultural, industrial and other purposes, with respect to registration of wells, regulation of depth, siting of wells, quantum of abstraction etc. An institutional framework to cover all ground water users-individual, institutional, public or private–shall be promoted in the form of water user associations or aquifer associations in rural and urban areas, and associations shall be made legally accountable to local regulatory authorities. The mobilization of ground water users shall be made utilizing existing legal procedures in relation to building/property construction, registration or property tax procedures or tenancy procedures. In rural areas, land titling and tax procedures, or in case of property less persons, procedures for provision of ration cards, election identity cards etc. shall be utilized for securing membership in ground water user institutions. The State Ground Water Board shall prepare guidelines for roles and responsibilities for different categories of ground water users and shall ensure that such guidelines are received and followed. Communication networks with user organizations shall be set up to monitor their response and actions.

8. Policy on Extraction, Recharge and Management

For the use and management of ground water, aquifer-based Ground Water Management Plans shall be prepared by the Local Elected Bodies in consultation and coordination with the State Ground Water Board, based on the scientific data base generated on ground water resources. Formulation of Ground water Management Plans will also require the cooperation and coordination of a number of government agencies and key stakeholder groups. Ground Water Management Plans shall contain strategies for ground water development taking into account equity issues geographically and community wise. They shall also contain strategies for preventing, stopping and curing ground water pollution, based on latest available scientific technologies within affordable costs. The precautionary principle shall be adopted in decision making on ground water development for future. Ground Water Management Plans so prepared shall be made available to the public for their information. Where necessary, experts and professionals in ground water use and management and related scientific disciplines shall be consulted by state agencies and local bodies. Where Land and Water Management Plans or Watershed Development Plans or Strategies exist or are being developed, Ground Water Management Plans will form a vital input into these. Rain Water Harvesting shall therefore be an integral component of all Ground Water Management Plans, with the exception of areas subject to water logging. It shall be made mandatory in all urban development plans through legislative measures. In rural areas, Rain Water Harvesting shall be mandated through policy directives and through sectoral legislation wherever possible. The State shall promote institutions that provide training in urban and rural rainwater harvesting and shall also fund programs through Local Elected Bodies and civil society organizations on a wide scale for implementation for Rain Water Harvesting projects. It shall be ensured that water quality is taken into account while recharging ground water reserves. If there is a risk of pollution in the catchments area of water bodies, ground water recharge techniques such as recharging shafts shall not be constructed. The Minor Irrigation Surface Water Bodies in the State have bee inventoried by the State Government. The traditional community and state constructed water bodies are critical for the purpose of ground water recharge. Their rejuvenation and maintenance needs to be taken up on an urgent basis. The State Ground Water Board shall coordinate with the concerned department to focus on development programs in areas identified with ground water depletion on a priority basis. At the same time, potential for ground water pollution from such water bodies shall be assessed and appropriate action taken. The Government of Uttar Pradesh shall enact a law to regulate ground water extraction. The identification and assessment of risk to ground water from agricultural fields shall be included in Ground Water Management Plans along with appropriate strategies to address these risks. Government of UP shall formulate guidelines and enact necessary legislation to control and regulate those agricultural practices that pose a risk to ground water. Organic farming practices shall be given a major policy emphasis. The State Ground water Board shall provide information to the relevant department of government on ground water bodies at risk, for focusing resources in promotion of organic farming. The State Ground Water Board shall prepare guidelines for other sectoral organizations for necessary policy and legal steps to be taken for protection of quantity and quality of ground water resources. Legislative measures shall be taken to ban on tapping age old reservoirs to protect historical reserves of water, to prevent structural changes in aquifers and consequent landslides or land subsidence.

9. Policy on Distribution and Pricing

Measures for efficiency improvement in ground water and surface water irrigation to reduce demand shall be implemented through the promotion of appropriate information and technologies. In collaboration between the energy and ground water departments and agencies, an appropriate economic incentive structure through power pricing and supply policy shall be introduced to promote ground water conservation.

10. Policy for Effective Implementation of Laws

The most common problem lies with the implementation of law. Therefore there is need for the laws and the law enforcing agencies to meet the issues and challenges of the ground water.

11. Policy on Inter Departmental Coordination

It shall be ensured that all developmental departments of government take into account ground water quality protection objectives in their policy and developmental planning. Information on identified pollutants and risk sites shall be shared with such departments to influence decision making on development planning. Coordination will be established with the State Land Use Board, Pollution Control Board, Irrigation Department, Drinking Water Agencies, Forest Department, Urban Development Department, Revenue Department, Industries Department, Geological Survey of India, GIS/Remote

Sensing agencies for the preparation of ground water aquifer maps in conjunction with land use and surface water uses and for the preparation of Ground Water Management Plans. Coordination will be established with Urban and Rural Local Bodies including Zilla Panchayats, Kshettra Panchayats and Village Panchayats for the collecting of direct evidence on ground water and related issues for the preparation of Ground water maps and for the preparation of Ground Water Management Plans. It shall be ensured through inter-sectoral coordination at state level that norms for including geological and hydro-geological factors and ground water quantity and quality in urban and rural planning policy and development planning are introduced. Such norms shall include the refusal of permission for development where there is significantly increased risk of ground water pollution or over-abstraction. Systems for monitoring and reporting compliance with such norms shall be introduced and operated. Coordination will be established with civil society organizations, environmental groups, urban residential associations, research institutions etc for mobilizing public opinion for ground water protection as well as for providing inputs into the planning process. Coordination will be established with representative institutions of industry for promoting ground water protection from pollution. Coordination will be established with electronic and print media in the state and private sector for state wide information campaign on ground water issues, and to give publicity to the state Ground Water Policy.

12. Policy on Institutional Arrangements

Presently the water resource planning for various uses is being done by the respective departments. Realising the scarcity value of this resource as also the need for co-ordinate planned development, the state government has constituted a "State Water Board" under the chairmanship of the Chief Secretary.

13. Policy on Implementation Agencies

The State Ground Water Board shall be the nodal agency to facilitate the implementation of this policy. All state departments and agencies to play a facilitative and regulatory role. The State Ground Water Board shall be responsible for preparation of ground water/surface water/land use maps in consultation and collaboration with local bodies. The State Ground Water Board shall implement programs systematically for capacity building of local elected body representatives and personnel attached to local bodies in all aspects of ground water use and management on a scientific basis.

Putting this type of approach into action requires a shift in a State's water policy and practice. Irrigation departments will have to move from supplying water only in the dry season to delivering adequate water during the monsoon, so that farmers can grow water-intensive crops, such as rice and sugarcane where the water irrigates fields and simultaneously recharges ground water aquifers. The water that is stored in the aquifers can then be pumped back up for a second post-monsoon cropping season. The advantages of this approach are that ground water levels are maintained, farmers' annual yields are increased, pumping costs are reduced and waterlogging is minimized.

14. Policy on Awareness of Information

The primary responsibility for ground water protection rests with any person, company or organization that is carrying out an activity that poses a threat to ground water. Education and guidance are, therefore, seen as essential mechanisms to improve ground water protection. A state wide educational campaign on ground water shall be conducted involving civil society organizations, industry, educational institutions, research institutions etc. The State Ground Water Board shall function as a nodal agency for this purpose. Based on information provided by district level Ground Water Information and Ground Water Quality Assessment and Monitoring Centers, prioritization norms shall be developed by the State Ground Water Board for selection of districts/blocks for development taking into account current levels of abstraction. Priority shall be given to blocks and districts that have low percentage of development to ensure equity in ground water access. These norms shall be given wide publicity.

15. Policy on Monitoring and Evaluation

Systematic, scientific and state wide study and monitoring of impacts of over abstraction of ground water on ecosystems, other water bodies and other wells shall be undertaken by the State Ground Water Board in collaboration with research institutions and other concerned departments and agencies of government.

16. Policy on Public Opinion

Coordination will be established with civil society organizations, environmental groups, urban residential associations, research institutions etc for mobilizing public opinion for ground water protection as well as for providing inputs into the planning process.

17. Policy on Private-Public Partnership

Appropriate technical and management techniques shall be identified and implemented through the aegis of Local Bodies in collaboration with State technical agencies and private sector institutions wherever possible. Coordination will be established with representative institutions of industry for promoting ground water protection from pollution. Coordination will be established with electronic and print media in the state and private sector for state wide information campaign on ground water issues, and to give publicity to the state Ground Water Policy.

18. Policy on Infrastructure and Development

Interdisciplinary diagnostic studies involving field studies of the plant environment, farm management practices, water supply and removal, and institutional linkages are necessary to diagnose site related problems and potential solutions. The Uttar Pradesh State Water Policy sets out several measures for the treatment of waterlogging and salinity. Modernization and upgradation of irrigation channels, improved 'conveyance management' using appropriate technology to enable quick adjustments of supplies as per requirements dictated by water availability and the priorities at the field, judicious use of canals, checking unauthorized use by cutting of canals and other means, leveling of fields, improvement and maintenance of water courses, improvement of drainage, adoption of improved irrigation and agriculture practices using appropriate technology to ensure optimal use of water for agriculture production, adoption of appropriate cropping pattern suitable for the area in question, strengthening of farmer institutional framework etc. Industrial pollution constitutes a major risk to ground water resources. It sets out that the agencies, departments, institutions and organizations that are required to be coordinated in order to achieve the objectives of this policy. The Government of Uttar Pradesh has set up a State Water Board to coordinate all activities relating to water. The overall responsibility for achieving the objectives of this policy shall vest in the State Water Board. The State Ground Water Board shall be vested with the responsibility of facilitating and implementing the Ground water Policy in coordination with all concerned institutions. The enactment of a Land Use Policy for Land Use Planning is very essential in the context of ground water protection. Coordination shall be established for this purpose with the concerned agency of Government.

19. Policy on Environmental Impact Assessment

Government shall ensure that risks to ground water reserves are addressed in all cases wherein Environmental Impact Assessments are legally required. The requirement of a State to conduct Environmental Impact Assessments in respect of activities that are likely to significantly affect the environment has been reflected in Principle 17 of Rio Declaration on Environment and Development and Article 5 of the Legal Principle for Environment Protection and Sustainable Development. The objective of the EIA is to ensure that environmental aspects are addressed and potential problems are foreseen at the appropriate stage of project design. A should be effectively implemented before taking up any project effecting ground water resources.

24

Possible Impact of Pharmaceutical Wastes on Aquatic Environment

A.K. Singh, Abubakar Ansari and Ruchi Singh

INTRODUCTION

Release of pharmaceuticals in the environment is associated with potential risks and becoming an increasingly important issue for environmental regulators and the pharmaceutical industry. Unplanned urbanization, discharge of untreated sewage and industrial effluents, destructive fishing practices are some of the reasons of increasing effluents (UNEP/GPA plan of action 2000). Pharmaceuticals and personal care products (PPCPs) are an emerging class of environmental pollutants (Fent *et al.* 2006).

Pharmaceuticals and personal care are more than 3000 different substances used as medicine for human therapy, including analgesics, antibiotics, contraceptives, beta blockers, lipid regulators, and impotence drugs (Ternes *et al.* 2004). These drugs are designed for a specific mode of action (Fent *et al.* 2006), and are absorbed and metabolized by the general by the body, and waste excreted in the faeces and urine as a mixture metabolites and unchanged substances (Richards *et al.* 2004). Once excreted, pharmaceutical compounds and their metabolites pass through treatment plants, sewage treatment plant (Quinn *et al.* 2008) and hence are released into surface waters, along with the treated effluent. This is reinforced by the fact that most people flush unused drugs in septic tanks or municipal treatment plant wastewater. Plant wastewater treatment plant (WWTP) which therefore has been identified as the main source of discharge of these compounds into the environment (Brun *et al.* 2006, Daughton and Ternes, 1999).

Pharmaceuticals and personal care products are also sometimes disposed of with household waste. Other sources include direct application in aquaculture farms, manure runoff or runoff from the application of sewage sludge and manure on farmland as fertilizer, through the hospital effluent (Kümmerer, 2001) or, finally, through leaching from landfills. Different sources of pharmaceutical compounds in the environment are showing in the Fig. 1. On many occasions, accidental or deliberate release of contaminants such as PPCPs, pose a serious threat to aquatic organisms, as their long-term effects are harder to identify and which are also present in small amounts ranging from ng/L to mg/L (Kolpin *et al.* 2002), therefore, exposure of aquatic organisms to these pollutants may pose a risk to human health through the food chain (Fent *et al.* 2006). It could also cause an ecological hazard that can lead to heritable mutations and loss of total genetic diversity (O'Brien and Dietrich, 2004).

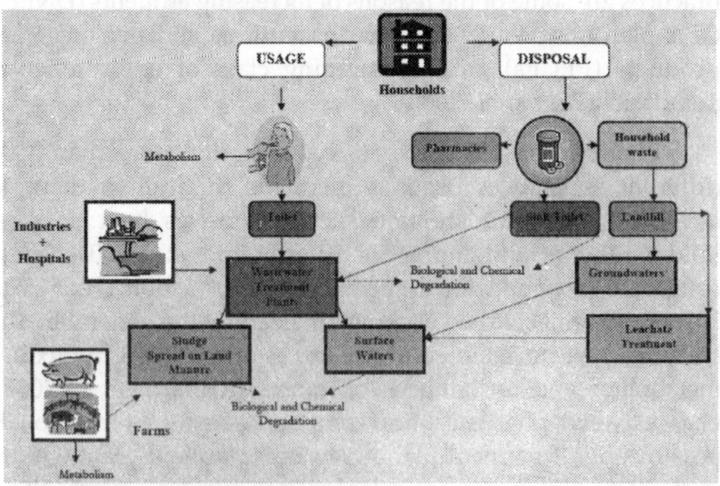

Fig. 1 : Source of pharmaceutical compounds into the environment

Most pharmaceutical studies have been conducted in northern European countries and central and in the United States, however, there is limited information and/or data available in India (Andreozzi *et al.* 2003, Carballa *et al.* 2004). This is due to the difficulty and costs associated with these micro-chemical analyses. This study aims to investigate the potential risks associated with the release of pharmaceuticals in the environment since it has become an important issue for environmental regulators and the pharmaceutical (Dietrich *et al.* 2002).

Effect of Pharmaceuticals on the Environment

Pharmaceuticals and personal care products are released to the environment and that are continuously used. However, its behaviour and fate and their metabolites are not known as indicated in a study on the fate of human pharmaceuticals in the environment (Fent *et al.* 2006). As other xenobiotics, pharmaceuticals also tend to be persistent (Sanderson *et al.* 2004), and can cross membranes because they are designed to maintain their chemical structure long enough for them to accomplish their work treatment (Ternes and Hirsch, 2000). Physical and chemical properties of pharmaceutical products for example, low volatility and high polarity makes it easy for distribution within the water column in aqueous form (Fent *et al.* 2006).

Therefore, pharmaceuticals are released into the environment, they are transported and distributed to the atmosphere, water, soil or sediment. Its distribution through the receiving aquatic bodies is determined by the compounds and the physico-chemical environments. For example, the transport of the pharmaceutical industry between different media depends mainly on the absorption behavior of the substance in soils, plants and water treatment system, water sediments, the absorption behavior varies with different drugs (Boxall, 2004). In addition, pharmaceuticals can also be distributed through food chains (Fent *et al.* 2006).

Once pharmaceuticals effluents come in the aquatic environment through various routes, for example, washing of unused medications by sewer systems, disposal of medications in landfills and excretion in humans in the urine or feces, which end in plants wastewater treatment plant (WWTP), surface water and / or in some cases in drinking water and ground water a few cases have been reported (Heberer and Stan, 1996). In wastewater systems, pharmaceuticals are primarily eliminated by the process of adsorption to suspended solids (sewage sludge) and biodegradation). The adsorption process depends on the hydrophobic and electrostatic property of the pharmaceutical industry. The adsorption of different drugs increases with a decrease in pH, but acidic drugs such as ibuprofen, naproxen and diclofenac has little tendency to adsorb to sludge (Fent *et al.* 2006) therefore absorption sewage sludge is an important method to remove acidic pharmaceuticals STPS (Fent *et al.* 2006). The levels of pharmaceuticals in digested sludge and sediments are suggested to be relatively low, as demonstrated in several controlled studies (Ternes *et al.* 2004; Uras and Kikuta, 2005). The removal of pharmaceuticals by wastewater treatment plants (WWTP) that vary according to the same complex and depends on the hydraulic retention time, technology / mechanism used by plants, other factors such as temperature and climate. In fact, some pharmaceutical products are not completely removed due to conventional treatment technology used in the STP does not seem to be sufficient to completely eliminate these specific compounds. Therefore, pharmaceuticals are discharged in varying amounts can reach surface waters, ground water (Farré *et al.* 2001) and sediments (Zuccato *et al.* 2000).

In surface waters, the drugs are eliminated by bio- or photodegradation (photolysis).

Photodegradation has been found that the process of elimination of pharmaceuticals such as diclofenac and carbamazepine clorofibric acid. Laboratory experiments have shown direct and indirect photolysis an important removal process (Andreozzi *et al.* 2003). The efficiency of photodegradation depends among other factors, the strength of solar radiation, and thus the latitude and season. Other factors are the properties of substances and water components that

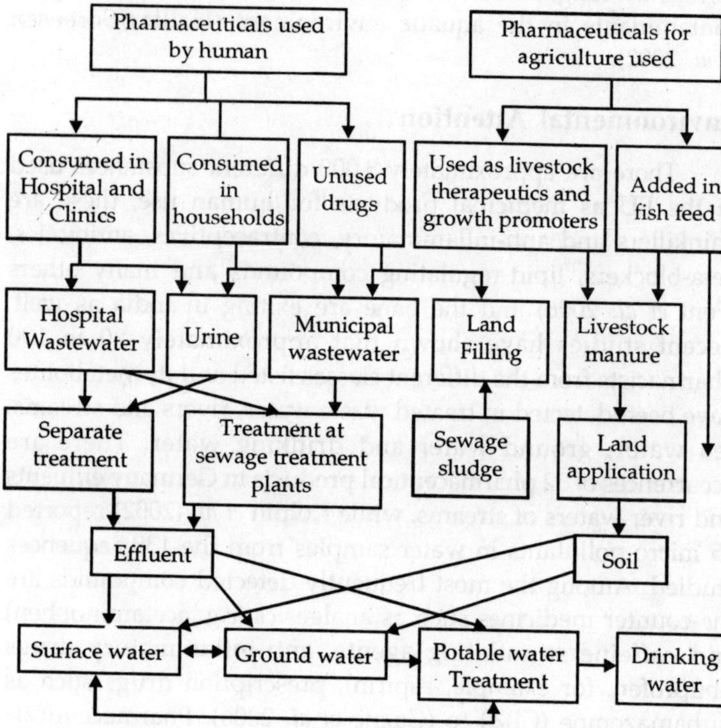

Fig. 2 : Pharmaceuticals pathway in the environment.

can act as photosensitizers. Pharmaceuticals also are removed from the medium through bioaccumulation in the food chain, but not enough information about it, except for studies of diclofenac in vultures is deposited. Some cases have reported the presence of pharmaceuticals in drinking water (Heberer and Stan, 1996) and ground water (Holm *et al.* 1995 and Ternes *et al.* 2001). Despite all the above procedures regarding the disposal and fate of pharmaceuticals in the environment, there is still lack of information in the literature about possible interactions of pharmaceutical compounds to aquatic organisms (Tixier *et al.* 2003). In general, the degradation processes reduce the power of drugs, or transform them into products that may have similar toxicity to the drug, or transformed into more toxic compounds and persistent. This therefore calls for more serious investigation on the possible effects of these emerging contaminants in the aquatic environment (Halling-Sorensen *et al.* 2002).

Environmental Attention

There are approximately 3,000 different substances used in the EU as medicinal products for human use, these are painkillers and anti-inflammatory, contraceptives, antibiotics, beta-blockers, lipid-regulating compounds and many others (Fent *et al.* 2006) and the bane are exiting in India as well. Recent studies have shown that approximately 80 to 100 pharmacists from the different classes listed and its metabolites have been detected in treated waste water, rivers and streams, sea water, ground water and drinking water. There are occurrences of 32 pharmaceutical products in Germany effluents and river waters of streams, while Kolpin *et al.* (2002) reported 95 micro-pollutants in water samples from the 139 sequences studied. Among the most frequently detected compounds are the-counter medicines such as analgesics (*e.g.* acetaminophen) and caffeine stimulating agents, anti-inflammatory drugs (ibuprofen, for example, aspirin), prescription drugs such as carbamazepine (CBZ) to (Gagne *et al.* 2006). Pharmaceuticals have been reported in India in the treated wastewater, where clofibric acid ranged from 0.8 to 2 g/L. Pharmaceuticals have also been detected in rivers in the US (Garrison *et al.* 1976), in concentrations ranging up to 1 mg/L (Richardson and Bowron, 1985). However, current knowledge on the environmental occurrence of pharmaceuticals has increased since the invention of new analytical techniques that are capable of determining polar compounds, even in minute quantities. Globally, the concentrations of pharmaceuticals have been found to fall within the range of nanograms per liter to micrograms per liter range of concentrations of the aquatic environment. Although it can be degraded in the environment by biotic (Winkler *et al.* 2001) or abiotic process (Andreozzi *et al.* 2002), it is assumed that products could act as persistent compounds simply by continuous infusion in the aquatic environment via STP effluents. The following table shows the different drug concentrations in the influent and effluent treatment plants, wastewater (STP) plants on the basis of a study by Fent *et al.* (2006).

Ecotoxicological Effect of Pharmaceuticals in the Environment

Pharmaceuticals can to cause some beneficial effects on the target organism, which is the treatment and prevention of disease in humans and animals, after which they are released into the aquatic environment in its native form or as metabolites. However, studies indicate that pharmaceuticals and personal care products are present in small amounts in surface waters, but their release continued in the aquatic system is expected to cause acute effects to non-target animals and microorganisms in environment (Boxall, 2004) because they are constantly exposed (Fent *et al.* 2006). The effect of pharmaceuticals in the environment is reinforced by the fact that pharmaceuticals are designed to perform certain biological effect in the target organism (Henschel *et al.* 1997). Therefore, its effect on non-objective can not be neglected. Pharmaceutical compounds are designed to react, either directly or interact with human and animal receptors, or to be toxic to pathogens such as bacteria, fungi and parasites. Compared with most of the lower animals have receptors similar to those of humans and higher animals. Because of this similarity, the pharmaceutical industry can have subtle effects on aquatic organisms and terrestrial non-target. Data on the presence and the complex behaviour of drugs have raised questions about how mixtures of human and veterinary drugs impact on the human health. Damage caused by pharmaceutical products range from one organism to another and from drug to drug and pharmaceutical products ability to cause effects on the environment does not depend on the rate of output, but on the persistence of environment and biological activity of high toxicity such high power that affect major biological functions such as reproduction (Fent *et al.* 2006). The specific reactivity of PPCPs redox forms the basis on which to fulfill their biological (therapeutic) effects, metabolism, elimination, and toxicity (Gagne *et al.* 2006). Exposure of aquatic organisms to these compounds can alter the oxidative status of cells by increasing oxidative stress that is because these bodies are not efficient as mammals in the elimination of lipophilic drugs and oxygen radicals. Although PPCP act in various tissues, most of them are metabolized in the liver (hepatocytes) (Gagne *et al.* 2006). The available data about the toxicity of pharmaceutical products are mostly related to acute lethal responses (Pascoe *et al.* 2003). Several studies have reported sublethal effects of pharmaceuticals, including the effects on reproductive success in aquatic organisms, affecting the maturation of oocytes and testes (O'Brien and Dietrich, 2004), reduced reproduction the cladoceran *Ceriodaphnia dubia* exposed to fluoxetine (Henry *et al.* 2004), the reduction potential for regeneration in cnidarians *Hydra vulgaris* exposed to diazepam (Pascoe *et al.* 2003). Henschel *et al.* (1997) Report on the pharmaceutical effects on Daphnia, and the reported effect of Diclofenac in vulture population in Asia. Several of these studies indicate that high concentrations and unrealistic environment of pharmaceutical products are required to cause acute toxicity. It has been shown that low concentration may cause mortality after long-term exposure. However, the lack of data clearly indicates that the ecotoxicity of pharmaceuticals in different concentrations on the environment is an important unresolved area at present (Daughton and Ternes, 1999).

Policy on the Usage of Pharmaceuticals

The government of India and the state governments as well have issued directives against pollution, which requires that the wastewater should pass through treatment plants, sewage treatment plant (WWTP) which will help to reduce the discharge of point sources of organic matter, phosphates and nitrates in aquatic ecosystems such as lakes, rivers and marine ecosystems. There by improving water quality, but now a new problem of pharmaceutical waste has emerged and the focus has shifted to pharmaceuticals and personal care. In 1995, the first requirement for ecotoxicity testing as a pre-requisite for registration of pharmaceutical products has been developed by the European Union (EU) Directive 92/18 EEC and the relevant Guidance Note "(EMEA, 1998) for veterinary pharmaceuticals. The European Commission released a draft Directive (Directive 2001/83/EC) which specifies that the authorization of a medicinal product shall be accompanied by an environmental risk assessment (EMEA, 2005).

The proposed EU project phase II should involve testing of all substances with the predicted environmental concentration (PEC) of 0.01 mg / L or more, or substances that have a specific mode of action (MOA) and the interaction directly and indirectly to a receiver. Specific pharmaceuticals to the Ministry of Agriculture have been described by the European Agency for the Evaluation of Medicinal Products (EMEA), indicating that the study of pharmaceutical compounds should take place, the process that involves animals and costly acute and chronic long-term studies. In general, the main purpose of in India guidelines is to protect the aquatic environment from any pharmaceutical substance are excreted and reach where STPs are either completely not existing or transformed into other compounds and ultimately discharged with the effluent treated water systems.

Various Regulatory Bodies that Oversee Pharmaceutical Waste Management :

- Environmental Protection Agency (EPA)
- Department of Transportation (DOT)
- Drug Enforcement Administration
- State Environmental Agencies,
- State Pharmacy Boards, and
- Local Publicity Owned Treatment Works (POTW)

Table 1 : Reported Sub Lethal Effects of Pharmaceuticals on Aquatic and Terrestrial Organisms

Substance(s)	*Medicine Class*	*Reported Effect*	*Reference*
Fenfluramine	Anorexic	Enhances release of serotonin (5-HT) in crayfish which in turn triggers the release of ovarysimulating hormone resulting in larger oocytes with enhances amounts of vitellin in fiddler crabs, stimulates the production of gonad-stimulating hormone accelerating testicular maturation	Doughton & Ternes, 1999
17a-Ethinylestradiaol	Synthetic steroid	Endocrine-disrupting effects on fish, reptoles and invertebrates	Sharma *et al.*, 2010
Methyltestosterone	Synthetic steroid	Intersex, reduced fecundity, oogenesis, spermatogenesis in snails	Sharma *et al.*, 2010
Avermectins	Parasiticide	Adult insects : loss of water balance, disruption of feeding and reduced fat accumulation, delayed ovarian development, decreased fecundity and impaired mating juvenile insects : delayed development, reduced growth rates, development of physical abnormalities, impairment of pupariation or emergence and a loss of developmental symmetry	Sharma *et al.*, 2010
Tetracyclines, macrolides and streptomycin	Antibacterials	Antibacterial resistance measured in soil bacteria obtained from sites treated with pig slurry	Sharma *et al.*, 2010

Cont...

Substance(s)	*Medicine Class*	*Reported Effect*	*Reference*
Cypermethrin	Ectoparaciticide	Impact on drug de-composition	Sharma *et al.*, 2010
Fenbendazole	Paraciticide	Impact on drug decomposition	Sharma *et al.*, 2010
Tylosin	Antibacterial	Impacts on the structure of soil microbial communities	Sharma *et al.*, 2010
Erythromycin	Antibacterial	Inhibition of growth cynobacteria and aquatic plants	Sharma *et al.*, 2010
Tetracycline	Antibacterial	Inhibition of growth cynobacteria and aquatic plants	Sharma *et al.*, 2010
Ibuprofen	Anti-inflammatory	Stimulation of growth of cynobacteria and inhibition of growth of aquatic plants	Sharma *et al.*, 2010
Fenofibrate	Lipid regulator	Inhibition of basal EROD activity in cultures of rainbow trout hepatocytes	Sharma *et al.*, 2010
Carbamazepine	Analgesic	Inhibition of basal EROD activity in cultures of rainbow trout hepatocytes Inhibition of emergence of Chironomus riparius	Sharma *et al.*, 2010
Diclofenac	Analgesic	Inhibition of basal EROD activity in cultures of rainbow trout hepatocytes	Sharma *et al.*, 2010
Propanolol	Beta blocker	Weak EROD inducer in cultures of rainbow trout hepatocytes	Sharma *et al.*, 2010
Sulphametha-zole	Antibacterial	Inhibition of basal EROD activity in cultures of rainbow trout hepatocytes	Sharma *et al.*, 2010
Clofibrate	Lipid regulator	Inhibition of basal EROD activity in cultures of rainbow trout hepatocytes	Sharma *et al.*, 2010

Cont...

Substance(s)	*Medicine Class*	*Reported Effect*	*Reference*
Diazepam	Antianxiety drug	Inhibition in the ability of dissected polyps from the cnidarian *Hydra vulgaris* to regenerate hypostome, tentacles and a foot	Pascoe *et al*, 2003
Digoxin	Cardiac glycoside	Inhibition in the ability of dissected polyps from the cnidarian *Hydra vulgaris* to regenerate hypostome, tentacles and a foot	Pascoe *et al*, 2003
Amlodipine	Calcium channel blocker	Inhibition in the ability of dissected polyps from the cnidarian Hydra vulgaris to regenerate hypostome, tentacles and a foot	Pascoe *et al*, 2003

CONCLUSION

In Indian rivers where more than 90 different pharmaceutical companies dump their waste have been reported to test positive for over 21 active drug ingredients. In some river alone, there was report of enough ciprofloxacin (a strong antibiotic) being dumped every day by drug companies. The drug contamination levels found in India's rivers were 150 times the detected levels found in the U.S (Sharma *et al.* 2010). These findings prove that drug companies have cared less about the consequences of drug residue they dump in water. People are not the only beings that are affected by pharmaceutical contamination; rather aquatic ecosystems are also being negatively impacted. According to an MSNBC report back in 2009, all kinds of drugs are being found in the bodies of fish near major U.S. cities. Researchers found drugs for high cholesterol, allergies, high blood pressure, bipolar disorder and depression in the livers and tissue of fish.

Researchers are in agreement that aquatic species of all types are being harmed by continuous exposure to water contaminated with pharmaceuticals. Many fish are experiencing reproductive problems as a result of exposure, pharmaceutical effluents. Beyond having their sperm damaged, some fish are actually changing sexes. Males are becoming females and females are becoming males as a result of drug exposure in the water. Since drug companies are failing to properly treat their wastewater before dumping it into rivers (even though they claim to be treating it), regulatory agencies need to step up and correct the problem. Regular monitoring of wastewater contaminant levels is the only way to halt the chemical contamination of waterways. State and local legislators would do well to put forth their own legislation to upgrade wastewater treatment facilities so they can properly filter out pharmaceuticals (and dispose of them safely).

The presence of pharmaceuticals in natural waters raises issues beyond the obvious concern about public health. Ecologically, pharmaceutical chemicals in waterways threaten wildlife with continuous exposures. Since exposures to the aquatic biodiversity through aquatic ecosystems are more intermittent, a higher concern than human health is recognised.

The ecological impacts – such as the likelihood of adverse effects on aquatic organisms and persistence of certain chemicals in the environment can be the basis for developing a priority list of pharmaceuticals. We need to develop a priority list of some of the most commonly detected pharmaceuticals that represent the greatest concern for environmental impact, accounting for these factors and the chemicals' recalcitrance to treatment.

REFERENCES

Andreozzi, R., Paxeus, N., Campanella, L., Lyberatos, G., Garric, J., Battilotti, M., (2002) : Ecotoxicological assessment and removal technologies for pharmaceuticals in wastewater. treatment. *Water Science, Technology*, 50 : 269–276.

Andreozzi, R., Raffaele, M., Nicklas, P., (2003) : Pharmaceuticals in STP effluents and their solar photodegradation in aquatic environment. *Chemosphere*, 50 : 1319–1330.

Boxall B.A., (2004) : The environmental side effects of medication. EMBO *reports* volume 5 (12).

Brun G.L., Bernier M, Losier R, Doe K, Jackman P, and Lee HB. (2006): Pharmaceutically active compounds in Atlantic Canadian sewage treatment plant effluents and receiving waters and potential for environmental effects as measured by acute and chronic aquatic toxicity. *Environmental Toxicology and Chemistry*, 25 (8) : 2163–76.

Carballa, M., Omil, F., Lema, J.M., Llompart, M., Garcia-Jares, C., Rodriguez, I., Gomez, M., and Ternes, T., (2004): Behavior of pharmaceuticals, cosmetics and hormones in a sewage treatment plant. *Water Res.* 38 (12): 2918–2926.

Daughton C.A., and Ternes T.A., (1999) : Pharmaceuticals and personal care products in the environment : agents of subtle change. *Environmental Health Perspectives,* 107 : 1-41.

Dietrich, D.R., Webb, S.F., and Petry, T., (2002) : Hot spot pollutants: pharmaceuticals in the environment. Toxicology Letters 131 : 1–3.

EMEA, (1998) : Note for Guidance : Environmental Risk Assessment for Veterinary Medicinal Products other than GMO containing and Immunological Products. EMEA, London (EMEA/CVMP/055/96).

EMEA, (2005) : Note for Guidance on Environmental Risk Assessment of Medicinal Products for Human Use, CMPC/SWP/4447/draft. The European Agency for the Evaluation of Medicinal Products (EMEA), London.

Farre, M.L., Ferrer, I., Ginebreda, A., Figueras, M., Olivella, L., Tirapu, L., Vilanova, M., and Barcelo, D., (2001) : Determination of drugs in surface water and wastewater samples by liquid chromatography–mass spectrometry : methods and preliminary results including toxicity studies with *Vibrio fischeri*. *J. Chromatogr.* A 938 (1/2), 187–197.

Fent K, Weston AA, and Caminada D. (2006) : Ecotoxicology of human pharmaceuticals. *Aquatic Toxicology* 76 : 122–59.

Gagné F, Blaise C, and Andre C. (2006) : Occurrence of pharmaceutical products in a municipal effluent and toxicity to rainbow trout (*Oncorhynchus mykiss*) hepatocytes. *Ecotoxicology and Environment Safety,* 64 : 329–336.

Garrison, A.W., Pope, J.D., and Allen, F.R., (1976) : Analysis of organic compounds in domestic wastewater. In : Keith, C.H. (Ed.), Identification and Analysis of Organic Pollutants inWater. Ann Arbor Science, Michigan, USA, page. 517–566.

Halling-Sorensen, B., Sengelov, G., and Tjornelund, J., (2002) : Toxicity of Tetracyline degradation products to environmentally relevant bacteria, including, selected tetracycline-resistant bacteria. Archives of Environmental Contamination and Toxicology 42 : 263-271.

Heberer, T., and Stan, H.J., (1996) : Occurrence of polar organic contaminants in Berlin drinking water. Vom Wasser 86, 19–31. Heberer, T., Stan, H.J., 1997. Determination of clofibric acid and *N*-(phenylsulfonyl)-sarcosine in sewage, river and drinking water. *Int. J. Environ. Anal. Chem.* 67 (1–4) : 113–123.

Henry, T.B., Kwon, J.-W., Armbrust, K.L., and Black, M.C., (2004) : Acute and chronic toxicity of five selective serotonin reuptake inhibitors in *Ceriodaphnia dubia. Environmental Toxicology and Chemistry,* 23 (9) : 2229–2233.

Henschel, K.P., Wenzel, A., Diedrich, M., and Fliedner, A., (1997): Environmental hazard assessment of pharmaceuticals. Regulatory Toxicology and Pharmacology 25 (3), 220–225.

Holm, J.V., Rugge, K., Bjerg, P.L., and Christensen, T.H., (1995): Occurrence and distribution of pharmaceutical organic-compounds in the ground water downgradient of a landfill (Grindsted, Denmark). *Environmental Science and Technology,* 29 : 1415–1420.

Kolpin, D.W., Furlong, E.T., Meyer, M.T., Thurman, E.M., Zaugg, S.D., Barber, L.B., and Buxton, H.T., (2002) : Pharmaceuticals, hormones and other organic wastewater contaminants in the US streams, 1999-2000: a national reconnaissance. *Environmental Science, Technology,* 36 : 1202-1211.

Kummerer, K., (2001) : Drugs in the environment: emission of drugs, diagnostic aids and disinfectants into wastewater by hospitals in relation to other sources a review. *Chemosphere,* 45 (6/7) : 957–969.

O'Brien, E. and Dietrich, D.R. (2004) : Hindsight rather than foresight: reality versus the EU draft guidelines on pharmaceuticals in the environment. TRENDS in Biotechnology 22 (7).

Pascoe, D., Karntanut, W., and Muller, C.T., (2003) : Do pharmaceuticals affect freshwater invertebrates? A study with the cnidarian *Hydra vulgaris. Chemosphere,* 51 : 521–528.

Quinn, B, Gagne, F and Blaise, C. (2008) : An investigation into the acute and chronic toxicity of eleven pharmaceuticals (and solvents) found in wastewater effluent on the cnidarian, *Hydra attenuata.* Science of the Total Environment, 389: 306-314.

Richards, S.M., Wilson, C.J., Johnson, D.J., Castle, D.M., Lam, M., Mabury, S.A., Sibley, P.K., and Solomon, K.R., (2004) : Effects of pharmaceutical mixtures in aquatic microcosms. Environmental Toxicology. *Chemosphere.* 23 : 1035–1042.

Richardson, M.L., and Bowron, J.M., (1985) : The fate of pharmaceutical chemicals in the aquatic environment. *J. Pharm. Pharmacol.* 37(1) : 1–12.

Sanderson, H., Brain, R.A., Johnson, D.J., Wilson, C.J., and Solomon, K.R., (2004a) : Toxicity classification and evaluation of four pharmaceuticals classes: antibiotics, antineoplastics, cardiovascular, and sex hormones. *Toxicology,* 203 : 27–40.

Sanderson, H., Johnson, D.J., Reitsma, T., Brain, R.A., Wilson, C.J., and Solomon, K.R., (2004b) : Ranking and prioritization of environmental risks of pharmaceuticals in surfacewaters. *Regulatory Toxicology and Pharmacology,* 39 (2) : 158–183.

Sharma N, Agarwal D, Khinchi M Gupta M.K, and Bisht S., (2010): Pharmaceutical waste management : A challlange to make environment ecofriendly. *International J. of Research in Ayurveda & pharmacy,* 1(2) : 332-338.

Ternes, T., Bonerz, M., and Schmidt, T., (2001) : Determination of neutral pharmaceuticals in wastewater and rivers by liquid chromatography–electrospray tandem mass spectrometry. *J. Chromatogr. A,* 938 (1/2): 175–185.

Ternes, T., and Hirsch, R., (2000) : Occurrence and behavior of X-ray contrast media in sewage facilities and the aquatic environment. *Environmental Science and Technology,* 34 : 2741–2748.

Ternes, T., Jos, A., and Siegrist, H., (2004) : Scrutinizing pharmaceutical and personal care products in wastewater treatment. *Environmental. Science and Technology,* 393–399.

Tixier, C., Singer, H.P., Oellers, S., and Muller, S.R., (2003) : Occurrence and Fate of carbamazepine, Ibuprofen, Ketoprofen and naproxen in surface waters. *Environ. Sci. Technol.* 37(6), 1061-1068.

Urase, T., and Kikuta, T., (2005) : Separate estimation of adsorption and degradation of pharmaceutical substances and estrogens in the activated sludge process. *Water Research.* 39 (7) : 1289–1300.

Winkler, M., Lawrence, J.R., and Neu, T.R., (2001) : Selective degradation of Ibuprofen and Clofibric acid in two model river biofilm systems. *Water Research,* 35: 3197–3205.

Zuccato, E., Calamari, D., Natangelo, M., and Fanelli, R., (2000) : Presence of therapeutic drugs in the environment. *Lancet,* 335 : 1789-1790.

25

Distribution of Heavy Metals in Water and Sediment of Lake Nainital

Rashmi Yadav and D.S. Malik

INTRODUCTION

Nainital lake is considered as the main attraction of the famous town of same name in the State of Uttarakhand. The lake has its importance in the socio-economic and cultural life of the people in the region. It is one of the second largest natural freshwater lake in the Kumaon region. The Nainital lake is extremely popular tourist destinations situated in the scenic Himalayas. The lake is also a source of drinking water, fisheries, irrigation etc. to the local people. Heavy metals have contributed a great ecological stresses due to their toxicity and accumulation in aquatic ecosystem. These elements, contrary to most pollutants are not biodegradable and undergo a global ecobiological cycle in which natural waters are the main pathways (Nurnberge, 1984). High concentration of heavy metals in the environment creates serious pollution problems and cause deterious effects in aquatic plants. Pollution of air, soil and water with heavy metals is a major environmental problem and metals cannot be easily degraded and the cleanup usually requires their removal (Wang, 1992 and Lasat, 2002).

The Nainital lake receives toxic metals, organic and inorganic pollutants from different sources, *viz.*, high soil erosion, illegal construction activities and tourism development, heavy litter inputs, automobile exhaust, domestic discharge and recreational use of lake water etc. Heavy metals today have a great ecological significance due to their toxicity and accumulation in water, sediments and biotic communities in lake ecosystem. The toxic substances could be degraded by biological treatment, but in the case of heavy metals, they remain in sediments and are slowly released in to the water body (Da La Noue and De Pawn, 1988). Metals do not remain soluble in water, gets adsorbed and accumulated on bottom sediments acting as a sink (Hakanson, 1980). Pollution by toxic metals can be much more serious and insidious problem than by organic substances, because these are intrinsic component of environment. At high concentration, all the metals are toxic to animals and plant both (Rai and Chandra, 1992, Sinha *et al.* 1997).

METHODOLOGY

Lake Nainital is a natural kidney shaped, warm monomictic type lake located at 29°24′ N latitude and 79°28′ E longitude at an altitude of 1,937 m above sea level in the State of Uttarakhand, India. The maximum length and width of the lake are 1.43 and 0.45 km while the maximum and mean depth of the lake is 27.3 and 16.5 m, respectively.

The following sampling locations were selected for the study :

Site-1 (Mallital Zone) : The upper part of lake is called Mallital and the beautiful Naina Devi temple is situated on the right side, after whom the town has been named.

Site-2 (Central Zone) : It is central part of the lake.

Site-3 (Tallital Zone) : The lower part of the lake is called Tallital. Bus Station of the city is situated in close vicinity of Tallital zone.

The samples of lake water were analyzed using standard methods (Wetzel, 1975; Trivedi and Goel, 1984; APHA, 1998). The metal fractionations in sediment samples were analyzed following Tessier sequential techniques (Tessier *et al.* 1979). The trace heavy metals were analyzed in the laboratory by Atomic absorption spectrophotometer AAS 4129 model (Singh *et al.* 1987; APHA, 1998 and EICL Methods Manual, 2004) using air-acetylene flame. Operational conditions were adjusted in accordance with the manufacturer's guidelines to yield optimal determination. Quantification of metals was based upon calibration curves of standard solutions of respective metals. These calibration curves were determined several times during the period of analysis. The detection limits for iron, manganese, copper, nickel, chromium, lead, cadmium and zinc are 0.003, 0.001, 0.001, 0.004, 0.002, 0.01, 0.0005 and 0.0008 mg/L respectively.

RESULTS AND DISCUSSIONS

The concentration of heavy metals in surface water samples of the lake is presented in Tables 1 and 2 and concentration of heavy metals in central zone is presented in Figures 1 and 2. Lake contains high concentration of toxic metals Pb, Fe, Ni and Mn. While there was not significant variations in other metals contents in lake, Surface lake. water have shown seasonal fluctuations, the levels of metals in water varied significantly throughout the year : Cr, 0.03-0.17; Cu, 0.02-0.13; Fe, 0.37-1.13; Mn, 0.10-0.44; Ni, 0.16-0.58; Pb, 0.15-0.53; Zn, 0.03-0.24; Cd, 0.03-0.12 (mg/l) respectively. The level of some these metals like Lead, Iron and Nickel are much higher than the limits recommended by WHO (1985) throughout the year.

All the metals concentration were recorded in high concentrations at the Mallital zone, which showed high quantity of metals seems due to differential magnitude of population clusters in their catchments area. The present results indicated that the pollution input in the lake is mainly through domestic waste and municipal sewage as compared to natural inputs reported by Singh *et al.* (1982), Purohit *et al.* (1986), Pande *et al.* (1980), Rekha and Singh, (1987). Bottom water samples also have shown seasonal fluctuation (Tables 3 and 4), the level of metals in water were recorded in the range of - Cr, 0.02-0.15; Cu, 0.01-0.10; Fe, 0.21-0.66; Mn, 0.02-0.27; Ni, 0.06-0.34; Pb, 0.09-0.37; Zn, 0.01-0.18; Cd, 0.01-0.09 (mg/l) respectively. The level of some these metals like Fe, Pb, and Ni are lower than the surface lake water samples. All the metals found in low concentration during monsoon season and maximum found in pre monsoon season at all the zones of the lake. All the metals concentration was recorded high values at Mallital zone of the lake. High metals content in water at Mallital zone may be due to different loading rates of pollutants at this site. Pande *et al.* (1980) also reported the high metallic pollution in water and bottom sediments of Nainital lake. The present observations indicated higher increasing trends of metallic pollution load in all zones of Nainital lake.

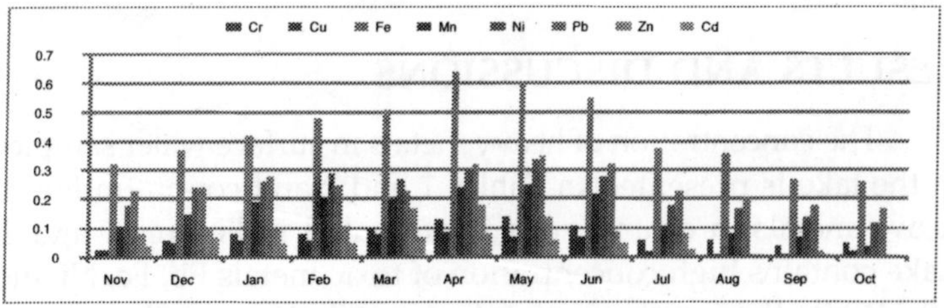

Fig. 1 : Distribution of Heavy metals in surface lake water (mg/l) in Central Zone of Nainital lake.

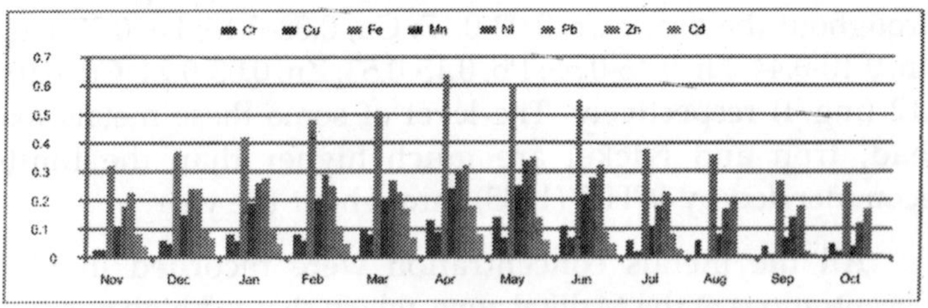

Fig. 2 : Distribution of Heavy metals in bottom lake water (mg/l) in Central Zone of Nainital lake.

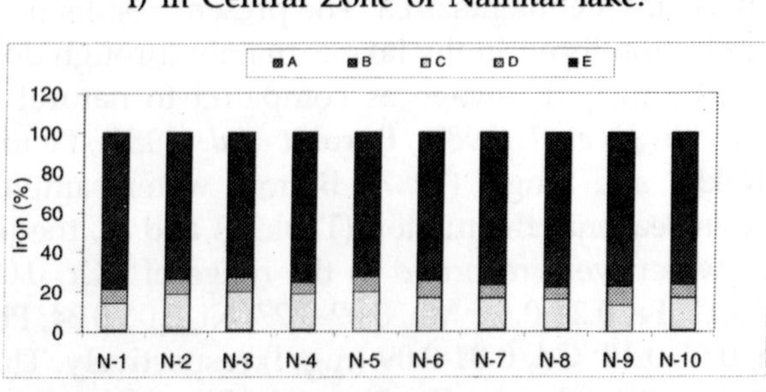

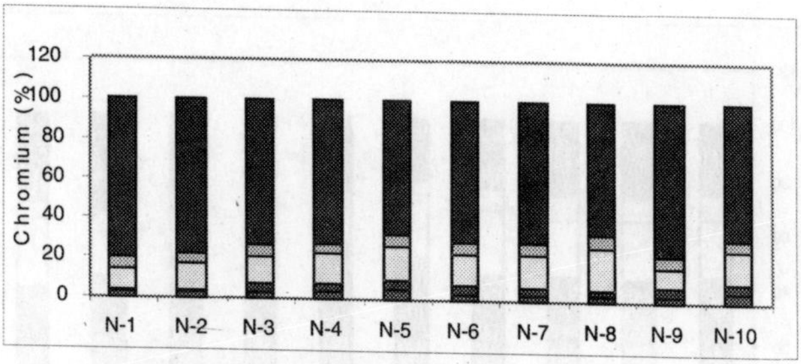

Chromium (%)
120
100
80
60
40
20
0
N-1 N-2 N-3 N-4 N-5 N-6 N-7 N-8 N-9 N-10

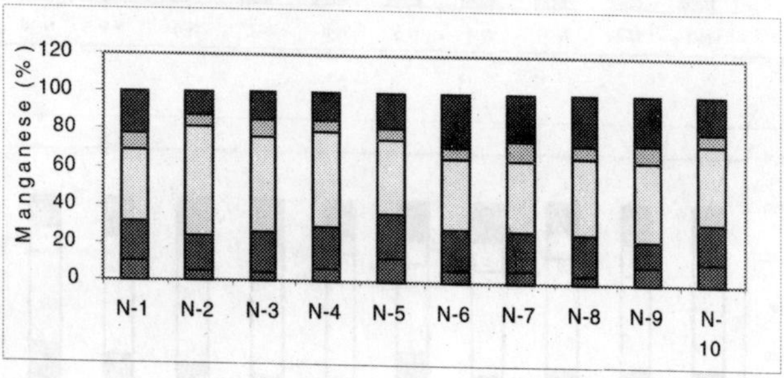

Manganese (%)
120
100
80
60
40
20
0
N-1 N-2 N-3 N-4 N-5 N-6 N-7 N-8 N-9 N-10

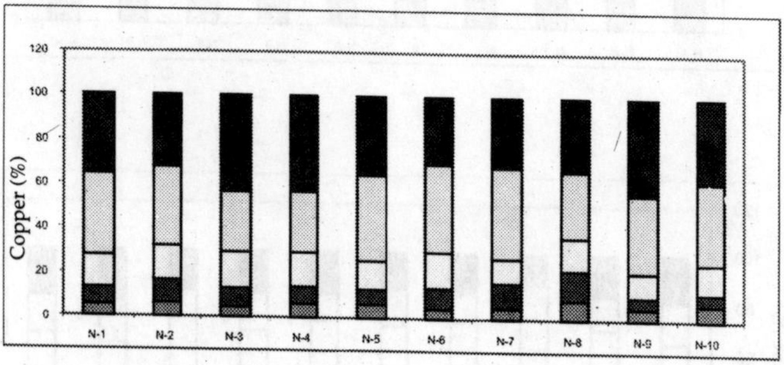

Copper (%)
120
100
80
60
40
20
0
N-1 N-2 N-3 N-4 N-5 N-6 N-7 N-8 N-9 N-10

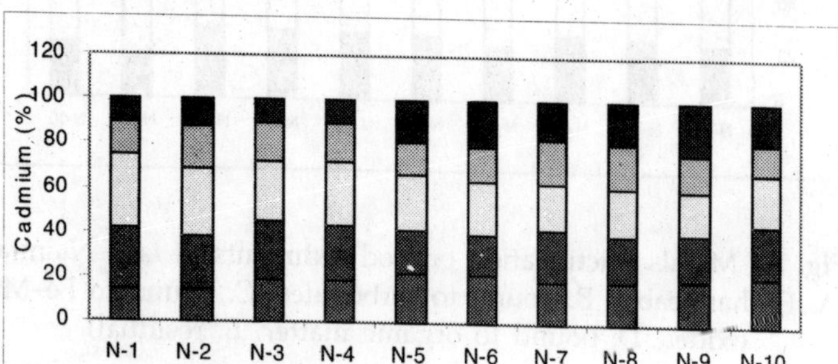

Cadmium (%)
120
100
80
60
40
20
0
N-1 N-2 N-3 N-4 N-5 N-6 N-7 N-8 N-9 N-10

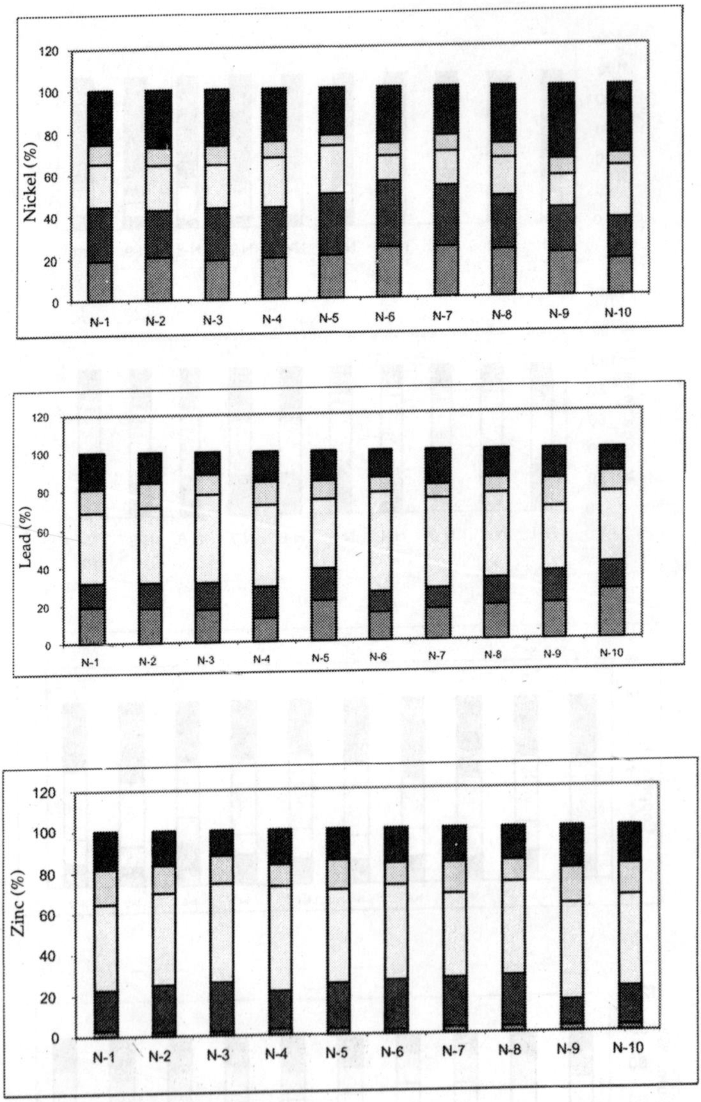

Fig. 3 : Metals fractionation on bed sediments of lake Nainital (A. Exchangeable, B. bound to carbonates, C. bound to Fe–Mn oxides, D. bound to organic matter, E. residual)

Table 1 : Distribution of Heavy Metals in Surface Lake Water (mg/l) in Mallital Zone of Nainital Lake

Months	*Cr*		*Cu*		*Fe*		*Mn*		*Ni*		*Pb*		*Zn*		*Cd*	
	Mean	*S.D*	*Mean*	*S.D*	*Mean*	*S.D*	*Mean*	*S.D*	*Mean*	*S.D*	*Mean*	*S.D*	*Mean*	*S.D*	*Mean*	*S.D*
November	0.11	0.03	0.06	0.02	0.64	0.03	0.2	0.03	0.31	0.04	0.41	0.04	0.11	0.04	0.06	0.03
December	0.14	0.04	0.08	0.02	0.77	0.04	0.23	0.04	0.37	0.02	0.49	0.06	0.12	0.02	0.09	0.04
January	0.09	0.01	0.08	0.01	0.82	0.03	0.27	0.06	0.42	0.04	0.52	0.05	0.16	0.04	0.1	0.04
February	0.09	0.04	0.1	0.01	0.88	0.04	0.3	0.02	0.45	0.05	0.52	0.03	0.19	0.05	0.13	0.05
March	0.14	0.02	0.12	0.02	1.04	0.06	0.35	0.04	0.48	0.04	0.45	0.04	0.2	0.03	0.08	0.04
April	0.17	0.01	0.11	0.03	1.05	0.04	0.4	0.06	0.49	0.02	0.48	0.04	0.22	0.04	0.12	0.06
May	0.16	0.02	0.13	0.02	1.13	0.04	0.44	0.04	0.58	0.05	0.53	0.04	0.24	0.02	0.12	0.04
June	0.15	0.04	0.12	0.03	1.1	0.03	0.4	0.03	0.54	0.04	0.53	0.06	0.18	0.04	0.11	0.05
July	0.1	0.02	0.06	0.03	0.78	0.04	0.32	0.04	0.36	0.03	0.38	0.04	0.12	0.02	0.07	0.04
August	0.07	0.03	0.06	0.01	0.69	0.05	0.29	0.06	0.31	0.05	0.35	0.05	0.11	0.04	0.04	0.03
September	0.08	0.02	0.06	0.02	0.66	0.05	0.26	0.05	0.25	0.03	0.32	0.05	0.09	0.04	0.03	0.01
October	0.08	0.01	0.05	0.03	0.61	0.03	0.18	0.03	0.21	0.04	0.27	0.04	0.08	0.02	0.03	0.02

Table 2 : Distribution of Heavy Metals in Surface Lake Water (mg/l) in Tallital Zone of Nainital Lake

Months	*Cr*		*Cu*		*Fe*		*Mn*		*Ni*		*Pb*		*Zn*		*Cd*	
	Mean	*S.D*	*Mean*	*S.D*	*Mean*	*S.D*	*Mean*	*S.D*	*Mean*	*S.D*	*Mean*	*S.D*	*Mean*	*S.D*	*Mean*	*S.D*
November	0.05	0.02	0.04	0.03	0.58	0.04	0.13	0.04	0.23	0.04	0.28	0.04	0.04	0.02	0.04	0.03
December	0.08	0.04	0.06	0.02	0.61	0.02	0.15	0.02	0.26	0.03	0.3	0.02	0.07	0.04	0.04	0.02
January	0.07	0.04	0.06	0.04	0.67	0.04	0.21	0.05	0.31	0.04	0.34	0.05	0.09	0.02	0.07	0.01
February	0.03	0.03	0.08	0.03	0.71	0.06	0.23	0.04	0.33	0.02	0.38	0.04	0.12	0.04	0.08	0.06
March	0.08	0.05	0.07	0.04	0.74	0.04	0.24	0.02	0.37	0.04	0.35	0.04	0.13	0.04	0.06	0.01
April	0.08	0.03	0.09	0.02	0.78	0.03	0.29	0.04	0.4	0.03	0.39	0.02	0.15	0.02	0.06	0.01
May	0.11	0.04	0.06	0.03	0.84	0.04	0.35	0.04	0.45	0.05	0.44	0.04	0.17	0.04	0.06	0.02
June	0.11	0.03	0.09	0.05	0.81	0.04	0.3	0.03	0.48	0.03	0.41	0.05	0.12	0.04	0.04	0.01
July	0.06	0.02	0.04	0.02	0.53	0.02	0.21	0.04	0.28	0.04	0.3	0.03	0.06	0.02	0.05	0.02
August	0.04	0.02	0.05	0.04	0.49	0.05	0.17	0.03	0.2	0.02	0.24	0.05	0.05	0.03	0.03	0.02
September	0.03	0.03	0.04	0.01	0.43	0.04	0.16	0.05	0.18	0.04	0.2	0.04	0.04	0.03	0.03	0.02
October	0.06	0.02	0.05	0.02	0.37	0.02	0.1	0.03	0.16	0.05	0.15	0.03	0.03	0.01	0.04	0.02

Table 3 : Distribution of Heavy Metals in Bottom Lake Water (mg/l) in Mallital Zone of Nainital Lake

Months	*Cr*		*Cu*		*Fe*		*Mn*		*Ni*		*Pb*		*Zn*		*Cd*	
	Mean	*S.D*	*Mean*	*S.D*	*Mean*	*S.D*	*Mean*	*S.D*	*Mean*	*S.D*	*Mean*	*S.D*	*Mean*	*S.D*	*Mean*	*S.D*
November	0.03	0.01	0.05	0.04	0.35	0.04	0.14	0.03	0.19	0.02	0.24	0.03	0.08	0.03	0.05	0.02
December	0.07	0.03	0.05	0.03	0.4	0.02	0.18	0.04	0.25	0.04	0.28	0.02	0.09	0.01	0.06	0.03
January	0.07	0.01	0.06	0.01	0.44	0.04	0.19	0.02	0.26	0.03	0.31	0.04	0.1	0.03	0.05	0.04
February	0.08	0.03	0.07	0.02	0.49	0.03	0.2	0.04	0.27	0.04	0.28	0.04	0.12	0.04	0.05	0.02
March	0.1	0.01	0.08	0.03	0.52	0.01	0.22	0.02	0.28	0.03	0.24	0.02	0.13	0.01	0.07	0.03
April	0.14	0.04	0.1	0.04	0.65	0.04	0.24	0.03	0.3	0.01	0.36	0.04	0.17	0.04	0.09	0.02
May	0.15	0.02	0.09	0.02	0.6	0.02	0.27	0.04	0.34	0.04	0.37	0.03	0.14	0.02	0.08	0.03
June	0.12	0.02	0.06	0.01	0.54	0.04	0.22	0.03	0.29	0.02	0.33	0.02	0.11	0.02	0.06	0.02
July	0.05	0.01	0.04	0.03	0.4	0.02	0.12	0.02	0.21	0.03	0.25	0.04	0.07	0.02	0.03	0.01
August	0.06	0.02	0.04	0.02	0.38	0.04	0.1	0.02	0.2	0.04	0.22	0.02	0.02	0.01	0.02	0.01
September	0.04	0.01	0.02	0.02	0.32	0.02	0.08	0.03	0.16	0.02	0.18	0.01	0.03	0.01	0.02	0.01
October	0.05	0.03	0.04	0.02	0.31	0.04	0.09	0.04	0.15	0.04	0.21	0.04	0.02	0.02	0.01	0.01

Table 4 : Distribution of Heavy Metals in Bottom Lake Water (mg/l) in Tallital Zone of Nainital Lake

Months	*Cr*		*Cu*		*Fe*		*Mn*		*Ni*		*Pb*		*Zn*		*Cd*	
	Mean	*S.D*	*Mean*	*S.D*	*Mean*	*S.D*	*Mean*	*S.D*	*Mean*	*S.D*	*Mean*	*S.D*	*Mean*	*S.D*	*Mean*	*S.D*
November	0.03	0.02	0.03	0.01	0.27	0.04	0.07	0.03	0.17	0.04	0.18	0.04	0.04	0.03	0.04	0.01
December	0.05	0.04	0.03	0.02	0.29	0.02	0.11	0.04	0.18	0.03	0.2	0.03	0.07	0.04	0.04	0.03
January	0.04	0.03	0.05	0.03	0.37	0.04	0.14	0.02	0.23	0.04	0.24	0.05	0.08	0.04	0.04	0.02
February	0.05	0.02	0.06	0.02	0.4	0.03	0.18	0.04	0.26	0.04	0.25	0.04	0.11	0.05	0.08	0.04
March	0.08	0.04	0.09	0.04	0.5	0.04	0.17	0.03	0.23	0.04	0.23	0.04	0.11	0.03	0.06	0.03
April	0.09	0.02	0.07	0.04	0.61	0.03	0.21	0.01	0.27	0.02	0.32	0.02	0.17	0.04	0.09	0.01
May	0.12	0.04	0.06	0.03	0.53	0.02	0.26	0.04	0.32	0.02	0.31	0.04	0.15	0.04	0.08	0.02
June	0.1	0.05	0.07	0.02	0.58	0.04	0.17	0.02	0.27	0.04	0.28	0.04	0.09	0.02	0.03	0.02
July	0.05	0.03	0.02	0.02	0.3	0.03	0.06	0.03	0.19	0.02	0.18	0.04	0.04	0.02	0.01	0.01
August	0.04	0.03	0.01	0.01	0.29	0.02	0.06	0.04	0.12	0.03	0.14	0.02	0.01	0.01	0.01	0
September	0.04	0.02	0.01	0	0.25	0.04	0.04	0.03	0.1	0.02	0.1	0.03	0.02	0.01	0.01	0.01
October	0.02	0.02	0.02	0.01	0.21	0.04	0.02	0.02	0.06	0.03	0.09	0.04	0.01	0	0.01	0.01

In the sediment samples of lake Nainital (Fig. 3), the concentration of chromium was found to vary from 59 to 87 mg/kg. The fractionation profile of chromium indicated that major portion of chromium is associated with residual fraction. A significant portion of chromium is also associated with the first three fractions (exchangeable, bound to carbonate and reducible) at Mallital and Tallital zone of the lake water and may enter into food chain. Fractionation profile of copper indicated that a major portion is bound to organic matter at most of the sites probably due to its more pronounced tendency for complexation with organic matter. A substantial percentage of copper is also found in reducible fraction (Fe-Mn oxide). As reported in earlier studies on polluted sediments (Samanidou and Flytianos, 1987; Pardo *et al.* 1993; Marin, *et al.* 1997),

Iron is the most abundant metal in all sediments because it is one of the most common elements in the earth's crust. Fractionation profile of iron in bed sediments of lake Nainital indicated that major portion (>70%) is associated with residual fraction characterizing stable compounds in sediments. Manganese, which is also abundant in nature, behaves in a different way. The fractionation profile of manganese indicated that it is mostly bound to reducible, carbonate and residual fractions, the rest being associated with exchangeable and organic matter fractions. More than 60% of the manganese was associated with first three fractions *i.e.* exchangeable, bound to carbonate and reducible fractions. The concentration of nickel was found to vary from 160 to 200 mg/kg in the sediments of the lake. More than 60% of the nickel was extracted in the first three fractions (exchangeable, carbonate and reducible. The concentration of lead varies from 88 to 125 mg/kg with maximum value at Mallital zone. The fractionation profile of lead indicates that major portion of lead is associated with reducible fraction at all the sites.

In the sediment samples of lake Nainital, the concentration of cadmium was found to be less than 20 mg/kg at most of the sites. More than 60% of the cadmium was associated in the first three fractions and may be easily remobilized by changes in environmental conditions. Nainital lake has a very significant importance for lake water utility, tourism, socio cultural and environmental ethics at national and international level. The present study reveals that there was heavy metal pollution in lake water and sediments. It is obvious that scientific treatment is necessary to minimize the pollution effects. The metal contents of various biotic components could be used for biomonitoring purpose and definitely contributing the remedial measures for restoration of the lake water quality.

REFERENCES

APHA (1998) : Standard methods for the examination of water and waste water, *American Public Health Association*, 19th edition, 1015, Fifteenth Street, NW, pp : (1-1) – 10-150.

Da La Noue, J., and De Paw, N. (1988) : The potential of microalgae biotechnology : A review of production and uses of microalgae. *Biotechnol. Adv.*, 6 : 725-770.

ECIL methods manual (2004) : Methods Manual, Atomic Absorption Spectrophotometer, AAS 4129, *Electronic Corporation of India Limited*, Hyderabad-500062, pp : 85.

Hakanson, L. (1980) : An ecological risk index for aquatic pollution control – a sedimentological approach. *Water Research*, 14 : 975–1001.

Lasat, M.M. (2002) : Phytoextraction of toxic metals : a review of biological mechanism. *J Environ Qual*, 31(1) : 109-20.

Marin, B., Valladon, M., Polve, M. and Monaco, A. (1997) : A reproducibility testing of a sequential extraction scheme for the determination of trace metal speciation in a marine reference sediment by inductively coupled plasma mass spectrometry. Analytic *Chimica Acta*, 342, 91–112.

Nurnberge, H.W. (1984) : The voltametric approach in trace metal chemistry of natural waters and atmospheric precipitation. *Anal Chem Acta*, 164 : 1-21.

Pande, H.K. and Singh, J.S. (1980) : Preliminary observations on phytoplankton productivity in Nainital and Bhimtal lakes. *Glimpses of Ecology*. Int. Scientific Publications, 375-382.

Pardo, R., Barrado, E., Castrillejo, Y., Velasco,M. A.and Vega, M. (1993): Study of the contents and speciation of heavy metals in river sediments by factor analysis. *Analytical Letters*, 26, 1719–1739.

Purohit, R., Sing, S.P. and Upreti, N. (1986), Effect of non-removal of the macrophytes biomass on the characteristics of water and community in Lake Nainital, U.P. India, int. Revue ges. *Hydrobiol* 71 (2), 245-257.

Rai, U.N. and Chandra, P. (1992) : Accumulation of copper, lead, manganese and iron by field population of Hydrodicty on reticulatum (linn.). Lagerheim. Sci. *Total Environ.* 116 : 203-211.

Samanidou, V., and Fytianos, K. (1987) : Partitioning of heavy metals into selective chemical fractions in sediments from rivers in northern Greece. *Science of the Total Environment*, 67, 279–285.

Singh, R. and Mahajan, I. (1982) : Phytoplankton and water chemistry of Rawalsar and Renuka lake. H.P. India. *J. Ecol.* 14(2), 273-277.

Sinha, S., Gupta, M. and Chandra, P. (1997) : Oxidative stress induced by iron in *Hydrilla verticillata* (D.F.) Royle : Response of Antioxidants Ecotoxicol. *Env. Safety* 38 : 286-291.

Tessier, A., Campbell, P.G.C., and Bisson, M. (1979) : Sequential extraction procedures for the speciation of particulate trace metals. *Analytical Chemistry*, 51, 844–851.

Trivedi, R.K. and Goel, P.K. (1984) : Chemical and biological methods for water pollution studies Karad, *Environmental publication*, pp : 1-251.

Wang, W. (1992) : Use of plants for the assessment of environmental contaminants. Rev. *Environ. Contam. Toxicol.* 126 : 87-127.

Wetzel, R.G. (1964) : A comparative study of the primary productivity of higher aquatic plants, periphyton and phytoplankton in a large, shallow lake. *Int. Rev. ges. Hydrobiol*, 49 : 1-64.

WHO (1985) : Guideline for drinking water quality, Recommendations, *World Health Organization, Geneva*, pp : 130.

26

Arsenic Contamination in Drinking Water and its Removal Strategies

Anamika Srivastava

ABSTRACT

Arsenic concentration is high in the drinking water. The greatest threat to public health from arsenic comes from drinking water. In this context of prevalence of high concentration of arsenic in drinking water, a wide range of technologies have been tried for the mitigation of arsenic from drinking water. The most common technologies utilized the conventional processes of oxidation, co-precipitation and adsorption onto coagulated flocs, adsorption onto sorptive media, ion exchange and membrane techniques for Arsenic mitigation. But the available conventional technologies for arsenic removal are moderately costly and need technical expertise therefore straight forward methods like Phytoremediation are inevitable for the people in developing countries. This chapter reviews the technologies used or can be used for the mitigation of arsenic from drinking water.

Keywords

Arsenic conatamination, Drinking water, Bio remediation technology.

INTRODUCTION

Albert Szent Gyorgyi, The Hungarian biochemist and Nobel Prize winner for medicine once said, "Water is life's matter and matrix, mother and medium. There is no life without water." "Third world war will not be over oil or energy but over water". The statement justify itself as much of the ill health in the underdeveloped countries is largely due to lack of safe drinking water. There can be no state of positive community health and well-being without safe water supply. Water is polluted with numerous pollutants among which arsenic poses most serious problem.

In Bangladesh and India, alluvial Ganges aquifers used for public water supply are polluted with naturally occurring arsenic, which adversely affects the health of millions of people. Excessive amounts of Arsenic can cause acute gastrointestinal (GI) and cardiac damage. Chronic doses can cause vascular disorders such as Blackfoot disease (Chen, S.L, *et al.* 1994) and epidemiological studies have linked arsenic to skin and lungs cancer (Tata and Arnold, 1990). Arsenic toxicity has no known effective medicine for treatment, but drinking of arsenic free water can help the arsenic affected people to get rid of arsenic toxicity.

In 1975, under the SAWA (Safe drinking water Act of 1994), EPA (Environmental Protection Agency) established a maximum contaminant level (MCL) for Arsenic at 0.05 ppm. On January 22, 2001 EPA adopted a new standard for arsenic in drinking water at 10 parts per billion (ppb), replacing the old standard of 50 parts per billion (0.05 ppm). The rule became effective on February 22, 2002 and this revised regulatory standard become active on January 23, 2006. The Arsenic and Clarifications to Compliance and New Source Contaminants Monitoring Final Rule was published in the Federal Register (66 FR 6976) on January 22, 2001.

The most commonly used processes of arsenic removal from water have been described by Chang and Others (1994), Hering and Others (1996), Hering and Others (1997), Kariten and Martin (1995), Shen (1973), and Joshi and Chaudhari (1996). A detailed review of arsenic removal technologies has been presented by Sorg and Logsdon (1978). Jekel (1994) has documented several advances in arsenic removal technologies.

In view of the lowering of the standard of the United States Environmental Protection Agency (EPA) for the maximum permissible levels of arsenic in drinking water, a review of arsenic removal technologies was carried out to consider the economic factors involved in implementing more stringent drinking water standards for arsenic (Chen and Others 1999). Many of the arsenic removal technologies have been discussed in detail in the AWWA (American Water Works Association) reference book (Pontius 1990). Comprehensive reviews of arsenic removal processes have been documented by Ahmed, Ali, and Adeel (2001), Johnston, Heijnen and Wurzel (2000) and Ahmed (2003). The AWWA conducted a comprehensive study on Arsenic treat ability options and evaluation of residuals management issue (AWWA 1999).There is several methods available for removal of arsenic from water in large conventional treatment plants. The most commonly used technologies include oxidation, co-precipitation and adsorption onto coagulated flocs, lime treatment, adsorption onto sorptive media, ion exchange resin and membrane techniques (Cheng *et. al.* 1994; Hering *et al.* 1996, 1997; Kartinen and Martin, 1995; Shen, 1973; Joshi and Chaudhari, 1996). The techniques for mitigation of arsenic from soil are solidification/stabilization, vitrification, pyrometallurgical recovery, *in-situ* soil flushing, and soil washing/acid extraction.

BACKGROUND

The Government of India and Bangladesh with the help of foreign aid agencies in the year 1970s and 1980s took an action in order to provide the masses with safe water. These agencies with the help of government dug over a million public pipe wells, providing the population with water that was believed to be clean and safe. The majority of the wells sunk by UNICEF. Ironically these wells led to the largest arsenic contamination crisis in the world. In the 1980's the health problem associated with arsenic poisoning were discovered in West Bengal. The first case of arsenic poisoning was identified in West Bengal in July 1983.

Bangladesh is facing the largest mass poisoning of a population in a history because of arsenic contamination. It is beyond the accident at Bhopal in 1984 and Chernobyl, Ukraine in 1986. (Smith). Mass majority of the rural population in Bangladesh does not have any alternatives. While those problems remain unsolved, thousands of villagers in Bangladesh will never be able to recover from a condition that has ruined their lives. The scale of this environmental disaster is greater than any seen before. Slowly, but severely, painful deaths are occurring and most of the rural population do not know why!! A poison that cannot be seen or even smell in the water is hardly thinkable for such simple-minded people. Another unfortunate and complicating fact about arsenic poisoning, Hiroki Hashizume adds, "is that it generally takes seven to ten years, some times longer, for the diseases to be recognized, when it finally is, it may be too late to treat."

The high concentration of arsenic contamination are found in various parts of the world including Argentina, USA, Vietnam, Cambodia, Hungry, Chile, China, Mexico, Thailand, Pakistan, Nepal, India and Bangladesh. The millennium summit in September 2000, the States of the United Nation reaffirmed their commitment to working towards a world in which sustaining development and eliminating poverty would have the highest priority. The eight millennium development goals comprise 18 targets and 48 indicators. The Goal 7 of millennium development goals indicate about environmental sustainability, Halve by 2015, the proportion of people without sustainable access to safe drinking water; Have achieve, by 2020, a significant improvement in the lives of at least 100 million slum dwellers.

ARSENIC

Arsenic is a major constituent of many minerals of the earth crust. The source of arsenic in the environment is the myriad anthropogenic activities. Arsenic are very much dependent on the chemical forms for its toxicity, availability and environmental ability reported by Tatken and Lewis 1983;

Deveral and Millard 1988. Further Wickstron *et al.* 1995 reported that trace concentration of arsenic can affect the physical and mechanical properties of metals and metal alloys.

Arsenic is very similar chemically to phosphorous. The similarity is so great that arsenic will partly substitute for phosphorus in biochemical reactions and is thus poisonous. As (III) exerts its toxic action by attacking SH groups of an enzyme, thereby inhibiting enzyme action. The enzymes which generate cellular energy in the citric acid cycle are adversely affected. The inhibitory action is based on inactivation of pyruvate dehydrgenase by complexation with As (III), whereby the generation of ATP is prevented. At high concentration it coagulates proteins. Arsenate in plants is taken via phosphate transporters at the plasma membrane of root cells and is rapidly reduced to arsenite once inside the cytoplasm. It is reported that arsenic can induce reactive oxygen species (ROS) in plants, which cause increases in lipid peroxidation and oxidative stress. (Hartley-Whitaker *et al.* 2001; Mascher *et al.* 2002; Srivastava *et al.* 2005). It also affects negatively chlorophyll levels in leaves (Mascher *et al.* 2002) though sometimes an increase has been reported (Paivoke & Simola, 2001). The pools of thiols in plants tissue play an important role on arsenic detoxification (Pickering *et al.* 2000; Vazquez *et al.* 2005). Arsenic also inhibits enzyme involved in DNA repair and thus responsible for DNA damage (Schaumloffel and Gebel, 1988). Arsenite is a week mutagen but potent comutagen (Hartwig and Beyersmann, 1989; Rossman *et al.* 1988).

Arsenic has been reported to replace phosphorous in the phosphate groups of DNA. (Dixon and Webb, 1958). Arsenic is effective mitotic poisons at particular concentration, due to their known affinity for thiol groups and induces various types of spindle disturbance. The clastogenic effects are S-dependent. The availability of cations affect the number of aberrations produced quantitatively. Effects of metallic salts are related directly to the dosage and duration of exposure. Studies on genotoxicity of arsenic showed that genotoxic effects could be in part responsible for metal phytotoxicity. (M.Patra, *et al.* 2004). Arsenic inhibits various physiological and biochemical processes resulting in consequent reduction in morphological characters and economic yield of agricultural and horticultural crops. Major characters affected are tillers (in cereals), plant height, leaf number and area, pod number and length (in legumes) and dry matter production (Bhowmik and Sharma, 1999).

In ground water it is present mainly in nonionic trivalent As (III) and ionic pentavalent As (V) inorganic form. The solubility of Arsenic in water is usually controlled by redox condition, pH, biological activity and adsorption reaction. The reducing condition at low Eh value converts arsenic into a more mobile As (III) form, whereas at high Eh value As (V) is the major arsenic species. As (III) is more toxic than As (V) and difficult to remove from water by most techniques.

The basic principals of arsenic removal from water are based on conventional techniques of oxidation, co-precipitation and adsorption onto coagulated flocs, adsorption onto sorptive media, ion exchange and membrane filtrations. Oxidation of As (III) to As (V) is needed for effective removal of arsenic from water by most treatment methods. These conventional technologies although available but costly and can not be afford by poor people and in absence of any choice either contaminated sources are abandoned or force to drink water from contaminated source. For these people a ray of light is innovative technology like phytoremediation; based on the simple principle of purification technology of the nature, *i.e.*, of the nature, by the nature and for the nature. This chapter presents a short review of these technologies along with innovative technologies especially *phytoremediation in detail* is intended to update the technological development in arsenic removal for successful implementation and adaptation to rural condition.

ARSENIC REMOVAL TECHNOLOGIES

A short review of Conventional technologies are as follows :

Soil and Waste Treatment Technologies : [Conventional Technologies]

Solidification and Stabilization Treatment : It is an

established treatment technology which is often used to reduce the mobility of arsenic in soil and waste. This process involves mixing a soil or waste with binders to create a slurry paste or other semi-liquid paste, which is then allowed time to cure into a solid form and thus reduces the mobility of contaminants in the environment. It physically binds contaminants and reduces them by converting into less soluble, mobile or toxic form. Frequently used binders are pozzolanic material, such as cement and lime. Others are fly ash, sulfur, phosphate. It generally produces a stabilized product that meets the regulatory threshold of 5mg/L. Limitation of this treatment is the use of proprietary binders or reagents which may be more expensive than the use of non-proprietary binders. Secondly, the needs to pre treat soil and waste prior to solidification/ stabilization may also advance management costs.

Vitrification

This technology for soil treatment can be applied either in *situ* or *ex situ*. It typically requires large amounts of energy to achieve vitrification temperature and therefore can be expensive to operate. Off gases may require further treatment to remove hazardous constituents. Vitrification is a high temperature aimed at reducing the mobility of metals by incorporating them into a chemically durable, leach resistant, vitreous mass. This process also may cause contaminants to volatilize or undergo thermal destruction, thereby reducing their concentration in the soil or waste.

Soil Washing/ Acid Extraction

It can be used in limited number of application. The process is limited to soil in which contaminants are preferentially adsorbed onto the fine fraction. The separated fine must be further treated to remove or immobilized arsenic.

But soils that vary widely and frequently in characteristic, such as soil type, contaminant type and concentration and where blending for homogeneity is not feasible, may not be suitable for soil washing. Secondly, cold weather can cause the washing solution to freeze and can affect leaching rates.

Pyrometallurgical Recovery

This technology use heat to convert an arsenic contaminated waste feed into a product with a high arsenic concentration that can be reused or sold. This process involves high temperature metals recovery and slag cleaning process. But the presence of impurities, such as other heavy metal may need to be removed, which increases the complexity of the treatment process.

In-Situ Soil Flushing

This technology extracts organic and inorganic contaminants from soil by using water, a solution of chemical in water without excavating the contaminant material itself. The solution is injected into or sprayed onto the area of contamination causing the contaminants to become mobilized by dissolution. After passing through the contaminant zone the contaminants bearing the flushing solution is collected by down gradient wells and pumped to the surface for removal. This technology works best when a single contaminant is targeted. Secondly, soil flushing may cause arsenic in the soil to precipitate and obstruct the soil pore structure and inhibit flow through the soil.

Water Treatment Technologies

It is reported that arsenic ubiquitously distributes in nature and is mainly transported by water in the environment. (Ferguson and Gavis, 1972). Arsenic concentration in the biota of aquatic ecosystem is much higher than those found in the organisms inhabiting terrestrial environment reported by Lunde, (1977). Arsenic is also very toxic to humans (Borum and Abernathy, 1994). The studies of risk assessment concerning chronic exposure to arsenic suggested that arsenic has a carcinogenic potential in humans (Thomas, 1994; Buchet *et al.* 1996). Chen *et al.* (1988, 1992) and Chiou *et al.* (1995) pointed out that arsenic levels in water are associated with cancer. In addition, Chen *et al.* (1980) indicated that high arsenic concentration in ground water is related to black foot disease (BFD).

The most common arsenic removal technologies can be grouped into the following four categories :

- Oxidation and Sedimentation.
- Coagulation and Filtration.
- Sorptive Filtration, and
- Membrane Filtration.

Oxidation and Sedimentation

It is inevitable to convert As (III) to As (V) for effective removal hence most treatment technologies include an oxidation step as pretreatment to convert arsenite to arsenate. Arsenite can be oxidized by oxygen, ozone, free chlorine, hypochlorite, permanganate, hydrogen peroxide and Fulton's reagent but atmospheric oxygen, hypochlorite and permanganate are commonly used for oxidation in developing countries. The oxidation processes convert noncharged arsenite to charged arsenate, which can be easily removed from water. *Passive sedimentation* has received considerable attention because of rural people's habit of drinking stored water from pitches. Ahmed *et al.* (2000) showed that more than 50 per cent reduction in arsenic content is possible by sedimentation of tube well water containing 380-480 mg/L of alkalinity as calcium carbonate and 8-12 mg/l of iron but cannot be relied to reduced arsenic to desired level. Most studies showed a reduction of zero to 25 per cent of the initial concentration of arsenic in water. In rapid assessment of technologies passive sedimentation failed to reduce arsenic to the desired level of 50 µg/l in any well (BAMWSP, DFID, WaterAid, 2001).

Coagulation and Filtration

In the process of coagulation and flocculation, arsenic is removed from solution through three mechanisms :

- **Precipitation :** The formation of insoluble compounds.
- **Co-precipitation :** The incorporation of soluble arsenic species into a growing metal hydroxide phase.

- **Adsorption :** The electrostatic binding of soluble arsenic to external surfaces of the insoluble metal hydroxide (Edwards 1994).

The Precipitation/Co-precipitation technology is costly as the cost generally increases with increased chemical addition and sludge produced from this process could be considered hazardous waste and requires additional treatment before disposal. In the coagulation-flocculation process aluminium sulfate, or ferric chloride, or ferric sulfate is added and dissolved in water under efficient stirring for one to few minutes. Aluminium or ferric hydroxide micro-flocs are formed rapidly. The water is then gently stirred for few minutes for agglomeration of micro-flocs into larger easily settable flocs. During this flocculation process all kinds of micro-particles and negatively charged ions are attached to the flocs by electrostatic attachment. Arsenic is also adsorbed onto coagulated flocs. As trivalent arsenic occurs in non-ionized form, it is not subject to significant removal. Oxidation of As (III) to As (V) is thus required as a pretreatment for efficient removal. This can be achieved by addition of bleaching powder (chlorine) or potassium permanganate.

Arsenic adsorbed on aluminium hydroxide flocs as Al-As complex is removed by sedimentation. Filtration may be required to ensure complete removal of all flocs. Similar reaction take place in case of ferric chloride and ferric sulfate with the formation of Fe-As complex as end product which is removed by the process of sedimentation and filtration. Arsenic removal is dependent on pH. In alum coagulation, the removal is most effective in the pH range 7.2-7.5 and in iron coagulation; efficient removal is achieved in a wider pH range usually between 6.0 and 8.5 (Ahmed and Raham, 2000). The effects of cations and anion are very important in arsenic removal by coagulation. Anions compete with arsenic and lower the removal rates.

The presence of more than one anion can have a synergistic effect on arsenic removal. The technologies developed based on coagulation–sedimentation-filtration-process include :

- Bucket Treatment Unit.

- Stevens Institute Technology.
- Fill and Draw Treatment unit.
- Tube well-attached arsenic treatment unit, and
- Iron-arsenic treatment unit.

Sorptive Filtration

Several sorptive media have been reported to remove arsenic from water. The efficiency of sorptive media depends on the use of an oxidizing agent as an aid to sorption of arsenic. The most commonly used media for arsenic removal in small treatment plants include :

- Activated alumina.
- Granular Ferric oxide and Hydroxide.
- Metallic Iron.
- Iron-coated sand or brick dust.
- Cerium oxide.
- Ion exchange media.

The activated alumina-based sorptive media used in Bangladesh and India include :

- BUET activated alumina.
- Alcan enhanced activated alumina.
- Apyron arsenic treatment unit.
- Oxide (India) Pvt. Ltd.
- RPM Marketing Pvt. Ltd.

Membrane Technique

It separates contaminants from water by passing it through a semi-permeable barrier or membrane. The membrane allows some constituents to pass through, while blocking others.

Synthetic membrane can remove many contaminants from water. They are of two main types :

- Low pressure membrane, used in micro filtration and ultra filtration, and
- High pressure membrane, used in nano-filtration and Reverse osmosis.

The types of membrane selected might affects the costs of the treatment and further treatment may be required prior to disposal or discharge, which is again expensive.

Field Test Kit

Field test kits have been developed for detection and measurement of arsenic by different institute. The detection and semi quantitative measurement of arsenic by all field kits is based on the Gutzeit procedure, which involves the conversion of all arsenic in water into arsenite by reduction and then formation of arsine gas by further reduction using nascent hydrogen in an acid solution in Gutzeit generator. The technique is also known as the mercuric bromide stain method. (APHA-AWWA-WEA-1985). The arsine gas, thus liberated produces a yellow to brown stain on a vertical paper strip impregnated with mercuric bromide. The amount of arsenic present in the water is directly related to the intensity of the colour. The colour developed on mercuric bromide-soaked paper is compared either with a standard colour chart or measured by a photometer to determine the arsenic concentration of the water samples. In some field test kits the generated arsine is passed through a column containing a roll of cotton moistened with lead acetate solution to absorb hydrogen sulfide gas, if any is present in the gas stream.

A number of researchers and organizations have evaluated the performance of arsenic field test kit. The NEERI evaluated the Asian Arsenic Network (AAN) kit (0.02-0.70mg/l), the National Institute of Preventive and Social Medicine kit (0.10-3.0 mg/l), the Aqua Kit and the AIIH and pH kit (NEERI and WHO 1998). The EPA issued verification reports and verification statements for these arsenic field test kits (Abbgy and Others, 2002; EPA 2003) (Ahmed, Minnatullah and Talbi, 2005).

A myriad research has been expended in the search for better practical methods for determination of arsenic species in water. Those include capillary zone electrophoresis (Li and Li, 1995), high performance liquid chromatography (HPLC) (Manning and Martens, 1997), ion exchange method (Ficklin 1983; Grabinski 1981), solvent extraction (Mok *et al.* 1986), and chemical precipitation (Van Elteren *et al.* 1991). The methods for the determination of arsenic in natural waters entail considerable problems. In most cases, the equipment for the separation is not portable, so it is difficult to perform the experiment in the field and to obtain the exact values of arsenic species.

SOIL, WASTE AND WATER TREATMENT TECHNOLOGIES [INNOVATIVE TECHNOLOGIES]

Biological Treatment

Biological treatment for arsenic is used primarily to treat water above-ground in processes that use microorganisms to enhance precipitation/co-precipitation. This treatment technology involves the use of microorganisms that acts directly on contaminant species. Microbes used in this treatment processes are Sulfate-reducing bacteria and Arsenic-reducing bacteria. It was found that arsenic was reduced to below 0.05 mg/l in one pilot scale application (Eleventh Edition: EPA-542-R-02-004).

It was reported that biological treatment of arsenic may produce less sludge than conventional ferric arsenic precipitation.

Phytoremediation : A detail review of Phytoremediation

An important tool of bioremediation has played a vital role to fight with the problem of environmental pollution, use of microorganism and plants are gaining popularity, but in case of microbial remediation genetically altered bacteria are useful to check pollutants. Whereas plants without much alteration could be used to mitigate pollutants in their natural environment which can also be utilized by layman as this technology is straightforward, not expensive and this method of mitigating pollutants are age-old practice. Being not expansive this method is necessary for the developing, countries.

Phytoremediation is the processes of restoration of the quality of environment by the application of plants. In parlance this processes is referred as "Green Clean Technology" which means cleaning up of environment with the help of plants. It may be stated that it is the purification technology of the nature, by the nature and for the nature. Phytoremediation is an emerging technology. This technology tends to have low-capital, operating and maintenance costs relative to other arsenic treatment technologies because it relies on the activity and growth of plants. The study conducted by McCutcheon and Schnoor, (2003) showed that phytoremediation is applicable to the remediation of arsenic because arsenic has been identified as suitable for phytoextraction and rhizofiltration.

The phytoremediation processes consist of the following components: Phytoextraction (by which plants can accumulate heavy metals in their stable plant parts *i.e.*, roots and shoots), Phytomining (extraction of precious metals which have high economic importance), Chaney *et al.*, 2000 reported that plant material can subsequently be used for nonfood purposes or ashed, followed by disposal in a landfill or, in the case of valuable metals, the accumulated elements can be recycled. Phytovolatilization (evaporation of certain metals from aerial parts of the plants), Phytostabilization (where the plants reduce the bioavailability of the heavy metals) and Rhizofiltration (where the plants roots absorb and concentrate heavy metals from polluted effluents).

Phytoextraction have two basic strategies: chelate assisted phytoextraction, also known as induced phytoextraction and long term continuous phytoextraction. Continuous phytoextraction is also being studied by several groups for the removal of oxianionic metals such as arsenic, selenium and chromium.

Salt *et al.* (1998) and Vassil *et al.* (1998) reported that metal phytoextraction relies on metal-accumulating plants to transport and concentrate polluting metals from soil into the harvestable aboveground shoots. Popular species for phytoextraction are Indian mustard (*Brassica juncea* L. *Czern.*) and Sunflower (*Helianthus annus*) because of their fast growth, high biomass and high tolerance and accumulation of metals and other inorganic (Blaylock and Huang, 2000; Salt *et al.* 1995b). The plants that accumulate several heavy metals in high concentration (> 1000 mg/g) at a rapid rate with high biomass production are referred to as hyper-accumulators. About this category of plants has been known since the late 1800s but the idea of using only appeared in the literature in the last 20 years. At present, at least 45 plant families are known to contain metal-accumulating species reported by Reeves and Baker in year 2000. Such plants can accumulate As, Cu, Co, Cd, Mn, Ni, Se, Pb or Zn upto levels that are hundreds to thousands times of those normally accumulated by plants growth under the same condition (Baker *et al.* 2000 ; Ma *et al.* 2001; Brooks,1998). Environment Canada has developed a database (PHYTOREM) of 775 plants with capabilities to accumulate or hyperaccumulate one or several key metallic elements.

Burlo *et al.* 1999; Carbonell-Barrachina *et al.* 1999a reported that the phytoavilability of arsenic is primarily determined by arsenic species and concentration in the medium. It was reported that plants readily takes up arsenite and arsenate, however it is greatly influenced by soil texture and competing phosphates. Low levels of phosphates displace arsenic from soil particle to advance uptake and phytotoxicity, whereas larger amounts of phosphates compete with arsenic at root surfaces to decrease uptake and phytotoxicity. (Peterson *et al.* 1981). Other factors such as plant species, types of compound and method of application influenced the accumulation of arsenic in plants. Arsenic accumulation in plants rarely reaches the levels that are harmful to animals and man reported by Bandyopadhyay, 1997; Dudka and Miller, 1999. The biochemical fate of arsenic in Indian mustard (*Brassica juncea* L. *Czern*) is studied by Pickering *et al.* (2000), showed that the root uptake the arsenate via the phosphate transport mechanism, a small fraction is exported to shoot via the xylem as the oxyanions (arsenate and arsenite). Once in the shoot, the arsenic is stored as an As (III)-tris thiolate complex. The majority of the arsenic remains in the root as an As (III)-tris thiolate complex which is indistinguishable from that found in the shoot and from As (III)-tris glutathione. Further, it was found that in plants arsenic uptake advances with the rate of plant growth. However, growth per unit of arsenic uptake was higher for plants in untreated soils than plants in arsenic treated soil (Onken and Hossner, 1995). Abedin *et al.* (2002) reported that arsenite uptake was active and was taken up at approximately the same rate as arsenate.

Phytovolatilization is a passive processes, it may be maximized by using plant species with high transpiration rates, by over expression of enzymes such as cystathionine-V-synthase that mediates Se volatilization (Van Huysen *et al.* 2003) and by transferring gene for Se volatilization from hyper accumulators to non accumulators (Le Duc *et al.* 2004). Volatilization of arsenic has been demonstrated for microorganisms, but these elements do not appear to be volatilized to significant levels by nontransgenic plants (Rugh *et al.* 1996).

Phytostabilization of metals may employ plants to reduce leaching, runoff and erosion via stabilization of soil by plant roots or root exudates may cause metals to precipitate, converting them to less bioavailable form suggested by Berti and Cunningham, 2000; Burken *et al.* 2000; Kramer and Chardonnens, 2001). Poplar are also known as arsenic accumulator. It was reported that Poplar maintain an upward flow to prevent downward leaching, while grasses prevent wind erosion and lateral runoff with thin dense root system. However, grasses do not accumulate as much metals in their shoots as dicot species, minimizing exposure of wildlife to toxic elements (Pilon Smits, 2005).

In grasses, ryegrass and *prairie* grasses are known for their phytoremediation capacity. Rhizofiltration provides a cheap method of clean-up for low concentration of contaminants in large volumes of water. A number of plant species including

Indian mustard (*Brassica juncea* (L) *Czern.*), rye, corn and sunflower can have rapidly growing roots and can be used for rhizofiltration. Rhizofiltration, the processes where plant roots grown in water absorb, concentrate and precipitate toxic metals and organic chemicals from polluted effluents (Schmoger *et al.*, 2000). The plants can be used as filters in constructed wetlands (Horne, 2000; Lytle *et al.* 1998) or in a hydroponic setup (Raskin *et al.* 1997). Water hyacinth (*Eichhornia crassipes* L.), an aquatic floating plants, is a high accumulator of metals, making it the most efficient phytoextraciting agent for wastewater polluted with heavy metals. It was reported by V. Hooda, 2007 that in constructed wetlands *Eichhornia crssipes* L., *Azolla* and *Lemna minor* (duckweed) are popular because they are good metal accumulators and can be harvested easily. Cattail and Poplar are also used as they are tolerant, grow fast and attain large biomass. Hydroponics system involves aeration and therefore is not limited to aquatic species; it often makes use of terrestrial species with large roots and good capacity to accumulate inorganic, such as, Sunflower (*Helianthus annuus*) (Dushenkov and Kapulnik, 2000) or Phaseolus vulgaris (Piechalak *et al.* 2002). Synder (2006) reported the highest arsenic concentration in the bladder of the water hyacinth, 162,000 parts per billion. It was found that water hyacinth may move the arsenic to their bladders to keep it from harming photosynthesizing portions of the plants. The stems and leaves had the second highest average arsenic concentration at 125,000 parts per billion. The roots had the lowest arsenic concentration, at 104,000 parts per billion. The bio-concentration factors for the bladders, stems and leaves and roots were: 538, 417 and 346 respectively. Because the bio-concentration factors were greater than one, water hyacinths are arsenic hyper accumulators by definition suggested by Synder (2006).

The basic principle of phytoremediation lies in the ability of plants to hyperaccumulate metals and metalloids. In arsenic contaminated soil the major component is arsenate. Due to microbial activity, transformation from arsenate to arsenite results in monomethylarsonic and dimethylarsinic acid. It was observed that the transformation induced by microorganisms in highly contaminated soil led to detoxification and thus made possible the use of microorganisms for phytoremediation. (Sarkar *et al.* 2003). It was reported by Meharg and Macnair in year 1991 that arsenic tolerance in non arsenic-hyperaccumulating plants involved a decreased uptake of arsenate due to suppression of the high affinity phosphate uptake system. This result confirms that arsenate is probably taken up via the phosphate uptake system.

Hitherto, Chinese brake fern (*Pteris vittata* L.) has been considered as efficient arsenic–hyper accumulator plant. (Ma *et al.* 2001; Zhu *et al.* 2005 ; Ma *et al.* 2006; Ma *et al.* 2007; Koller *et al.* 2007). *Pteris vittata* L. is a first known arsenic hyperaccumulator. It is reported that age greatly affects the physiological activity of a plants, especially its roots (Taiz and Eduardo, 1998). Generally roots of a young plant display greater ability to absorb ions than those of old plants when they are similar in size (Marschner, 1995). The arsenic-depletion rate by a 12-month old fern plant was just 42-52 per cent of that observed for 3-month old fern plants at the end of three days. This result suggests that it is important to use healthy young plants for more efficient plant arsenic removal. This was reported by Lena Q. Ma and Co-workers in year 2004. Similar results were reported in rice (Anti *et al.* 2001) and pumpkin (*Cucurbita moschata* Poir) (Swiader *et al.* 1986), where they found that N uptake rates for rice and nutrient uptake for pumpkin were slower with increasing plant age. It was found that Chinese brake fern has the extraordinary ability to hyperaccumulate arsenic in its fronds up to 27000 mg arsenic per kg dry weight (Wang *et al.* 2002).

It was reported that arsenic uptake by the corn plant roots for the As (V) and As (III) treatments were 95 and 112 ppm, respectively. However, in the shoots of the As (V) treatment had 18 ppm whereas As (III) treatments had 12 ppm. (J.G.Parsons *et al.* 2007). A hyper accumulating Chinese brake fern (*Pteris vittata* L.) has been used in the remediation of arsenic contaminated soil, waste and water. The fern can tolerate as much as 1500 ppm of arsenic in soil and can have a bio-concentration factor up to 265. The arsenic concentration in the plant can be as high as 2 per cent (dry weight). [Eleventh Edition: EPA-542-R-02-004].The study conducted by Zhang *et al.* 2002 reported that Chinese brake fern showed 60-74 per cent of the arsenic in the fronds was present as arsenite compared to only 8 per cent in the roots. Similar results were reported by Francesconi *et al.* 2002 for *Pityrogramma calomelanos*, another arsenic hyperaccumulating fern. The bioconcentration factors in *Pteris vittata* fronds from all arsenic contaminated soils were significantly greater than one, ranging from 4.7 to 48, suggesting the capacity of plant to bioconcentration of arsenic from all soils. (Ma *et al.* 2008).

Robinson *et al.* (2006) suggested that aquatic plants have the potential to accumulate arsenic and thus involved in arsenic phytoremediation. An aquatic weed has the potential to save thousands of lives from arsenic, according to an article published by the Royal Society of Chemistry in the United Kingdom. The Water hyacinth (*Eichhornia crassipes* L.) has been show to be capable of up taking both arsenite and arsenate (Misbahuddin, 2002) as well as a number of heavy metals including chromium, lead and zinc (Saha *et al.* 2001). Robinson, 2005 suggested that water hyacinth reduce the arsenic concentration in the water to 10 µg/l or less as per the US EPA or WHO standards for drinking water.

The dried hyacinth root has the potential to remove arsenite and arsenate. It is reported that over 80 per cent of the arsenic was taken up within 30 minutes exposure and 96 per cent of both chemicals species of arsenic were removed within two hours, leaving a solution containing less than 10 mg/l the WHO guideline value [Harrington and Harris, 2006]. *Hydrilla verticillata* plants accumulates higher concentration of arsenite and arsenate in laboratory condition however, further research is needed to analyze the same result in field condition. (Tripathi *et al.* 2007). It was found that the water lettuce (*Pistia stratiotes* L.) could effectively absorb arsenic between ranges of 0.25 to 5.0 mg/l to the extent of 82.0 to 28.8 per cent for a biomass of 20g/l at pH 7.0 after 144 hours. The removal efficiency was, however, noted to be maximum 87.5 per cent at pH 6.0. (Basu, Kumar and Mukherjee, 2003). Salt *et al.* (2000) suggest through their research on uptake of arsenic by Indian mustard (*Brassica juncea* (L.) *Czern*) that arsenate enters the roots as a phosphate analog and is promptly reduced to arsenite. Little arsenic is thus transported to the aboveground tissues. The addition of dimercaptosuccinate to the hydroponics growth solution caused significant amounts of arsenic to move into the shoots, perhaps offering a way of mitigating arsenate from contaminated soil. Hitherto the studies on arsenic phytoremediation showed that this method is the cheapest and safe among all other arsenic treatment methods.

Electrokinetics

This is an *in situ* technology intended to be applicable to soil, waste & water. This technology is most applicable to fine-grained soil, such as clays. It is based on the theory that a low-density current applied to soil will mobilize contaminants in the form of charged species. A current passed between electrodes inserted into the subsurface intended to cause water, ions and particulates to move through the soil. Contaminants arriving at the electrodes can be removed by means of electroplating or electrodeposition, precipitation or co-precipitation, adsorption or by pumping of water near the electrode.

CONCLUSION

A remarkable technologies development in arsenic removal based on conventional arsenic removal processes has been taken place during last 2-3 years. All the technologies described in this chapter have their merits and demerits and are being refined to make suitable in all conditions. Arsenic removal technologies (conventional and innovative) have to compete with other technologies in which costs appear to a major determinant in the selection of a treatment option by the users. Therefore, straightforward methods are inevitable for the people residing in developing countries, particularly technologies utilizing materials that are widely available and cheap; henceforth the need is to shift to innovative technology like Phytoremediation.

REFERENCES

Ahmed M.F. and Rahaman M.M. (2000) : Water Supply and Sanitation-Low Income Urban Communities, International Training Network (ITN) Center, BUET.

Ahmed. M.F., (2003) : Treatment of arsenic contaminated water. In: M.F.Ahmed, Ed, arsenic contamination: Bangladesh Perspective. 354-403. Dhaka, Bangladesh : ITN-Bangladesh.

BAMWSP, DFID and WaterAid Bangladesh (2001) : Rapid Assessment of Household level Arsenic Removal Technologies, Phase-I and Phase-II, Final Report, WS Atkins International Limited.

Cheng, C.R., Liang, S. Wang, H.C. and Beuhler, M.D. (1994) : "Enhanced coagulation for arsenic removal, "J. American Water Works Association, 86 (9), pp : 79-90.

Chen H.W, Frey M.M., Clifford D, McNeill L.S. and Edward M. 1999. Arsenic treatment consideration. J. American water works association 91 (3) : 74-85.

Choong T.S.Y., Chauh T.G., Robiah Y, Koay G and Azni I, 2(007) : Arsenic toxicity: health hazards and removal technique from water: Overview. Desalination, Vol–217, Issue-1-3, November, Pg : 139-166.

Das H.K., Mitra A.K., Sengupta P.K., Hossain A, Islam F and Rabbani, (2004) : Arsenic contamination in rice, vegetable and fish in Bangladesh : a preliminary study. Environmental International, Vol. - 30, Issue 3, May, pg : 383-387.

Fayiga A.O., Ma L.Q., Rathinasabapathi B, (2008) : Effect of nutrients on arsenic accumulation by arsenic hyper-accumulator *Pteris vittata* L. Environmental and Experimental Botany 62, 231-237.

Gupta D.K., Tripathi R.D., Mishra S, Srivastava S, Rai U.N., Yang X.E., Huang H and Inouhe M., (2008): Arsenic accumulation in root and shoot vis-à-vis its effects on growth and level of phytocheletins in seedling of *Cicer arietinum* L. J. Environ. Biol. 29(3), 281-286.

Gonzaga M.I.S, Santos J.A.G, Ma L.Q., (2008) : Phytoextraction by arsenic hyper-accumulator *Pteris vittata* L from six arsenic contaminated soils: Repeated harvests and arsenic redistribution. Environmental Pollution 154 : 212-218.

Harrington and Harris, (2005) : The use of dried hyacinth roots to remove arsenic from surface water. Journal of Environmental Monitoring, Issue-7, pg : 279-282.

Hooda V., (2007) : Phytoremediation of toxic metals from soil and waste water. J. of Environmental Biol. 28 (2) : 367-376.

Jekel M.R., (1994) : Removal of Arsenic in drinking water treatment. In. J.O. Nriagu ed. Arsenic in the environment. Part-1 : Cycling and Characterization. New York : John Willy and Sons, Inc.

Kariten, E.O. and Martin C.J. (1995) : An overview of Arsenic removal processes, J. Desalination , 103, pp. 79-88.

Lombi E., Zhao F, Fuhrmann M, Ma L.Q, McGrath, S.P. (2002) : Arsenic distribution and speciation in the fronds of the hyperaccumulator *Pteris vittata*. New Phytol. (156), 195-203.

Meyer D.E.R., Auchterlonie G.J., Webb R.I and Wood B, (2007) : Uptake and localization of lead in the root system of *Brassica juncea*. Environmental Pollution 153, 323-332.

Mukherjee S and Kumar S, (2005) : Arsenic uptake potential of water lettuce (*Pistia stratiotes* L). International Journal of Environmental Studies, Vol 62 (2), April, Pg : 249-258.

Nriagu J.O., Bhattacharya P, Mukharjee A.B., Bundschuh J, Zevenhoven R and Loeppert R.H., (2007) : Arsenic in soil and ground water: an overview. Trace metal and other contaminants in the environment. Vol 9, pg : 3-60.

Pasha C and Narayan B., (2008) : Determination of arsenic in environmental and biological samples using toluidine blue or saffranine by simple spectrophotometer method. *Bull Environ. Contam. Toxicol.* 81, 47-51.

Patra M, Bhowmik N., Bandopadhyay B and Sharma A. (2004): Comparison of mercury, lead and arsenic with respect to gene toxic effect on plant system and the development of genetic tolerance. Environmental and Experimental Botany 52, pg : 199-223.

Pierce M.L and Moore C.B. (1982) : Absorption of arsenite and arsenate on amorphous iron hydroxide. Water Resources 16, 1247-1253.

Porter E.K and Peterson P.J. (1977) : Arsenic tolerance in grasses on mine waste. *Environmental Pollution.*,Vol. 14, Issue 4, December, Pg : 255-265.

Salt D.E., Smith R.D., Raskin I., (1998) : Phytoremediation. Annu. Rev. Plant Physio. *Plant Mol. Biol.* 49, 643-683.

Singh A.K. (2007): Approaches for removal of arsenic from ground water of north eastern India. Current Science. Vol. 92, No. 11, June (2007).

Sorg T.J. and Logsdon G.S., (1978) : Treatment technology to meet the Interim Primary drinking water regulation for inorganic: Part 2. J. *American water works Association* 70 (7), 379-393.

Tu S, Ma L.Q., Macdonald G.E. and Bondada (2004) : Effects of arsenic species and Phosphorous on arsenic absorption, arsenic reduction and thiol formation in excised parts of *Pteris vittata* L. Environmental and Experimental Botany 51, Pp : 121-131.

Xia S, Dong B, Zhang Q, Xu B, Gau N and Causseranda C. (2007): Study of arsenic removal by Nanofiltration and its application in China. Desalination Vol. 204, Issue 1-3, Febrary, 374-379.

WEB REFERENCE

EPA (United States Environmental Protection Agency). (2002) : Arsenic treatment technologies for soil waste and water. www.epa.gov/tioclu-in.org/arsenic.

EPA (United States Environmental Protection Agency). (2003) : Environmental technologies verification program. www.epa.gov/etv/verifications/vcenter1-21.

Johnston, R, Heijnen. H and Wurzel P. (2000) : Safe Water technology. www.who.int/water_sanitation_health/arsenic/Arsenic UNRep7.html.

Rasmussen. L, and Anderson K, J. (2002) : Environmental health and human exposure. www.who.int/water_sanitation_health/arsenic/ArsenicUNRep2.html.

Wegelin, M, Gechter D, Hug S, Mahmud A, and Motaleb. (2000) : SORAS-A simple arsenic removal processes. phys4.harvard.edu/-wilson/mitigation/SORAS_Paper.

27

Assessment of Drinking Water Quality in the Vicinity of Industrial Area at Gajraula (U.P.), India

Pawan Kumar and D.S. Malik

ABSTRACT

The present study aims at identifying the ground water contamination in terms of physico-chemical characteristics of villages located in vicinity of Gajraula industrial area. Ground water samples were collected by hand pumps and bore well, based on the depth of 40 and 120 feet of water table respectively. Analytical techniques were adopted in the standard methods for examination of drinking water quality in ground water samples and results compared with the standards given by WHO and BIS guidelines for drinking water quality. Water quality index was calculated for quality standard of ground water for drinking purposes. The present results revealed that the water quality is moderately degraded due to high range of seven water quality parameters such as Temperature, Conductivity, TDS, Alkalinity, Ca-Hardness, Mg-Hardness and COD. The water quality index (WQI) also indicated the all the water quality ratings comes under the marginal category (45-64) *i.e.* water quality is frequently threatened or impaired and conditions usually depart from natural or desirable levels.

Keywords

Physico-chemical characteristics, Ground water, Industrial pollution, Water quality index.

INTRODUCTION

Gajraula is a prominent industrial area of western Uttar Pradesh, owes its significance to diverse group of industries, which includes distillery and its associated chemical units, pulp & paper, phosphate fertilizer plant, textiles, pharmaceuticals, dairy and other units. The industrial effluents contain toxic chemicals, hazardous compounds, suspended solids and non-biodegradable materials. The major source of surface and ground water pollution is injudicious discharge of untreated industrial effluents directly into the surface water bodies resulting surface and ground water pollution described by Nasrullah, *et al.* (2006) and Malik *et al.* (2009). The effluents from the industries and sewage of industrial towns have drained collectively through a local drain known as Bagad nallah, which assumes the shape of seasonal river, especially during monsoon. The quality of Bagad river water is a great concern to deteriorate the quality of the surface water and ground water due to releasing their effluent without any adequate treatment and polluting the upper water table through the leaching process in and around industrial area. The ground water quality of selected four village located around industrial area have been affected a lot and causes serious disease among the human beings and other livestock population. Most of the labour class society generally depends on ground water of hand pumps for drinking as well as other domestic utilities. Hence, some scientific remedial measures should be taken to improve the drinking water quality in and around of industrial area. The industrial effluents if not treated properly controlled, can pollute and cause serious damage to the ground water resources (Olayinka, 2004).

Water quality index (WQI) has provided an assessment of water quality trends for management purpose even though it is not meant especially as an absolute measure of the degree of pollution or the actual water quality (Anonymous, 1997). Horton, (1965) proposed the first time water quality index; a great deal of consideration has been given to bio-monitoring of pollution load in water bodies. The water quality index system is well known method of expressing water quality that offers a simple, stable and reproducing unit of measure, which responds to change in physical and chemical characteristics of water (Brown, *et al.* 1972). WQI is defined as a rating reflecting the composite influence of different water quality parameters and calculated from the point of view of the suitability of ground water for human consumption (Ramakrishnaiah, *et al.* 2009).

The present study is aimed for an impact assessment of industrialization on ground water quality in and around Gajraula industrial area and measured the WQI to find out the suitability of drinking quality of ground water for human consumption.

MATERIALS AND METHODS

Gajraula city is approximately 115 Km. away from Delhi on national highway NH-24, Delhi-Lucknow road. Gajraula is well known and one of the oldest industrial area of district J.P. Nagar (Uttar Pradesh) situated on globe at a longitude 78° 13′48.75′′ E and latitude 28° 50′59.26′′ N at 679 feet above sea level (207 msl). Ground water samples were collected by grab sampling method from hand pumps and bore well having the water table at depth of 40 feet and 120 feet respectively, from sampling sites selected at four villages located nearby Gajraula industrial area namely, Naipur **(S_1)**, Bhanpur **(S_2)**, Shahbazpur **(S_3)**, and Tigaria **(S_4)**, during summer, monsoon and winter seasons (April 2008 to March-09). The distances of villages from the outgoing industrial effluent drains were as **S_1**-(0.5-1.0 Km), **S_2**-(2.0-2.4 Km), **S_3**-(2.5-2.9 Km), **S_4**-(3.4-4.0 Km), approximately, located in vicinity of industrial area.

The ground water samples (n = 05) were collected from each of hand pumps **(HP no.-5)** and bore wells **(BW no.-5)** from each sampling sites during the particular seasons. The water samples were analyzed by standard methods (APHA, 1998). The results were compared with standard given by WHO (2004) and BIS (1991). Water quality index was calculated on the basis of fifteen physico-chemical parameters according to CCME water quality indexing criterion (2001).

The CCME Water Quality Index :

$$\text{CCMEWQI} = 100 - \left[\frac{\sqrt{(F_i)^2 + (F_2)^2 + (F_3)^2}}{1.732}\right]$$

Where,

F_1 = Number of failed variables/ Total number of variables × 100

F_2 = Number of failed test/Total number of test × 100

F_3 = (nse/001nse + 0.01)

The divisor 1.732 normalizes the resultant values range between 0 and 100, where 0 represents the "worst" water quality and 100 represents the "best" water quality.

Once WQI value has been determined, water quality is ranked by relating it to one of the following categories :

Rank	*CCME-WQI*	*Description*
Excellent	95-100	Water quality is protected with a virtual absence of threat or impairment; conditions very close to natural or pristine levels; these index values can only be obtained if all measurements are within objectives virtually all of the time.
Good	80-94	Water quality is protected with only a minor degree of threat or impairment; conditions rarely depart from natural or desirable levels.
Fair	65-79	Water quality is usually protected but occasionally threatened or impaired; conditions sometimes depart from natural or desirable levels.
Marginal	45-64	Water quality is frequently threatened or impaired; conditions often depart from natural or desirable levels.
Poor	0-44	Water quality is almost always threatened or impaired; conditions usually depart from natural or desirable levels.

RESULTS AND DISCUSSION

Ground water is mostly uses for drinking purpose. So it is important that the quality of ground water should be monitored and assessed the suitability of drinking water for consumption by human and livestock. The present study showed that the how much acceptable ground water quality of hand pumps and bore wells of four villages compared with drinking water standards.

The water of low temperature containing higher value of dissolved oxygen and low value of BOD and COD. Temperature of ground water ranged from 18.33-32.36°C in bore well and hand pump water at S_4 and S_1 respectively during winter and summer season. The influence in temperature may be due to different seasonal climatic variations and different sampling time (Parashar, *et al.*, 2008). pH value varied between range as 7.25-8.05, minimum and maximum pH were recorded in hand pump and bore well water at S_1 and S_2 during winter and monsoon season respectively. pH values of all water samples were within the permissible limit given by WHO. Electrical conductivity with regards to water quality referred to the amount of salts in the water and is a numerical expression of the ability of an aqueous solution to convey an electric current. It is also an approximate indicator of total dissolved ions such as heavy metals and widely used for pollution monitoring (Nasrullah, *et al.* 2006). The values of conductivity of ground water of bore well and hand pumps ranged from 925.45-1399.59 μmho/cm at S_3 during summer and winter season, respectively. Total dissolved solids exhibited turbid condition of water pollution strength. The values of total dissolved solids in ground water ranged from 610.80-923.73 mg/l, lowest and highest values of TDS were recorded in bore well and hand pump water at S_3 during summer and winter season respectively. Drinking water with high total dissolved solids generally showed inferior potable quality and induced an unfavorable physiological reaction in the transient consumer and gastrointestinal infections. (Kulshrestha, *et al.*, 2002) also observed the similar range (840-1050 mg/l) of total dissolved solids in tube well

Table 1 : Physico-Chemical Characteristics of Ground Water of Hand Pumps (HP)

Parameters	*Unit*	S_1			S_2			S_3			S_4			
Winter Standards		*Summer*	*Monsoon*	*Winter*	*Summer*	*Monsoon*	*Winter*	*Summer*	*Monsoon*	*Winter*	*Summer*	*Monsoon*	*Winter*	*Standard*
Temperature	°C	32.35	27.95	23.92	23.41	24.48	18.61	23.43	24.46	19.39	24.30	23.73	19.42	12*
pH	—	7.63	7.57	7.27	7.38	7.61	7.40	7.76	7.55	7.63	7.41	7.68	7.41	6.5-8.5*
Conductivity	µmho/cm	1227.00	1267.27	1278.11	1157.66	1221.26	1174.98	1206.70	1310.56	1399.59	1169.42	1287.19	1220.27	400*
TDS	mg/l	781.88	836.08	832.99	764.06	806.03	775.48	796.42	864.97	923.73	771.82	849.55	805.38	500*
Free-CO_2	mg/l	0.840	0.191	0.318	0.843	0.152	0.401	0.766	0.247	0.254	0.680	0.193	0.570	1.2
Alkalinity	mg/l	329.79	335.95	333.52	325.97	329.97	332.11	334.89	329.03	338.80	329.54	339.83	337.30	200
T. Hardness	mg/l	277.73	271.86	282.35	279.51	268.99	276.03	274.90	253.62	269.52	282.87	290.18	278.40	300
Ca	mg/l	181.17	178.27	183.78	175.33	164.73	164.63	172.16	152.67	163.20	170.67	176.68	174.39	75
Mg	mg/l	96.97	94.07	99.10	104.18	104.26	111.41	102.74	100.95	106.32	112.20	113.50	104.01	30
DO	mg/l	4.53	4.24	3.51	4.09	5.23	5.08	5.29	5.37	6.36	5.20	4.25	5.55	3*
BOD	mg/l	10.64	10.88	12.22	10.48	10.63	12.17	9.99	10.84	10.36	9.56	7.56	11.44	
COD	mg/l	22.43	22.42	25.62	22.55	22.62	25.29	22.56	23.71	24.54	23.17	23.04	25.26	10*
Chloride	mg/l	32.10	35.95	43.83	33.20	36.99	36.73	33.19	32.07	38.77	33.96	33.89	40.92	250
Potassium	mg/l	5.40	5.56	5.28	5.08	5.54	4.75	5.05	6.08	5.57	4.94	5.16	5.32	12*
Nitrate	mg/l	5.01	6.05	6.95	4.34	5.95	7.24	5.07	5.83	6.85	4.46	5.81	6.43	45
Nitrite	mg/l	1.06	1.75	1.93	1.07	1.66	1.87	1.10	1.55	1.88	1.13	1.65	1.89	45

Note : Each site (HP-no.5) showed mean value of five samples (n = 5 × 5). Standard value with *indicates WHO standards and without * indicates BIS.

Table 2 : Physico-Chemical Characteristics of Ground Water of Bore Wells (BW)

Parameters	*Unit*	S_1			S_2			S_3			S_4			
Winter Standards		*Summer*	*Monsoon*	*Winter*	*Summer*	*Monsoon*	*Winter*	*Summer*	*Monsoon*	*Winter*	*Summer*	*Monsoon*	*Winter*	*Standard*
Temperature	°C	23.05	26.58	20.76	24.30	24.26	19.93	27.69	24.77	19.58	21.78	3.74	18.33	12*
pH	—	7.61	7.38	7.96	7.44	8.05	7.71	7.87	7.49	8.04	7.49	7.90	7.26	6.5-8.5*
Conductivity	µmho/cm	1039.99	1006.69	1041.25	987.10	924.97	989.37	925.45	932.43	955.72	965.15	955.87	949.97	400*
TDS	mg/l	686.39	664.41	587.23	651.48	610.48	652.99	610.80	615.40	630.77	637.00	630.87	626.98	500*
Free-CO_2	mg/l	0.541	0.340	0.113	0.526	0.345	0.117	0.406	0.332	0.122	0.669	0.218	0.123	1.2
Alkalinity	mg/l	282.40	301.30	276.31	272.21	289.84	260.17	265.58	295.39	272.48	277.29	277.57	276.15	200
T. Hardness	mg/l	261.34	269.01	258.42	259.83	264.35	259.36	230.64	238.07	248.35	240.76	238.22	39.02	300
Ca	mg/l	161.13	166.43	159.64	159.78	164.17	156.43	129.68	135.20	131.54	139.32	142.79	138.22	75
Mg	mg/l	100.20	102.57	98.78	100.05	100.18	102.93	100.96	102.88	116.81	101.44	95.43	100.79	30
DO	mg/l	4.78	6.29	5.28	5.02	5.44	4.71	5.25	4.83	6.21	4.87	7.27	6.23	3*
BOD	mg/l	8.07	8.63	8.20	7.80	7.47	6.76	6.80	5.61	5.35	6.68	6.93	6.26	
COD	mg/l	16.09	16.16	16.17	15.53	15.19	13.99	15.39	15.44	14.21	14.84	15.52	14.85	10*
Chloride	mg/l-	20.98	22.43	20.23	19.91	22.76	21.37	20.12	23.48	22.69	24.41	23.02	23.76	250
Potassium	mg/l	4.04	3.26	3.27	3.86	3.25	3.85	3.85	3.56	3.99	3.83	3.15	3.35	12*
Nitrate	mg/l	2.47	3.10	2.44	2.35	2.30	2.79	2.10	2.62	2.16	1.85	2.61	1.94	45
Nitrite	mg/l	0.97	0.99	1.02	0.79	0.97	1.07	1.11	1.12	1.05	1.03	1.01	1.07	45

Note : Each site (BW-No. 5) showed mean value of five samples (n = 5 × 5). Standard value with *indicates WHO standards and without *indicates BIS.

water during summer season. Free CO_2 in ground water ranged from 0.113-0.843 mg/l, while the lowest and highest value of free CO_2 in bore well and hand pumps water was recorded at S_1 and S_2 during winter and summer season respectively. All the values of free CO_2 were within the permissible limit given by BIS. Alkalinity of water is strongly correlated with pH value, if pH of water sample is also high than alkalinity, if pH value low than alkalinity, it may shown lower trend. Alkalinity in ground water was recorded in the ranged between 260.17-339.83 mg/l in bore well and hand pump water at S_2 and S_4 during winter and monsoon season respectively. Water with more than 200mg/l (BIS) value of alkalinity, hence values of alkalinity have shown higher mark on permissible limit for potability of drinking water quality and contributed for degradation of ground water resources.

The high degree of water hardness can definitely be attributed to the disposal of untreated and improperly treated sewage and industrial wastes (Shanker, *et al.*, 2008). Total hardness of ground water samples were recorded in the ranged between 230.64-290.18 mg/l in bore well and hand pump water at S_3 and S_4 during summer and monsoon season, respectively. The values of Ca-Hardness in ground water varied between 129.68-181.17 mg/l in bore well and hand pump water at S_3 and S_1 during summer seasons. The similar trend of calcium hardness occurred in ground water also reported by Ramakrishnaiah, *et al.*, (2009) and significantly correlated hardness of drinking water with health of human beings. Mg-Hardness in ground water was recorded in the ranged between 94.07-113.50 mg/l in hand pump water at S_1 and S_4 during monsoon season, respectively.

Dissolved oxygen in ground water samples were recorded in the ranged between 3.51-6.38 mg/l in hand pump water at S_4 and S_3 during winter season respectively. The lower value of DO (3.5) at site-S_1 during winter season may be due to the lowest distance (500 meter) from the outgoing industrial drain which indicates organic pollution in water due to percolation of effluent containing soluble organic compounds and enhanced value 5.23 in monsoon due to the rainfall *i.e.* percolation of rain water towards ground water affects the ground water quality. Similarly enhanced value of DO 6.3 at S_3 and 5.5 at S_4 may be due to the higher distance 2-3 and 3-4 Km far from the outgoing industrial effluent drains. The difference between DO values in hand pump water and bore well water were found due to the depth level of ground water, bore wells having a least pollution in comparison to hand pump. In case of S_3 shown fluctuations in DO values in different seasons may be due to the depth of hand pump and bore well water. All the values of dissolved oxygen recorded under the permissible limit 3.0-14.0 mg/l given by WHO. Biochemical oxygen demand (BOD) is inter-related with the levels of dissolved oxygen in the water. Organic compounds such as carbohydrates, proteins and fats, which can be a result of urban run-off, domestic sewage and industrial effluent are broken down by the micro-organisms present in the water and exerted an oxygen demand in aquatic system (Efe, *et al.*, 2005). BOD values in water were recorded in the ranged between 6.26-12.22 mg/l in bore well and hand pump water at S_4 and S_1 during winter season respectively. Chemical oxygen demand (COD) represents the amount of oxygen required for chemical decomposition of organic matter in water. Amount of oxygen required for chemical decomposition of organic matter were recorded in the ranged between 13.99-25.62 mg/l in bore well and hand pump water at S_2 and S_1 during winter season respectively. All the values of COD recorded high comparison than standard value (10 mg/l) given by WHO. The dissolved oxygen, BOD and COD have not favored the permissible value of drinking water quality, which existed in ground water of villages located near to industrial area. Hence, the existed drinking water has been contaminated by the leaching industrial effluents and intermixing with ground water table.

Chloride values of ground water were varied from 19.91-43.83 mg/l in bore well and hand pump water at S_2 and S_1 during summer and winter season respectively. Excessive chloride imparts bitter taste to water and corrode steel and may cause cardio-vascular problem. Potassium in ground water samples shown ranged from 3.19-6.08 mg/l in bore well and hand pump water at S_4 and S_3 during monsoon season respectively. The values of potassium were recorded under the permissible limit given by WHO. Nitrate values in ground water samples ranged from 1.85-7.24 mg/l in bore well and hand pump water at S_4 at S_2 during summer and winter season, respectively. Water with more than 45 mg/l of nitrate is not permissible for drinking as per BIS and limit is mandatory. Beyond this, it may cause metheamoglobinemia or blue baby disease in infants. It may also be carcinogenic in adults (Basappa, 2003). Nitrite values in ground water samples were recorded in the ranged between 0.79-1.93 mg/l in bore well and hand pump water at S_2 and S_1 during summer and winter season respectively. All the values of nitrite come under the standard value given by BIS.

Water quality index calculated on the basis of existed seasonal physico-chemical characteristics of ground water in all sampling sites. WQI ratings were measured in the range of 54.97-57.58 on S_1 site, lowest and highest value of WQI in hand pump and bore well water recorded during monsoon and winter season respectively. WQI ratings were measured in the ranged from 55.93-57.94 on S_2 site, the lowest and highest value was recorded in hand pump and bore well water during monsoon and winter season respectively. Water quality ratings were calculated in the ranged from 55.55-57.88 on S_3, the minimum and maximum value of WQI recorded in hand pump and bore well water during winter seasons respectively. WQI ratings were recorded in the range from 55.40-58.28 on S_4, minimum and maximum WQI values recorded in hand pump and bore well water during monsoon and winter season respectively. Water quality indexing values come under the marginal category in drinking water of hand pump and bore well at all sampling sites. Line diagram (Fig. 1) clearly indicated that quality of hand pumps water at the depth 40 feet is slightly critical quality of bore well water at 120 feet depth from earth surface.

The deterioration of ground water quality occurred due to leeching processes of surface industrial waste water in water aquifers. At present most of the industries lacking of either independent waste water treatment facilities or common treatment plant for purification of industrial waste water. The ground water resources must be non-contaminated, purified and treated efficiently on sustainable basis at point and non point sources of pollution level. The drinking water quality should be properly maintained and be available for human society to get healthy longer life.

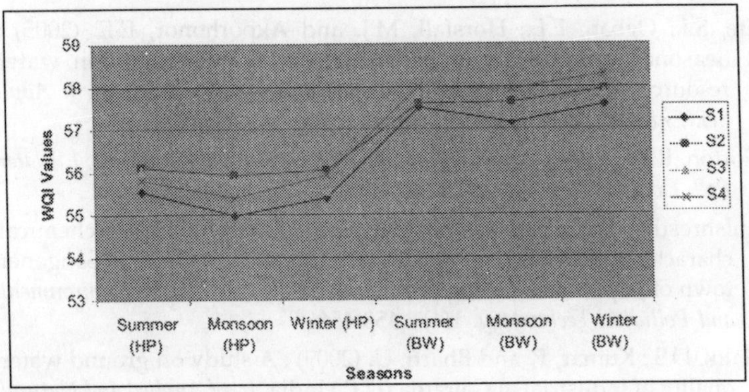

Figure 1 : WQI rating of ground water quality of hand pump and bore well.

ACKNOWLEDGEMENT

The authors are thankful to Prof. A.K Chopra, Registrar, Prof. B.D. Joshi, Co-ordinator of UGC (SAP) and Prof. D.R. Khanna, Head, Department of Zoology and Environmental Science for the facilities provided to carry out the work successfully. The author (Pawan Kumar) is gratefully acknowledged the financial support from UGC, Delhi.

REFERENCES

Anonymous (1997) : A Hach, Water Analysis Handbook, 3rd *Edn USA*: Hach Company pp : 1-45.

APHA (1998) : Standard methods for the examination of wastewater, *American Public Health Association, Washington, D.C.* pp : 1190.

Basappa, R. M. (2003): Status of ground water quality in Bangalore and its environment. Report, Department of Mines and Geology, Bangalore. pp. 156

BIS (1991) : Bureau of Indian Standards IS : 10500, Manak Bhavan, New Delhi, India.

Brown, R.M.; McLelland, N.; Deininger, R.A. and O'Connor, M.F. (1972) : A water quality index, crashing the psychological barrier. *Indicators of Environmental Quality*, pp : 455

CCME (2001) : Drinking water quality index manual, Canadian Water Quality Index, 1.0., pp : 4.

Efe, S.I.; Ogban, F.E.; Horsfall, M.J. and Akporhonor, E.E. (2005) : Seasonal variation of physico-chemical characteristics in water resources quality in western Niger Delta Region, Nigeria, *J. Appl. Sci. Environ. Mgt.* 9(1) : 191-195.

Horton, R.K. (1965) : An index system for rating water quality, *J. Water Poll. Contam. Res. Fed.*, 37 : 300-305.

Kulshrestha, S., Dhindsa, S.S. and Singh (2002) : Physico-chemical characteristics of underground water and effluent water in Sanganer town of Jaipur City during pre-monsoon season, *Nature Environment and Pollution Technology*. 1(4) : 453-456.

Malik, D.S.; Kumar, P. and Bharti, U. (2009) : A study on ground water quality of industrial at Gajraula (U.P.) India, *Jr. of Applied and Natural Science*, 1(2) : 275-279.

Nasrullah., Naz, R.; Bibi, H.; Iqbal, M. and Durrani, M.I. (2006) : Pollution load in industrial effluent and ground water of Gadoon Amazai Industrial Estate (GAIE) Swabi, NWFP, *Journal of Agriculture and Biological Science*, 1(3) : 18-24.

Olayinka, K.O. (2004) : Studies on industrial pollution in Nigeria- the effect of textile effluents on the quality of ground water, *Nigerian Journal of Health and Medical Sciences*, 3 : 44-50.

Parashar, C.; Verma, N.; Dixit, S. and Shrivastava, R. (2007) : Multivariate analysis of drinking water quality parameters in Bhopal, India, *Environ. Monit. Assess.*, 140 : 119-122.

Ramakrishnaiah, C.R.; Sadashivaiah, C. and Ranganna, G. (2009): Assessment of water quality index for the ground water in Tumukpur taluk Karnataka state, India, *E.J. of Chem.*, 6(2) : 523-530.

Sastry, K.V. and Rathee, P. (1999) : Ground water quality in three villages of Rohtak district. *J. Natcon.*, 11(2) : 175-182.

Shanker, B.S.; Balasubramanya, N. and Reddy, M.T.M. (2008) : Impact of industrialization on ground water quality- a case study of Peenya industrial area, Bangalore, India, *Environ. Monit. Assess.*, 142 : 263-268.

WHO (2004) : Guidelines for Drinking Water Quality. Vol.1-4, Geneva, pp : 515.

28

Water Quality Assessment of River Ganga for Irrigation in Garhmukteshwar (U.P.)

Praveen Chauhan

ABSTRACT

This chapter is an attempt to analyze the water quality of river Ganga at Garhmukteshwar for irrigation purpose. Water samples were collected from five sampling stations. The study area has been divided into three seasons. Winter (Nov-Feb), summer (Mar-June) and rainy (July-Oct). Water quality variables were measured in the river over a period of one year (2007). The samples were analyzed for Electrical Conductivity (EC), total dissolved solids (TDS), Magnesium Content (MC), Sodium Absorption Ratio (SAR), Sodium Percent (SP) and Permeability Index (PI). Study of all these characteristics indicates that river water in rainy season is not suitable for irrigation purpose because of high values of total dissolved solids, electrical conductivity and sodium percent.

Keywords

Ganga water, Garhmukteshwar, Irrigation Water Quality, SAR, PI.

INTRODUCTION

The nature of water described in the ancient Indian Scripture Rig-Veda translates as : 'The water which are from heaven and which flow after being dug, and even that spring by themselves, the bright pure water which lead to the sea, may those divine waters protect me here'. The universe is constituted of five basic elements: earth, water, light, air and ether. According to Rig Veda all life is evolved from water. Water is the first need of living creatures in all ages and it is regarded as one of the essential factors for human existence. Water is one of the most abundantly available resources in nature. Water is distributed in nature in different forms such as rain water, spring water, mineral water and ground water. Water is one of the widely distributed and abundant substance found in the nature, it covers about 75% of the earth surface therefore earth is sometimes called a "Water Planet" and it occurred in sea, river, ocean, pond, streams and even in the atmosphere in the form of humidity. Most of the earth surface water is in the sea or ocean contain about excessive quantities of salts. Fresh water occured in lakes, ponds, rivers and steams and is also available in the forms of rain, snow, dew etc.

Rivers play a major role in assimilation or carrying off the municipal and industrial waste water and run off from agricultural land. The municipal and industrial waste water discharge constitutes the constant polluting source, whereas, the surface run off is a seasonal phenomenon, largely affected by climate in the basin. Seasonal variations in precipitation, surface run off, interflow, ground water flow and pumped in and outflows have a strong effect on river discharge and subsequently on the concentration of pollutants in river water (Vega, *et al.* 1998). Rivers are the most important water resources in the world in general and India in particular, great civilization developed along the bank of rivers and even today most of the development has taken place in the cities or in the location near the rivers. The rivers provide water for industry, agriculture commercial aquaculture and domestic purposes. Unfortunately the same rivers are being polluted by indiscriminate disposal of sewage and industrial wastes as a plethora of human activities. River pollution has already acquired a serious dimension in India, with most of its fourteen rivers being grossly polluted. The situations are more or less same in small rivers and tributaries. The available natural freshwater resources today are threatened by hazards of pollution; particularly rivers are greatly polluted due to release of untreated effluents and waste materials from agriculture concerns and industries located around rivers. The poor living conditions of people in settlements around rivers, non availability of treatment from urban areas and negligence of industries for treatment of effluents before release in natural water bodies are the major reason for pollution of Indian rivers (Chavan and Wagh, 2005).

River Ganga, the holiest river of all rivers and lifeline of the north India originates from Goumukh glacier. It emerges after confluence of main streams of the two important rivers of hills, river Bhagirathi and river Alaknanda at Devprayag. Beside this Ganga's head streams Mandakini, Nandakini, Bhilangana, Dhauliganga and Pinder all originate from northern Himalayas. After descending 2,827 meters at Haridwar, the river Ganga cuts across the Shivalik hills and for the first time it enters in great plain of the Uttarakhand state. Therefore, Haridwar is known as the gate way of God or Devbhumi. Ganga river flows down from Haridwar towards south and then south-east touching many important cities and towns like Kanpur, Allahabad, Varanasi and lastly terminates in the Bay of Bengal covering about 2,506 km in India.

The main stream of the river Ganga, the Bhagirathi originates from the ice cave of Goumukh (33Ú 55′ N, 79Ú 07′E) at the snout of Gangotri glaciers in the Garhwal portion of western Himalyas, at an elevation of about 4100 meters. Mana pass, 8 Km. away from Badrinath joints it at Devprayag. It is below this confluence, the united stream is known as Ganga river. After a run of some 267 Km, it emerges into Indo-gangetic plains at Haridwar where it swells into a mighty system, 750 m wide (approximately). The water quality of river is deteriorated not only by the encouraging waste water discharges but also due to excessive abstraction of water from river mainly for irrigation rendered then dry or with meager flow in non-monsoon periods followed by increasing waste water. Load has left many river stretches in alarming condition.

One of the important problems arising out of improper disposal of untreated industrial waste is river pollution. The pollution of river water adversely affects the various water uses to which river may be put to the bad effect of pollution of water courses by wastes are many and far reaching. Water pollution ultimately leads to national loss directly or indirectly (Sastry *et al.* 1972). The increasing demand on water arising from fast growth of industries has put pressure on limited water resources. While most people in urban cities of developing countries have access to piped water, several others still rely on borehole and river water for domestic use. Most of the rivers in the urban areas of the developing world are the end point of effluents, if not treated and properly controlled can also pollute surface and ground water (Phiri, *et al.* 2005).

Ganga river passes down through district Ghaziabad of Uttar Pradesh state of India, where it touches several towns which are situated at its banks. It has been studied that no more work done on Ganga has been conducted at district Ghaziabad. Therefore, it is supposed to carryout investigations on river Ganga in the region of district Ghaziabad where no such studies have been done. Keeping the above in view, the present study has been under taken to determine the existing pollution load in terms of physico-chemical, persistent heavy metals and microbiological characteristics are great concern for deterioration of the water quality in river Ganga at Garhmukteshwar in the District Ghaziabad (Uttar Pradesh) River Ganga passes through the towns Sahpur Kotla, Garhmukteshwar Ghat, BrijGhat, Mohiuddinpur and its down streams Pooth in district Ghaziabad where main bathing festivals of Hindus are organized.

The District Ghaziabad (including Noida and Greater Noida now part of District Gautam Budh Nagar) covering an area of 2590 sq. Km lies between 28° 26' and 28°54' North latitudes and 77°12' and 78°13' East longitude. The shape of the District resembles an uprooted tree with its canopy being towards west and trunk towards east touching Ganga river. The District is located in the extreme western part of the Uttar Pradesh State and is bounded by District Meerut in the north, and in the south by District Bulandshar, River Ganga forms the natural boundary in the east, separating it from District Moradabad. River Yamuna forms the western boundary in parts.

The district of Ghaziabad lies in the doab of famous rivers of northern India, the Ganga and the Yamuna. It is bounded on the east by district Moradabad and on the west by the Union Territory of Delhi and are separated by the rivers Ganga and Yamuna in the east and west respectively. In the north and south, the district boundary extends to the boundaries of Meerut and Bulandshar districts respectively. Total area of the district is 2594 sq. Kms.

This study was performed at five stations starting from sampling station A (Shahpur Kotla-situated at the right bank of river Ganga flowing as entry point through the boundary line at dist. Ghaziabad) sampling station B (Garhmukteshwar Ghat-situated in the west direction on the right bank of river Ganga) sampling station C (Brijghat-situated on the right bank of river river Ganga), sampling station D (Mohiduinpur- At this station small cottages tanneries drain their waste water directly in river Ganga), sampling station E (Pooth-this station has historical cultural, mythological and archeological significance). This sampling station has some small industrial units manufacturing steel, brass and leather items. The industrial waste cum sewage from small houses is also flowed into the river. Throughout the year, faecal matter, decaying leaves and flowers, wooden parts, wasteclothes could be observed in the Ganga at this point. Few seasonal nullahs also open into the river.

Irrigated agriculture is dependent on an adequate water supply of usable quality. In irrigation water, quality characteristics were mentioned on the chemical and physical characteristics of the water and only rarely are any other factors considered important. Here attempt has been made to assess the irrigation water quality of river Ganga at Garhmukteshwar. The quality characteristics studied in the present investigations are as follows: Total dissolved solids (TDS), Electrical Conductivity (EC), Magnesium Content (MC), Sodium percent (SP), Sodium absorption ratio (SAR), and Permeability Index (PI).

EXPERIMENTAL METHODOLOGY

Water samples were collected for one year (2007) for three seasons *i.e.* summer, rainy and winter. 90 samples were collected from five sampling stations located within a radius of 25Km. Water samples were collected in plastic jerry canes. Grab sampling was generally applied during the sampling. Water samples were analyzed by standard methods. MC, SP, SAR and PI were calculated as follows :

Magnesium Content (MC)

$$\text{Mg Content} = [Mg^{2+}/(Mg^{2+} + Ca^{2+})]$$

Sodium Percentage (SP)

$$Na\% = [(Na^{+} + K^{+}) / (Ca^{2+} + Mg^{2+} + Na^{+} + K)]\ 100$$

Sodium Absorption Ratio (SAR)

$$SAR = Na^{+}/[(Ca^{2+} + Mg^{2+})/2]$$

Residual Sodium Carbonate (RSC)

$$RSC = (CO_3^{2-} + HCO_3^{-}) - (Ca^{2+} + Mg^{2+})$$

Permeability Index (PI)

$$PI = [(Na^{+} + HCO_3^{-})/(Ca^{2+} + Mg^{2+} + Na^{+})]\ 100$$

RESULT AND DISCUSSION

The results obtained from analysis of water samples of river Ganga are shown in Tables 1-3.

Table 1 : Irrigation Water Characteristics in Winter Season 2007

Parameter	*S-A*	*S-B*	*S-C*	*S-D*	*S-E*
EC	211.66	223.33	305.00	268.00	389.00
TDS	170.66	200.00	199.66	176.80	216.70
MC	29.41	30.70	34.86	29.14	25.01
SP	30.11	29.24	27.02	37.13	42.66
SAR	71.00	57.57	68.26	74.40	66.84
PI	110.51	75.16	65.51	90.01	100.53

Table 2 : Irrigation Water Characteristics in Summer Season 2007

Parameter	*S-A*	*S-B*	*S-C*	*S-D*	*S-E*
EC	272.66	265.66	283.00	316.00	323.00
TDS	173.33	250.00	186.66	208.50	230.10
MC	28.81	30.01	35.61	37.52	32.05
SP	24.45	23.02	24.52	27.20	33.16
SAR	50.00	53.71	51.35	55.47	47.72

Table 3 : Irrigation Water Characteristics in Rainy Season 2007

Parameter	*S-A*	*S-B*	*S-C*	*S-D*	*S-E*
EC	311.66	325.00	343.00	343.66	324.33
TDS	205.33	268.66	226.33	226.00	243.00
MC	30.70	34.06	36.60	35.70	33.05
SP	41.52	41.21	42.80	50.56	51.24
SAR	67.28	60.83	85.68	81.45	70.32
PI	78.48	74.28	72.42	88.22	94.38

Total Dissolved Solids (TDS)

Water used for irrigation can vary greatly in quality depending upon type and quantity of dissolved salts. Salts are present in irrigation water is relatively small but significant amounts. They originate from dissolution or weathering of the rocks and soil, including dissolution of lime, gypsum and other slowly dissolved soil minerals. These salts are carried with the water to wherever, it is used. In the case of irrigation, the salts are applied with the water and remain behind in the soil as water evaporates or is used by the crops. A salinity problem exists if salt accumulates in the crops root zone to a concentration that causes a less in yield. Water with TDS less than 450mg/l is considered good and that with greater than 2000mg/l is unsuitable for irrigation purpose. In the resent study, the minimum value of TDS was found in winter season with a range of 70.66 to 216.00ppm. The maximum value was found in rainy season with a range of 205.33 to 268.66. Same trend of temperature was observed by Khanna (2001) in river Ganga at Haridwar. More or less a similar status of temperature has been reported by Badola and Singh (1981) in river Alaknanda and in river Kallayi by John (1976). Similar variations were also observed by Singh, *et al.* (1989) in river Ganga, Yamuna and Sangam at Allahabad. More beyond or less of temperature may affect the physico-chemical and other biotic characteristics of running riverine system.

Electrical Conductivity (EC)

Electrical conductivity determines the total amount of ionisable salts in water. It is due to ionization of dissolved inorganic solids. The seasonal fluctuations in this factor were caused mainly by variations in the ionic precipitations and the diluting effect of rains (Welch, 1948). The primary effect of high electrical conductivity water on crop productivity is the inability of the plants to compete with ions in the soil solution for water. The higher the electrical conductivity, the less water is available to plants, even though the soil may appear wet. Because plants can only transpire "pure" water, usable plant water in the soil solution decreases dramatically as electrical conductivity increases. The least value of conductivity was during winter season as compared to other seasons. The maximum conductivity was recorded in rainy season which may be due to run off of surface water from agricultural lands, industrial effluents and sewage drains etc. Similar trends were recorded by Singh, *et al.*, (1989) in river Ganga at Allahabad. CPCB (2003) also reported conductivity from 0.517 Mmhos to 0.641 Mmhos in the stretch of Bithur, Kanpur to Allahabad.

Magnesium Content

Magnesium content of water is considered as one of the most important qualitative criteria in determining the quality of water for irrigation. Generally, calcium and magnesium maintain a state of equilibrium in most waters. More magnesium in water will adversely affect crop yield as the soils become more alkaline. In the present study, the magnesium content of the water of river varies from 25.01mg/l to 37.52mg/l. So, water is suitable for irrigation purpose in terms of magnesium content. Chugh (2000) reported the same trend of Mg-Hardness in river Ganga.

Sodium Percent

Sodium percent is another important factor to study sodium hazard. This parameter judge the quality of water for the use of agricultural purpose. Use of high percentage sodium water for irrigation purpose stunts the plant growth. Sodium reacts with soil to reduce its permeability. Sodium percent in water is a parameter computed to evaluate the suitability for irrigation. When SP > 15%, reduced permeability will occur. The finer the soil texture and the greater the organic matter content, the greater the impact of sodium on water infiltration and aeration. The sodium percent values of river were 23.02 to 51.24.

Gypsum can be added to the soil to reduce the effect of high percentage of sodium in irrigation water.

Sodium Absorption Ratio (SAR)

Excess sodium in water produces the undesirable effects of changing soil properties and reducing soil permeability. Hence the assessment of sodium concentration is necessary while considering the suitability for irrigation. Sodium absorption ratio is an important parameter for the determination of suitability of irrigation water because it is responsible for the sodium hazard. The degree to which irrigation water tends to enter into cation-exchange reactions in soil can be indicated by the sodium adsorption ratio. Sodium replacing adsorbed calcium and magnesium is a hazard as it causes damage to the soil structure. In the present study SAR was found to be 47.72 to 85.68.

Permeability Index (PI)

The soil permeability is affected by long term use of irrigation water. Sodium, Calcium, Magnesium and Carbonate content of the soil influence it. Doneen evolved a criterion for assessing the suitability of water for irrigation based on the permeability index. Accordingly, water can be classified as class I, class II and class III orders. Class I and Class II water are categorized as good for irrigation with 75% or more maximum permeability. Class II water are unsuitable with 25% of maximum permeability. In the present study the minimum value of PI is 65.51. Hence water of River is of good irrigation quality. In the present study, TDS, MC, SP, SAR and PI were found to be within permissible range except in rainy season. In rainy season TDS and PS were a little higher. Hence irrigation water quality of river was found to be good except in rainy season.

CONCLUSION

In the present study MC, SAR and PI was found in the permissible range for Ganga river water. But EC, TDS are higher in rainy season. Hence, Ganga river water is suitable for irrigation purpose except in rainy season.

REFERENCES

APHA (1998) : Standard methods for examination of water and waste water, American Public Health Association, Washington D.C.

Chavan, T.P. and Wagh, S.B. (2005) : Physico-chemical characteristics of industrial effluent near Sukhna river at midc area, Chikalthana, Aurangabad. *J. of Industrial Pollution Control* 21(1) pp 23-26.

Chopra, A.K. and Rehman, A. (1995) : A study of self purification of physico-chemical properties of Ganga canal at Jawalapur, Haridwar. *Him. J. Env. Zool,* **9** : 11-13.

Chugh, Tarun (2000) : Seasonal variation in the microbial Ecology of river Ganga at Haridwar. Ph. D. Thesis, G.K.V. Hardwar. pp : 49-147.

D.R. Khanna, S. Singh, A. Gautam and J.P. Singh, (2003). India, *J. Nat. Con.* 15, 165.

D.R. Khanna (1993). Ecology and pollution of Ganga River, Ashish Publishing House, Delhi.

H.M. Raghunath (1987). Groundwater Wiley Eastern Ltd, Delhi India, 563.

Joshi, D.; Kumar A. and Agrawal, N. (2009) : Assessment of the irrigation water quality of river Ganga: *Rasayan J. Chem.* Vo. 2 : 285-292.

Manual on water and waste water analysis, NEERI Publications (1988).

N. Manivaskam (1996). Physico-chemical examination, sewage and industrial, Pragati Prakashan Meerut.

P.K. Trivedi and P.K. Goel (1986) : Chemical and biological methods for water pollution studies, Env. Publication, Karad.

S.N. Biswas, H. Mohabey and M.L. Malik (2002). Asian J. Chem., 16.

S.S. Ahmed, Q.H. Mazumder, C.S. Jahan, M. Ahmed and S. Islam (2002). Journal, Geological society of India, 60, 411.

S.S. Pathak., K.K. Upadhyaya and S.K. Joshi (2002). Him. J.Env. Zool, 151.

Sundaray, S.K.; Nayak, B.B. and Bhatta, D. (2008) : Environmental studies on river water quality with reference to suitability for agricultural purposes : Mahanadi river estuarine system, India-a case study. *Environ Monit Assess.* DOI 10.1007/s 10661-008-0431-2.

Vega, M.; Pardo, R.; Barrado, E. and Deban, L. (1998) : Assessment of seasonal and polluting effects on the quality of river water. Water Res. **32** : 3581-3592.

29

Studies on Impact of Urban and Industrial Wastewater on Water and Sediment Quality of River Gomti at Lucknow, India

Sanjay Kumar Gupta, Virendra Mishra, Krishna Gopal and Jaswant Singh

ABSTRACT

The surface water pollution issues have been enlisted as one of the most serious problems in developing countries. The river Gomti one of the major tributary of Ganga, received pollution load both from the point and non point sources. In the present study, the impact of urban and industrial activities on the Gomti river water and sediment quality was assessed at seven different locations thoughout the river from upstream to downstream at Lucknow. The comprehensive evaluation of the river water quality was done by analyzing its physico-chemical parameters such as pH, turbidity, conductivity, TH, TA, TSS, TDS, TS, DO, BOD, COD, TOC and major ions such as fluoride, chloride, sulphate, nitrate and phosphate. Heavy metals such as Cd, Cr, Pb, Ni, Mn, Fe and Zn were analysed in water and sediment both. Assessment of heavy metal contamination in river sediment was done by calculation of the iron normalized metal enrichment. To assess the anthropogenic influence in the study area, the evaluation of sediments contamination with heavy metals was made using the index of accumulation, which permits assessment of the extent of sediment contamination with respect to the global standards. The concentration of Cd,

Cr, Cu and Fe in the river water, were in the range of ND – 0.006, 0.033-0.336, 0.01-0.041 and 0.228-2.845 mg/l respectively and Mn, Ni, Pb and Zn were in the range of 0.009-0.050, 0.031-0.06, 0.003-0.048 and 0.056-0.090 mg/l respectively. In the monitoring of river sediment, the Cd was recorded in the range of ND -0.006 mg/l and detected in river sediment of only at two sites *i.e.* Mohan Meakins and Martyrs Memorial. The Cr, Cu, Fe, Mn, Ni, Pb and Zn were recorded in the range of 0.033-0.336, 0.010-0.041, 0.228-2.845, 0.009-0.050, 0.031-0.060, 0.003-0.048 and 0.056-0.071 µg/g respectively.

Keywords

Gomti river, Heavy metals, Water quality, Geo- accumulation index.

INTRODUCTION

Today water resources have been the most exploited natural system since man strode the earth. A considerable amount of wastewater from majority of urban and industrial activities, released in to water resources either untreated or adequately treated. This has created a problem of surface and subsoil water pollution (Chapman, 1996). Rapid population growths, land development along river basin, urbanization and industrialization have subjected the rivers to increase stress, giving rise to water pollution and environmental deterioration (Sumok, 2001). India's finite and fragile water resources are stressed and depleting due to mainly anthropogenic factors on one hand and demand from various sectors ranging from agriculture to industry which are growing rapidly, on the other hand. Water is rapidly becoming a scarce resource in India, yet continues to be used inefficiently (Rao, 2004). The geographical area of India is 3,287,590 sq km. The length of its coastline is about 7500 km. There are thirteen major river basins (area more than 20,000 sq km) in the country, which occupy 82.4% of total drainage basins, contribute 85% of total surface flow and house 80% of the country's population. Most of the rivers in the urban areas of the developing world are the end point of effluents discharged from the industries (Phiri *et al.* 2005).

The river Gomti one of the major tributary of Ganga, originates from a natural reservoir in the swampy and forested area (elevation of about 200 m; North latitude 28° 34′ and east longitude 80° 07′) about 3 km east of Pilibhit in Uttar Pradesh and about 50 km south of the Himalaya foot-hills. The catchment area of river Gomti is about 25,800 Km^2 and during its whole journey of about 730 Kms. it is flowing through the districts of Pilibhit, Shahjahanpur, Sitapur, Lucknow, Barabanki, Sultanpur, Jaunpur and Ghazipur in Uttar Pradesh and finaly confluence's with the Ganga river in Ghazipur about 30 km north of Varanasi. River Gomti receives urban wastewater, agricultural runoff and industrial influents both from the point and non point sources, directly or through its tributaries and 42 wastewater drains mainly from four major townships Lucknow, Jagdishpur, Sultanpur and Jaunpur. A barrage is also constructed on Gomti river in Lucknow City to restricts its water flow with a view to maintain the water level to facilitate water intake at Gaughat for domestic supplies.

In the past two decades much attention has been paid to the status of water as well as sediment quality of the river Gomti (Singh, *et al.* 1997, Gupta and Subramanian, 1998, Singh *et al.* 2004, Singh *et al.* 2005, Gaur *et al.* 2005). In the present study, the impact of urban and industrial activities on the Gomti river quality was assessed at Lucknow by evaluating its physico-chemical parameters such as temperature, pH, turbidity, conductivity, total hardness (TH), total alkalinity (TA), total suspended solids (TSS), total dissolved solids (TDS), total solids (TS), dissolved oxygen (DO), biological oxygen demand (BOD), chemical oxygen demand (COD) and total organic carbon (TOC) and major ions such as fluoride (F), chloride (Cl), sulphate (SO_4), nitrate (NO_3), phosphate (PO_4) of the river water as well as heavy metals such as cadmium (Cd), chromium (Cr), lead (Pb), nickel (Ni), manganese (Mn), iron (Fe) and zinc (Zn) at six different locations. The measurement of pollution in water is not conclusive due to water discharge fluctuations and low residence time. The same holds true for the suspended material, therefore the river sediments are the very important sources for the assessment of natural as well as anthropogenic contaminations (Forstner and Wittmann, 1983; Muller, 1989). Heavy metals constitute a special group of contaminants of aquatic systems and deserve special attention. Since, metals are not removed by natural degradation processes; they may become enriched in sediments over time (Phillips, 1977; Lee *et al.* 2006). Metal contamination of sediments is an issue of growing concern worldwide (NRC, 1989). Both natural processes and anthropogenic activities are responsible for introducing metals in to the aquatic systems (Nriagu, 1990; Wong *et al.* 2006). Metals in aquatic systems become part of the water-sediment system and their distribution is controlled by a dynamic set of physical-chemical interactions and equilibria, largely governed by pH and type of legends and chelating agents, oxidation state of the mineral components and the redox conditions of the system (Sorme and Lagerkvist, 2002). Metal contaminated sediments may release heavy metals back to the overlying water column and thus, pose risk to aquatic life and ecosystems. Due to their particle reactivity, heavy metals tend to accumulate in sediment and as a result may persist in the environment long after their primary sources has been removed (Frostner and Wittman, 1981; Peakall and Berger, 2003; Tang *et al.* 2008). Metal accumulation in sediments provides a record of the spatial and temporal history of pollution (Jardo and Nickless, 1989; Heyden and New, 2004). Hence, sediment monitoring can provide important information on various pollution events. In view of the increasing pollution load and consequent deterioration of the Gomti river water quality, the present study was undertaken with a view to assess the extent and rate of metal pollution with special reference to the geochemical distribution patterns of different metals in water and sediments.

MATERIALS AND METHODS

Study Area

The Gomti river, one of the major tributaries of the Ganga river system in northern India has been selected for this study. In the present study, total seven sites (Figure), namely Gaughat

(S-1), Mohan Meakins (S-2), Martyr's Memorial (S-3), Hanuman Setu (S-4), Nishat Ganj (S-5), Pipraghat, (S-6) and Malhaur D/S (S-7), were selected to assess the impact of localized domestic and industrial effluents in the water and sediments of the Gomti river. Gaughat was the first sampling sites because from here river water is used as raw water for drinking purposes. Second site was Mohan Meakin at this site effluent comes from Brewery Industry and city. Third site was selected due to its importance as Martyr's Memorial, which is also used for recreation. Fourth site was selected due to that place is used for relegious purposes (Immersion of idols etc.) and learning of swimming two swimming clubs are situated here. The fifth site *i.e.* Nishatganj Bridge was selected due to fall of two big nala on the both side of Nishatganj bridge. These nalas length is too much and carrying the domestic sewage and waste water from both sides of city. The sixth site was Pipraghat, at this site a crematorium is situated and here domestic waste comes from cantonment area. The seventh site was Malhaur, it is the outer part of the city and population is very less and like other site no such big nalas are meeting to the river Gomti. The first site (S-1) is located in the area of relatively low river pollution and is upstream of the Lucknow city. Other three sites (S-2 to S-5) are located in the region of high river pollution as there are a number of wastewater drains (24 nos.) and two highly polluted tributaries emptying into the river in this stretch. The last two sites (S-6 to S-7) are in the downstream region of moderate pollution as the river considerably recovers in the course. The description of sites chosen for study :

1. Gaughat, U/S Lucknow (S-1) : This location is just upstream of the Lucknow city near the water supply intake point from the river. It is about 5 km upstream of Indian Institute of Toxicology Research (IITR) formerly ITRC. All the three locations are relatively pollution free as there is no major point pollution source discharging into the river in the upstream catchment, except the Sarayan river and seasonal sugar mill effluents drain.

2. D/S Mohan Meakins, Mid Lucknow (S-2) : This location on the river is downstream of the Mohan Meakins distillary in the central city. It is about 1 km upstream of IITR. Between Gaughat and Mohan Meakins (D/S) locations, there are several outfalls/drains discharging into the river directly.

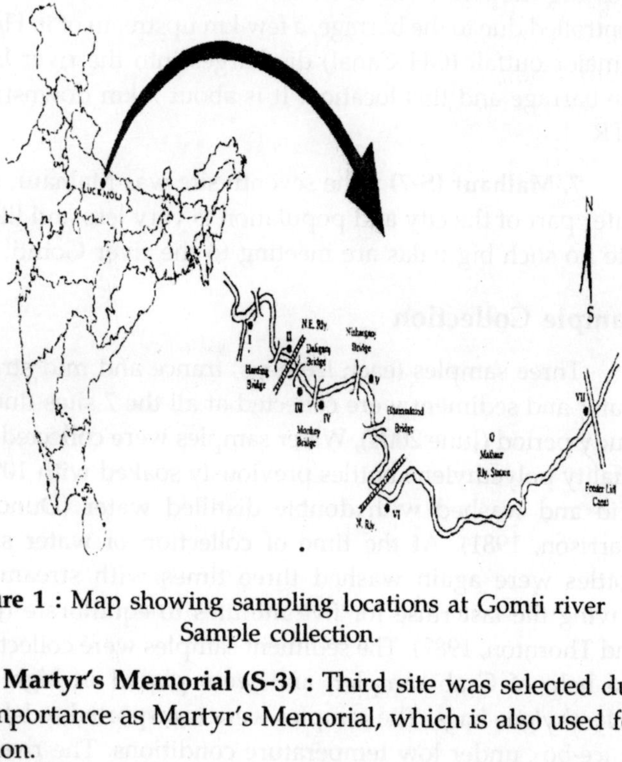

Figure 1 : Map showing sampling locations at Gomti river Sample collection.

3. Martyr's Memorial (S-3) : Third site was selected due to its importance as Martyr's Memorial, which is also used for recreation.

4. Hanuman Setu (S-4) : Fourth site was selected due to that place is used for religious purposes (Immersion of idols etc.) and learning of swimming, two swimming clubs are situated here.

5. Nishatganj Bridge (S-5) : The fifth site *i.e.* Nishatganj Bridge was selected due to fall of two big nala on the both side of Nishatganj bridge. These nalas length is too much and carrying the domestic sewage and waste water from both sides of city.

6. Pipraghat, D/S of Lucknow (S-6) : This location is situated in the Cantonment area, the downstream of Lucknow city and is just after the Lucknow-Sultanpur Railway line crossing the river. The river flow at this location is reduced/ controlled due to the barrage, a few km upstream of it. However, a major outfall (GH Canal) discharges into the river between the barrage and this location. It is about 7 km downstream of IITR.

7. Malhaur (S-7) : The seventh site was Malhaur, it is the outer part of the city and population is very less and like other site no such big nalas are meeting to the river Gomti.

Sample Collection

Three samples (each from sis, trance and mid stream) of water and sediment were collected at all the 7 sites during the study period (June 2004). Water samples were collected in high quality polyethylene bottles previously soaked with 10% nitric acid and washed with double distilled water (Duncan and Harrison, 1981). At the time of collection of water samples, bottles were again washed three times with stream water, leaving the last rinse for five minutes to equilibrate (Johnson and Thornton, 1987). The sediment samples were collected with the help of Grab sampler, and were placed in high density polyethylene bags. The samples were transported to laboratory in ice-box under low temperature conditions. The river water temperature was measured at the time of sampling using Mercury Thermometer and other parameters were measured in the laboratory.

Physico-chemical Analysis

All the physico-chemical parameters of the river water were done as per the standard methods of water and waste water analysis (APHA, 1998). Different physico-chemical parameters and their analytical protocols are given in the table 1.

Table 1 : Physico-chemical Parameters and Their Analytical Protocols (APHA, 1998)

Parameters	Method
Temperature	Mercury Thermometer
pH	Electrometric method
Dissolved oxygen	Winkler's method
Total solid	Gravimetric Method (105°C)
Total dissolved solid	Gravimetric Method (180°C)
Conductivity	Electrometric method
Turbidity	Nephelometric method
Total hardness	Titrimetric (EDTA) method
Total alkalinity	Titrimetric (EDTA) Method
COD	Dichromate Reflux method
BOD	Winkler's method (Azide Modification)
Fluoride	Ion Analyser (ORION)
Chloride	Argentometric method
Nitrate	Ion Analyser (ORION)
Sulphate	Turbiditymetric method
Phosphate	Spectrophotometric method (Stannous chloride method)
TOC	TOC Analyser (Shimadzu)
Heavy metals	Acid digestion method

Sampling and Analysis of Metals in Water

For the 'total metal' estimation, river water samples were acidified at the time of collection in the field with concentrated HNO_3 (5 ml/l of water sample, to reduce the pH of the sample, pH > 2.0). The total metal content was determined by digesting 100 ml of sample with of concentrated nitric acid upto dryness, cooled and filtered through whatman no. 42 filter paper. The final volume was made upto 20 ml with 0.1N nitric acid and selected heavy metals were analysed by ICP-AES (Model Labtam- 8440).

Sampling and Analysis of Metals in River Bed Sediments

The samples were air dried for removal of water contents associated with the sediments. The dried samples were then ground with pestle mortar, homogenized, sieved to below 100 mesh size, sealed in clean polythene bags and stored in refrigerator till the final analysis. The total metal content was determined by digesting 1.0 g of homogenized and powdered sediment sample from each site with a mixture (4:1 ml) of concentrated nitric and perchloric acid (Moon *et al.* 1994). The digested samples were filtered through Whatman filter No. 42 and finally volume were made 20 ml with 0.1 N HNO_3 and analyzed for heavy metals using ICP-AES (Model Labtam- 8440).

ASSESSMENT OF HEAVY METAL CONTAMINATION IN RIVER SEDIMENT

Enrichment Factor

The primary objectives of the sediment sampling study is to estimate how sediment chemistry changes throughout a watershed and to relate these changes to natural and manmade sources of pollution. The iron normalized metal enrichment in the bed sediments was calculated following the method given by Jain (2004).

$$\text{Metal Enrichment} = \left[\frac{M}{Fe}\right] + 1000$$

Where *M* is the mean metal concentration in the sediment (mg/g), and *Fe* is the mean concentration of iron in the sediments (mg/g).

Geo-Accumulation Index

To assess the anthropogenic influence in the study area, the evaluation of sediments contamination with heavy metals was made using the index of accumulation, *I*geo which permits assessment of the extent of sediment contamination with respect to the global standards. *I*geo was calculated by the method of Muller (1979) using the following mathematical relation :

Where, *Cn* is the mean measured total concentration of the examined element *'n'* in the studied bottom sediment, *Bn* is the average (crustal) geochemical background value for concentration of the element 'n' in fossil argillaceous sediment ("average shale") and 1.5 is the factor compensating the background data (correction factor) due to lithogenic effect (Taylor, 1964). The *I*geo can be classified into seven grades (0 - 6). The highest grade, 6 reflect 100 - fold enrichment and 0 reflects the background concentration. The following description classification is given for the index of geo-accumulation by Forstner *et al.* (1990) as follows :

< 0	Practically unpolluted
0-1	Unpolluted to moderately polluted
1-2	Moderately polluted
2-3	Moderately to strongly polluted
3-4	Strongly polluted
4-5	Strong to very strongly polluted
> 5	Very strongly polluted

Reagents and Instrumentation

All chemicals and reagents were used of analytical reagent grade (obtained from Merck, India Ltd.). Double distilled water was used throughout the study. Before use all glass and plastic ware was soaked in 10% HNO_3 for 24 h. The washing was completed with a double distilled water rinse. Stock standard solutions of metals were prepared by dissolving ultrapure metals/compounds (99.99% pure) obtained from Merck, India Ltd. Heavy metals were estimated by using ICP-AES (Inductively coupled plasma atomic emission spectrometer) Model Labtam- 8440.

Quality Assurance/Quality Control

All the chemicals used in these experiments were supplied by Merck and of analytical grade. High purity deionized water (Milli-Q) was used for preparation of reagents and standards. Analytical grade concentrated nitric acid (65% v/v) and perchloric acid (70% v/v) was used. The standard solutions were prepared from stock solution of 1000 mg/l (Merck) by successive dilutions with Milli-Q water. After each five samples, one blank was run to check the possible contamination. The same digestion procedures were applied to the inter-laboratory quality control samples and the recover values were ranged from 98–102%.

RESULTS AND DISCUSSIONS

Physico-Chemical Parameters of River Water

The water samples of river Gomti from seven selected sites were monitored for all physico-chemical parameters. The temperature, pH, DO, TSS, TDS, TS, conductivity, Turbidity, Hardness, Alkalinity, COD, BOD and TOC were measured and results are shown in Table 3. Four ions such as chloride, fluoride, nitrate and phosphate were also monitored at all selected sites of the river and the results are presented in Table 4. The temperature ranged from 31 to 33°C throughout the river whereas conductivity and turbidity were in the range of 425.60-645.00 ms and 1.3-1.5 NTU respectively. The water pH was noticed slightly alkaline at all sites and it was in the range of 7.85-8.57. The range of pH and conductivity was in group A, CPCB (1995) classification, as the water is being taken as raw water supply for drinking purposes at first sampling site Gaughat. The total solids, dissolved solids and suspended solids were in the range of 336-489, 260-445 and 26-99 mg/l respectively. The total hardness and total alkalinity was in the range of 118-266 and 194-298 mg/l. Dissolved oxygen, BOD, COD and TOC were in the range of 1.50-8.33, 0.93-13.67, 3.93-42.40 and 9.31-21.09 mg/l respectively. The conductivity, turbidity, total solids, dissolved solids, suspended solids, hardness, alkalinity and COD were noticed highest at Mohan

Meakins site and lowest at Gaughat site. The dissolved oxygen was noticed lowest at Mohan Meakins due to high organics resulted from discharge of distillery effluent from Mohan Meakins and highest at Gaughat.

Table 2 : Classification of Inland Surface Waters (CPCB, 1995)

Characteristics	*A*	*B*	*C*	*D*
DO (mg l^{-1}) min.	6	5	4	-
BOD (mg l^{-1}) max.	2	3	3	-
TDS (mg l^{-1}) max.	500	-	1500	2100
pH value	66.5-8.5	66.5-8.5	66.5-8.5	66.5-8.5
Conductivity	–	–	–	2.250
SO_4^{2-} (mg l^{-1}) max.	400	–	400	–
NO_3^- (mg l^{-1}) max.	20	–	50	–
Cl^- (mg l^{-1}) max.	250	–	600	600

A – Drinking water source without conventional treatment but after disinfiection.

B – Outdoor bathing (organized).

C – Drinking water source with conventional treatment followed by disinfection.

D – Irrigation, industrial, cooling, controlled waste disposal.

Temperature is the most important parameters in natural surface water systems for assessing the water quality as it plays a major role in the aquatic environment through influencing a variety of physico-chemical and biological processes/ parameters. Most chemical reactions involving dissolution of solids are accelerated by increased temperatures. The solubility of gases, on the other hand, decreases at elevated temperatures. During the study period, temperature was recorded ranged 30 -33°C. The different temperature at different locations during the same seasons was noticed due to the different sampling time on the same day.

The hydrogen (H^+) and hydroxyl (OH^-) ions are controlling variables in aqueous systems as they influence both physico-chemical and biological processes in the aquatic environment.

The equilibrium between these two ionic species is influenced by reaction with acids and bases introduced into the aqueous system. The river water pH was recorded in the range of 7.85-8.57. The river water was noticed at alkaline side during the study period at all sites, this could be due to high carbonate, bicarbonate salts of calcium and magnesium contributed by urban wastewater.

Conductivity of water is mainly contributed by dissolved solids mainly mineral salts, therefore it's a very good indicator of ionic strength of water. In this study conductivity were recorded in the range of 425-645 μmho/cm. The higher conductivity values could be attributed to high mineral salt concentration which comes from effluent discharges into the stream.

The total dissolved solids (TDS) in water comprise inorganic salts and small amounts of organic matter. The principal ions contributing to TDS are carbonate, bicarbonate, chloride, sulphate, nitrate, sodium, potassium, calcium and magnesium. The study revealed high loads of TDS at all six sites except Gaughat up stream, which could be contributing to the high level of BOD. The higher TDS increases the chemical and biological oxygen demands and ultimately depletes the dissolved oxygen level. The TDS was recorded in the range of 261-445 mg/l. The TDS is related to conductivity in that if they are high, the conductivity will also be high. Inorganic solids such as clay and silt bring about total suspended solids (TSS). This TSS value was noted in the range of 34-99 mg/l. The total solids (TS) were found in the range of 295- 544 mg/l.

The biochemical oxygen demand (BOD) is an approximate measure of the amount of degradable organic matter in water. In this monitoring the values were higher at almost all sites and were in the range of 4.37-18.67 mg/l. Only one site was recorded with the lower BOD value which was upstream Gaughat, because of less or no source of organic load at that site. Other (sites were either receiving the sewage from the city or wastewater and effluent from the industry. At Lucknow the higher BOD of different sites was mainly due to waste water

Table 3 : Physico-Chemical Parameters of Gomti River Water

Sampling Sites	*Temp. (°C)*	*pH*	*Turbidity (NTU)*	*Cond. (µmho/ cm)*	*TSS (mg/l)*	*TDS (mg/l)*	*TS (mg/l)*
Gaughat	33	8.57	1.3	425	34	261	295
M. Meakin	33	8.09	1.5	645	99	445	544
M. Memorial	32	7.99	1.5	549	85	390	475
H. Setu	31	7.85	1.5	534	73	367	440
N. Bridge	31	8.38	1.3	484	52	345	397
Pipraghat	33	8.20	1.5	471	76	322	398
Malhaur	31	7.85	1.3	425	34	261	295
Min.	31	7.85	1.3	425	34	261	295
Max.	33	8.57	1.5	645	99	445	544
Average	32	8.13	1.41	504.71	64.71	341.57	406.29
Gaughat	166	194	6.33	4.37	13.92	9.31	
M. Meakin	218	298	2.58	18.67	26.4	18.47	
M. Memorial	184	241	1.90	15.85	23.73	15.95	
H. Setu	189	252	2.60	13.47	20.7	21.09	
N. Bridge	178	222	3.70	12.80	19.37	10.47	
Pipraghat	196	274	3.60	14.10	21.77	11.29	
Malhaur	166	194	5.90	11.30	17.98	9.31	
Min.	166	194	1.90	4.37	13.92	9.31	
Max.	218	298	6.33	18.67	26.4	21.09	
Average	**185.29**	**239.29**	**3.80**	**12.94**	**20.55**	**13.70**	

drains which still owe the heavy load of organic as well as the inorganic to the river Gomti. The highest value of BOD was recorded at Mohan Meakins site only because of the wastewater of Mohan Meakins factory. BOD and DO directly indicate the total organic load in aquatic system. In our study the DO was noticed in the range of 1.90–6.33 mg/l whereas the chemicals oxygen demand and total organic carbon were recorded in the range of 12.93-26.40 mg/l and 9.31–21.09 mg/l.

The alkalinity and hardness of water both are the very important parameters. The higher values of these parameters directly indicate the wastewater influence on surface water quality. The total hardness and total alkalinity was in the range of 118-266 and 194-298 mg/l. The conductivity, turbidity, total solids, dissolved solids, suspended solids, hardness, alkalinity and COD were noticed highest at Mohan Meakins site and lowest at Gaughat site. The dissolved oxygen was noticed lowest at Mohan Meakins due to high organics resulted from discharge of distillery effluent from Mohan Meakins and highest at Gaughat.

Major Ions of the Water

Nitrates and nitrites are considered together because of conversion from one form to the other in the environment. Very high nitrate and nitrite levels are associated with sewage contamination. Nitrate levels in natural waters seldom exceed a daily load of 0.00622 tones (0.1 mg/l) according to Chapman (1996). When influenced by man's activities, surface waters attain nitrate concentration of up to 5 mg/l. They can also be reduced to nitrites by denitrification process. The maximum contaminant level (MCL) is 10 mg/l. Nitrites levels in natural waters are usually very low (0.001 mg/l) and are rarely higher than 1 mg/l (Chapman, 1996). The values found in the study 2.27 – 8.03 mg/l could be linked to contamination from human activities. The magnitude of the nitrate concentration were almost in the same range at three site *i.e.* Mohan Meakins, Martyr's. Memorial and Hanuman Setu whereas at Nishatganj Bridge and Piparaghat it was on higher level.

Domestic wastewaters, particularly those containing detergents, industrial effluents and fertilizer run-off contribute to elevated levels of phosphates in surface waters. In most natural surface waters, phosphorus ranges from 0.005 to 0.20 mg/l. In this study phosphorus was recorded 0.05-0.84 mg/l. High concentrations of phosphates can indicate the presence of pollutants that are largely responsible for atrophic conditions. This could be attributed to the people's activities. The high phosphate level at four sites could be due to higher detergents concentrations released in the wastewater from the city and also resulting from the washing of cars. Its also a good indication of fecal contamination from the city. The values at other two sites *i.e.* Gaughat and Martyr's. Memorial were recorded 0.05 and 0.40 mg/l. All the values recorded, were found to be below 1 mg/l and therefore within acceptable level.

Table 4 : Major Ions in Gomti River Water

Sampling Sites	*F (mg/l)*	*Cl (mg/l)*	*SO_4 (mg/l)*	*NO_3 (mg/l)*	*PO_4 (mg/l)*
Gaughat	0.41	9.57	10.24	2.27	0.051
M. Meakin	0.47	13.25	14.31	4.70	0.713
M. Memorial	0.49	13.82	15.80	4.60	0.405
H. Setu	0.50	17.37	15.35	4.43	0.548
N. Bridge	0.47	18.46	16.37	6.43	0.392
Pipraghat	0.51	22.38	16.88	8.03	0.701
Malhaur	0.52	21.72	17.72	4.63	0.839
Min.	0.41	9.57	10.24	2.27	0.051
Max.	0.52	22.38	17.72	8.03	0.839
Average	**0.48**	**16.65**	**15.24**	**5.01**	**0.520**

Sulphate is considered to be a permanent solute of water (H_2O) that reacts with organic matter to produce sulphur (S), water and carbon dioxide (CO_2). It may however, be reduced to sulphides or volatilised as hydrogen sulphide (H_2S), in the presence of hydrogen ions (H^+), which has a foul smell. The most common mineral of sulphates are iron sulphide, lead sulphide, zinc sulphide, calcium sulphate and magnesium sulphate. The majority of sulphates are soluble in water. The quality of stream water is affected by sulphates.

Chloride is one of the major anions found in water and sewage. Its presence in large amounts may be due to natural processes such as the passage of water through natural salt formations in the earth or it may be an indication of pollution from sewage or industrial wastes. Any change in normal chloride content of natural water should be reason for suspecting pollution from one of these sources. The concentration of chloride was noticed in the range of 9.57 – 22.38 mg/l. At Piparaghat and Malhaur it was noted higher in comparison to other five sites. The higher concentration of chloride at

Piparaghat was due to industrial waste water as well as the higher concentration at Malhaur could be due to high sewage contamination. Only Gaughat site was monitored with the low concentration of chloride and also due the not a single waste water opening was there. The fluoride was found in the range of 0.41 – 0.52 mg/l. The minimum concentration of fluoride was recorded at Gaughat whereas the higher concentration of fluoride was at Piparaghat.

Heavy Metals in River Water

The spatial distribution of heavy metals in Gomti river water was studied. Eight heavy metals *i.e.* Cd, Cr, Cu, Fe, Mn, Ni, Pb and Zn were analysed in water samples of the river Gomti. The concentration of heavy metals in Gomti river water were compared with the BIS limits for surface water (Inland surface water, 1982), and in most of the cases such as in case of Cd, Cu, Mn, Ni and Pb the heavy metal concentration were found well below the BIS limit. In case of Cr and Fe the concentration were found above the BIS limit at most of the places. The concentration of Cd, Cr, Cu and Fe were in the range of ND – 0.006, 0.033 - 0.336, 0.01 - 0.041 and 0.228 - 2.845 mg/l respectively and Mn, Ni, Pb and Zn were in the range of 0.009 - 0.050, 0.031 - 0.06, 0.003 - 0.048 and 0.056 - 0.090 mg/l respectively. Cadmium was found below the detection limit at most of the sites except Mohan Meakins. Chromium was noticed high at Malhaur whereas copper was at Gaughat and lower at Martyrs Memorial and Nishatganj sites respectively. The concentration of iron was higher at Malhaur, manganese was at Mohan Meakins and nickel was also at Malhaur whereas these three metals were individually lower at Martyrs Memorial, Piparaghat and Nishatganj. Lead was noticed higher at Martyrs Memorial and lower at Nishatganj whereas zinc was noticed Martyres Memorial and lower at Malhaur.

The concentration of iron and manganese in well-aerated water is seldom high but under reducing conditions, which may exist in some groundwater, lakes or reservoirs and in the absence of sulphide and carbonate, high concentrations of soluble state II iron and manganese metals may be found. Both iron and manganese promote the growth of bacteria. Results reveals the concentration of iron 0.228-2.845mg/l and manganese 0.009-0.050 mg/l in river Gomti at Lucknow. Iron bacteria derive their energy from the oxidation of ferrous iron (Fe^{2+}) to ferric iron (Fe^{3+}). The above problems usually arise in water distribution systems and streams when the iron concentration approaches 0.3 mg/l. The resulting slime growth may change the taste and odour of water.

Studies elsewhere reported elevated concentrations of cadmium, copper, selenium including zinc in stream waters and sediments reduced species diversity and abundance in aquatic communities. At Lucknow the cadmium was ND – 0.006 mg/l, copper was Cu 0.01-0.041 mg/l and zinc were recorded as 0.056-0.090 mg/l.

The natural lead content of lake and river water worldwide has been estimated to be between 1 and 10 µg/l (Chapman, 1996). Higher values indicate contamination from industrial sources. In this monitoring the concentration of lead was recorded 0.003-0.048 mg/l. The concentration of chromium and nickel were recorded in summer 0.033-0.336 and 0.031-0.06 mg/l respectively. Cadmium was found below the detection limit at most of the sites except Mohan Meakins. Chromium was noticed high at Malhaur whereas copper was at Gaughat and lower at Martyrs Memorial and Nishatganj sites respectiverly. The concentration of iron was higher at Malhaur, manganese was at Mohan Meakins and nickel was also at Malhaur whereas these three metals were individually lower at Martyrs Memorial, Piparaghat and Nishatganj. Lead was noticed higher at Martyrs Memorial and lower at Nishatganj whereas zinc was noticed Martyrs Memorial and lower at Malhaur.

The distribution of metals in water was not uniform over the six sampling sites of the river Gomti at Lucknow. This variation in the concentrations may be due to differences in the sources of the metals and prevailing physico-chemical conditions and complex reactions such as adsorption, flocculation and redox condition taking place between water and sediments. This is because of in reverine system sediment act as reservoir of pollutants where all persistent toxicants resulted from anthropogenic activities, were settle down and absorbed.

Heavy Metals in River Sediments

Metals differ from other toxic substances in that they are neither created nor destroyed by humans. Humans have used heavy metals in many different areas for thousands of years; this use influences their potential for health effects in at least two major ways: first, by environmental transport, that is, by human or anthropogenic contributions to air, water, soil, and food, and second by altering the speciation or biochemical form of the element. Although several adverse health effects of heavy metals have been known for a long time, the exposure to these elements continues; moreover, it is even increasing in some parts of world, in particular in the less developed countries, though emissions have declined in most developed countries over the last 100 years. Owing to their toxicity, persistence and tendency to accumulate in water and sediment, heavy metals and metalloids, when occurring in higher concentrations, become severe poisons for all living organisms.

The spatial distribution of heavy metals in Gomti river sediment was also monitored. In the monitoring of river sediment at Lucknow, the cadmium was recorded in the range of ND-0.006 mg/l and detected in river sediment of only at two sites *i.e.* Mohan Meakins and Martyrs Memorial. The Cr, Cu, Fe, Mn, Ni, Pb and Zn were recorded in the range of 0.033-0.336, 0.010-0.041, 0.228-2.845, 0.009-0.050, 0.031-0.060, 0.003-0.048 and 0.056-0.071 µg/g respectively. Chromium and copper was noticed higher at Martyrs Memorial and Gaughat and lower concentration were on Mohan Meakins and Nishatganj. Iron was found higher at Malhaur and lower at Mohan Meakins. Manganese and Nickel were recorded high at Mohan Meakin and Malhaur whereas lead and zinc were at Martyr's Memorial as well as the lower concentration of manganese and nickel were at Pipraghat and Nishatganj, lead and zinc were at Nishatganj and Gaughat respectively.

Trace metals are distributed globally with a wide range of concentrations in sediments and biota. High concentrations of heavy metals in sediments from estuarine and coastal areas in Asian countries have also been reported (Chen *et al.*, 2007). Trace metals can be recirculated in the aquatic environments via natural or anthropogenic processes and then back to the water bodies, resulting in deterioration of the water quality and longterm implication of human health and ecosystem. Chen *et al.* (2007) investigated the distribution and accumulation of trace metals in sediments of Kaohsiung Harbor, Taiwan and found that the metal concentrations at the vicinity of river mouths were higher than those at other locations and metal concentrations correlated closely to the physico-chemical properties of sediments. These results clearly show that the physical- and chemical properties of river basins and anthropogenic processes have significant influence on the fate and transport of trace metals in sediments of river Gomti at Lucknow.

Enrichment Factor

Iron normalized heavy metal concentrations (M/Fe) at different sampling sites in river sediments are presented in Table 5. The metal pair ratios clearly reflect maximum enrichment of Cd, (2.11) at Mohan Meakin followed by (1.57) at Martyrs Memorial and (1.15) at Hanuman Setu. The Cr enrichment were found maximum (6.61) at Hanuman Setu followed by (6.23) at Mohan Meakin and (5.72) at Nishatganj Bridge. The Cu enrichment were found maximum (10.07) at Martyrs Memorial followed by (7.77) at Mohan Meakin and (6.12) at Hanuman Setu. The Mn enrichment were found maximum (48.25) at Gaughat followed by (33.92) at Mohan Meakin and (31.62) at Nishatganj Bridge. The Ni enrichment were found maximum (4.14) at Martyrs Memorial followed by (4.10) at Gaughat and (4.00) at Mohan Meakin. The Pb enrichment were found maximum (7.23) at Martyrs Memorial followed by (6.29) at Hanuman Setu and (5.72) at Nishatganj Bridge. The Zn enrichment were found maximum (22.17) at Martyrs Memorial followed by (17.07) at Hanuman Setu and (16.99) at Mohan Meakin.

In natural river systems, elements as well as metals exist together in relative proportion to each other. These ratios depend on a large number of processes in a geochemical cycle including weathering, transport and deposition (Jain, 2004). The metal pair ratio clearly reflects maximum enrichment of all studied metals (Cd, Cr, Cu, Mn, Ni, Pb and Zn). The spatial trends of metal enrichment in bed sediments of the Gomti river are in agreement with the data reflecting sources/activities of the corresponding catchments regions.

Table 5 : Iron-Normalized Mean Metal Concentration [(M/Fe) × 10^3] Values for Gomti River Sediments at Different Sites

Sampling Sites	*Cd*	*Cr*	*Cu*	*Mn*	*Ni*	*Pb*	*Zn*
Gaughat	0.555	065	17	48.25	4.105	63	7.75
Mohan Meakin	2.11	6.237	77	33.92	4.004	13	16.99
Martyrs Memorial	1.575	68	10.07	30.00	4.147	23	22.17
Hanuman Setu	1.156	616	12	27.25	3.816	29	17.07
Nishatganj Bridge	0.525	724	02	31.62	3.835	72	11.31
Pipraghat	0.694	033	55	19.11	2.333	02	15.37
Malhaur	0.514	603	79	23.90	3.665	43	10.26

Geo-accumulation Index

Index of geoaccumulation (*I*geo) calculated for all heavy metals in river sediments are given in Table 6. The geo-accumulation index (*I*geo) consists of seven grades ranging from unpolluted to very highly polluted. Class 6 indicates a 64 - fold enrichment above the background value. The *I*geo class for Gomti river sediments varies from metal to metal and place to place. Chromium, copper, iron, manganese, nickel, and zinc remain in class 0. The *I*geo class for cadmium increases from 3 to 5 while lead attains the value of 1. The *I*geo class for Cd was found maximum at Mohan Meakin and Martyrs Memorial followed by Hanuman Setu and so on, whereas, *I*geo class for Pb was found 1.0 at Gaughat, Martyrs Memorial, Hanuman Setu and Nishatganj Bridge.

The geo-accumulation index comprised of seven grades (0-6), indicates various degrees of enrichment above the background values ranging from unpolluted to very strongly polluted sediment. Based on the geo-accumulation indices, the Gomti river sediments from Gaughat to Malhaur are considered to be unpolluted with respect to Cr, Cu, Fe, Mn, Ni and Zn. It is unpolluted to moderately polluted with Pb. In case of cadmium it varies from moderately polluted to strongly polluted.

Table 6 : Geoaccumulation Index (Igeo) of Gomti River Sediments

Sampling Sites	*Cd*	*Cr*	*Cu*	*Fe*	*Mn*	*Ni*	*Pb*	*Zn*
Gaughat	3	0	0	0	0	0	1	0
Mohan Meakin	5	0	0	0	0	0	0	0
Martyrs Memorial	5	0	0	0	0	0	1	0
Hanuman Setu	4	0	0	0	0	0	1	0
Nishatganj Bridge	3	0	0	0	0	0	1	0
Pipraghat	3	0	0	0	0	0	0	0
Malhaur	3	0	0	0	0	0	0	0

CONCLUSIONS

Based on the results it is confirmed that out of seven site monitored, pollution load increased just of first site i.e Gaughat. Higher BOD, COD, TDS and metals were observed throughout the all sampling sites, which showed that the water quality of river Gomti is highly influenced by heavy load of urban waste water. The humid subtropical conditions as well as the hydrology of the river Gomti plays an important role in transportation and deposition process. The higher concentration of hyavy metals from Mohan Meakins (Second site) to Piparaghat (last site) may be due to low flow of the river as at the Gaughat there is a barrage which controls the flow of the water. Another reason would be the high discharge of untreated wastewater coming through 42 drains to the river. The urban runoff at Lucknow is the main source of pollution for river Gomti. The total metal concentration in sediment or water does not show the available part of a metal and is not useful for the biological impact evaluation. Therefore, the assessment of bioaccumulated metals should be made together with determination of metal concentrations in water and sediment.

A number of control measures can be recommended to mitigate the situation in Gomti. Hygiene awareness and education, change in the behaviors patterns of people in order to make them aware of the diseases related use of contaminated river use either for bathing or other recreational practices. Enforcing the Environmental Protection and Pollution Control legislation will ensure that all industries to keep their effluent pollutant concentrations to minimum levels. The reduction of discharges from water treatment plants and sewage works would also improve the water quality in the stream.

REFERENCES

APHA (1998) : Standard Method for the Examination of water and waste water 20th ed. American Public Health Association, New York.

Central Pollution Control Board, CPC (1995) : Classification of Inland surface waters (CPCB Standards). *Water Quality Parivesh*, **1** (4), 6.

Chapman, P.M. (1996) : Presentation and interpretation of sediment quality triad data. *Ecotoxicology*. **5** : 327-339.

Chen, C.W., Kao, C.M., Chen, C.F. and Dong, C.D. (2007) : Distribution and accumulation of heavy metals in the sediments of Kaohsiung Harbor, Taiwan. *Chemosphere*. **66** : 1431-1440.

Duncan, P.H.L. and Harrison, R.M. (1981) : Cleaning methods for polythene container prior to determination of trace metals in freshwater samples. *Analytical Chemistry*. **53**; 345.

Forstner, U. and Wittmann G. (1983): Metal pollution in aquatic environment. Berlin : Springer-Verlag, 484pp.

Frostner, U. and Wittman, G.T.W. (1981): Metal pollution in the aquatic environment. Springer verlag, Berlin, Germany 1981.

Forstner, U., Ahalf, W., Calmano,W. and Kersten, M. (1990): Sediment Criteria Development-Contribution from Environmental Geochemistry to Water Quality Management, in: D. Heling, P. Rothe, U. Forstner and P. Stoffers (eds), Sediments and Environmental Geochemistry: Selected Aspects and Case Histories, Springer-Verlag, Berlin Hidelberg, pp. 311-338.

Gaur, V.K., Gupta, S.K., Pandey, S.D., Gopal, K. and Mishra V. (2005): Distribution of heavy metals in sediment and water of river Gomti. *Environmental Monitoring Assessment*. **102** : 419-433.

Gupta, L.P. and Subramanian, V. (1998) : Geochemical factors controlling the chemical nature of water and sediments in the Gomti River, India. *Environmental Geology*. **36** (1-2) : 102-108.

Heyden, C.J.V. and New, M.G. (2004) : Groundwater pollution on the Zambian Copper belt. deciphering the source and the risk. *The Science of the Total Environment*. **327** : 17-30.

ISI (1982) : Indian Standard. Methods of sampling and microbiological examination of water (First Revision). IS 1622. 1981, Indian Standards Institute, New Delhi, pp. 7-13.

Jain, C.K. (2004): Metal fractionation study on bed sediments of River Yamuna, India. *Water Research*. **38**: 569-578.

Jardo, C.P. and Nickless, G. (1989) : Chemical association of Zn, Cd, Pb and Cu in soils and sediments determined by the sequential extraction technique. *Environment Technology Letter*. **10**: 743-52.

Johnson, C.A. and Thornton, I. (1987) : Hydrological and chemical factors controlling the concentrations of Fe, Cu, Zn and As in a river system contaminated by acid mine drainage. Water Research 21 : 359-365.

Lee, C.S., Li, X., Shi, W., Cheung, S.C. and Thornton, I. (2006) : Metal contamination in urban, suburban, and country park soils of Hong Kong. A study based on GIS and multivariate statistics. *The Science of the Total Environment*. **356** : 45-61.

Moller, A., Muller, H. W., Abdullah, A., Abdelgawad, G. and Utermann, J. (2005) : Urban soil pollution in Damascus, Syria. concentrations and patterns of heavy metals in the soils of the Damascus Ghouta. *Geoderma*. **124** : 63-71.

Moon, C.H., Lee,Y.S. and Yoon, T.H. (1994) : Variation of trace Cr, Pb and Zn in sediment and water of an urban stream resulting from domestic effluents. *Water Research*. **28** (4) : 985-991.

National Research Council (1989) : Contaminated marine systems-assessment and remediation. National Academy Press, Washington, D.C.

Nriagu, J.O. (1990) : Global metal pollution. Poisoning the biosphere ? *Environment*. **32**, 7-11; 28-33.

Peakall, D. and Burger, J. (2003) : Methodologies for assessing exposure to metals. speciation, bioavailability of metals, and ecological host factors. *Ecotoxicology and Environmental Safety*. **56**: 110-121.

Phillips, D.J.H. (1977) : The use of biological indicator organisms to monitor trace metal pollution in marine and estuarine environments, a review. *Environment Pollution*. **13** : 281-317.

Phiri, O., Mumba, P., Moyo, B.H.Z. and Kadewa, W. (2005) : Assessment of the impact of industrial effluents on water quality of receiving rivers in urban areas of Malawi. *International J. Environmental Science and Technology*. **2** (3), 237-244.

Rao, J.V. (2004) : Effects of monocrotophos and its analogs in acetylcholinesterase activity's inhibition and its pattern of recovery on euryhaline fish, *Oreochromis mossambicus*. *Ecotoxicology and Environmental Safety*. **59** : 217-222.

Singh, K.P., Malik, A., Mohan, D. and Sinha, S. (2004) : Multivariate statistical techniques for the evaluation of spatial and temporal variations in water quality of Gomti River (India) – a case study. *Water Research*. **38** : 3980-3992.

Singh, K.P., Malik, A., Mohan, D. and Takroo, R. (2005) : Distribution of persistent organochlorine pesticide residues in Gomti River, India. *Bulletin of Environmental Contamination Toxicology*. **74** : 146-154.

Singh, M., Ansari, A.A, Muller, G. and Singh I.B. (1997): Heavy metals in freshly deposited sediments of the Gomti River (a tributary of Ganga River): effects of human activities. *Environmental Geology*. **29** (3/4) : 246-252.

Singh, V.K., Singh, K.P. and Mohan, D. (2005) : Status of heavy metals in water and bed sediments of river Gomti – a tributary of the Ganga river, India. *Environmental Monitoring Assessment*. **105** : 43-67.

Sorme, L. and Lagerkvist, R. (2002): Sources of heavy metals in urban wastewater in Stockholm. *The Science of the Total Environment*. **298** : 131-145.

Sumok, P. (2001) : River water quality monitoring, Sharing Sarawak experience. In Proceedings of 6th SITE research seminar (pp. 1-4), 13-14 September.

Tang, C.W., Ip, C.C., Zhang, G., PaulShin, K.S., Qian, P., Li, X. (2008): The spatial and temporal distribution of heavy metals in sediments of Victoria Harbour, Hong Kong. *Marine Pollution Bulletin*. **57** : 816-825.

Vikaskumar, G.H., Dunstan, R.H., Geary, P.M., Coombes, P., Roberts, T.K. and Rothkirch, T. (2007). : Comparison of water quality parameters from diverse catchments during dry periods and following rain events. *Water Research*. **41**: 3655-3666.

Wong, C.S.C., Li, X. and Thornton, I. (2006) : Urban environmental geochemistry of trace metals. *Environmental Pollution*.**142**, 1-16.

30

Sediment Contamination Analysis and the Trend of Elemental Enrichment at an Industrial Complex During a Dredging Campaign

Madhurima Pandey, Sunita Gupta and Piyush Kant Pandey

ABSTRACT

The ecological health of many areas is threatened by inputs of toxic compounds and the accumulation of these contaminants in sediments. The sediment associated contaminants can affect biota or result in high concentrations in the tissues of indigenous organisms. Particularly the water bodies associated with mega-industrial complex such as steel industry are more prone to such contamination due to the release of various air and liquid pollutants. In this study the pollution level in the sediment settled over the years due to the Bhilai Steel Plant and associated power plant on the quality characteristics of sediments. The large water tank named as Maroda Tank is used for the providing water supply for Bhilai city and process requirement of the Bhilai Steel Plant. The other pond studied is recycling pond used for temporary storage of recycled water. Total 40 sediment samples were taken across the cross-section of both the water tanks to analyse the sediment pollution. Analysis shows the accumulation of trace metals in the sediments of water tanks of the Bhilai Steel Plant industrial area. The build up of contaminants in the Recycle pond is a result of large scale industrial activities taking place where this water is used as coolant, solvent and washing fluid. It is also found that the industrial contaminants inputs have sharply increased recently due to reduction in water consumption per ton of crude steel production.

Keywords

Contaminants, Sediment, Steel plant, Trace metals, Toxic compound, Water body.

INTRODUCTION

The investigation of the sediment has become a major topic in the research of aquatic system, as they reflect the current quality of the system and provide information about the impact of human activities because since the industrial revolution human activities have led to an increased input of heavy metals into the environment also accumulation in nearby areas. The heavy metal contaminants are absorbed at the surface of the sediment. Many investigations (Banat *et al.* 1972; Forstner & Muller, 1997; Muller, 1997, 1980, 1991; Pilz, 1999; Chopra *et al.* 2009; Jumbe *et al.* 2009; Phuong *et al.* 1998; Sarika *et al.* 2008; Abhishek *et al.* 2008; Hejabi *et al.* 2009; Shukla *et al.* 2009; Lokeshwari *et al.* 2006; Sudhira *et al.* 2007; Pilotte *et al.* 1978; Singh *et al.* 2002; Pascual, 2004; Knight, 1997; Anglin, 1995; Sammak, 2001) have been carried out in order to assess the pollution level of heavy metals in the sediments.

Metals are ubiquitous in the environment. Study of the trace metal pollution of the environment concerns mainly with the "toxic metals" The word "toxic" stands for relatively mobile metals (viz. As, Be, Cd, Co, Cr, Cu, Hg, Mn, Ni, Pb, Sb, Sn, Te, Tl, V, and Zn) for which values are reported at levels considered as harmful to the nature. Although it is well known that metals such as Cr, Co, Fe, Mn and Zn among others are essential for many biological systems including humans. Other elements such as As, Cd, Pb, Hg do not posses any beneficial biological effect and are toxic at certain levels. However, all metals including micronutrient elements can exert toxic effects at high concentrations (Peters *et al.*, 1986; Domingo, 1994; Christensen, 1995). Heavy metals are the most serious pollutants within the natural environment due to their toxicity, persistence and bioaccumulation problems (MacFarlane and Burchett, 2000).

The problems associated with trace metals in water entering the natural urban aquatic ecosystems have been well documented and studied. Heavy metals such as Cd, Hg, Pb, Cu, and Zn are regarded as serious aquatic pollutants because of their toxicity, tendency to be incorporated into food chains, and ability to remain in an environment for a long time (Kishe *et al.* 2003). Heavy metals are widespread pollutants of great environmental concern as they are nondegradable, toxic and persistent with serious ecological ramifications on aquatic ecology (Chopra *et al.* 2009; Jumbe *et al.* 2009 and Jumbe *et al.* 2009).

Sediments are integral and inseparable parts of the aquatic environments because they help to determine the overall assessment of heavy metals in water vis-a-vis aquatic life and survivability (Shukla *et al.* 2009). Since sediments play a very important role in physicochemical and ecological dynamics, any change in toxic concentrations of heavy metal residues on the sediments will affect the natural aquatic life support systems (Shukla *et al.* 2009). Particularly the heavy metals exhibit very short residence time in the mobile form and tend to settle in the ultimate sink: the soil and the sediment. Tracking the entire pathway is difficult owing to the inherent complexities of the environmental processes. However, the chemical analysis of the sediment provides valuable information on the spatial distribution of the heavy metals. It has been recognized since eighties that the sediment analysis of the lakes shows a good correlation with increasing rates of anthropogenic activities and the industrialization (Hamilton *et al.* 1979; Kansanen *et al.* 1985; Galloway *et al.* 1979).

Therefore, the main objective of this work was to see weather the sediment of water bodies located in an industrial complex are also contaminated by other toxic trace metals. The second aim was to find if the values of pollution can pose any serious hazard to the people of the township who depend on such water body for the supply of the drinking water. The studied water bodies are sufficiently large and it is presumed that it can serve as stable sediment sink so that its analysis can provide a useful baseline for the study of future depositional trends. The dredging operation which was carried out to increase the water holding capacity of the studied tanks provided a good opportunity for the collection of freshly exposed depths.

The environment of Bhilai city is expected to be influenced by the wide ranging industrial practices, population increase and long range transport from the other adjoining areas. Therefore, the investigations on the chemical composition and the spatial and vertical distribution of the trace metal content could provide a sight on the total deposition, its trend and probable impacts.

MATERIAL AND METHODS

About the Site

There are two water tanks recycle tank (Maroda tank I) upper maroda tank (Maroda tank II) having with a dimension of 2.45km^2 and 5.33 km^2 respectively. These tanks are used to provide fresh water for the use of Bhilai steel plant and its township. The approximate water holding capacity of the upper Maroda tank is 36.0 MM^3. Second tank, recycle tank, is used by Bhilai steel plant for the temporary holding and ultimate recycling of water, used for the purpose of cooling, washing, etc. The water holding capacity of the tank is 9.0 MM^3. Samples were collected during the ongoing dredging operations in both the tanks. Each pond was divided into grid of approximate ..... Duplicate samples were collected from each sampling site. Three major depth were targeted for analysis *i.e.* top layer of the sediment, 0.15 meters and 0.3 meters. These depth were arbitrary taken to avoid any contamination by mechanical dredger.

Table 1 : Analysis of Trace Elemental Concentration (mg/kg) in Sediment of Upper Maroda Tank (2008)

	Ca	*Mg*	*Fe*	*Ni*	*Cr*	*Zn*	*Mn*	*Pb*	*Cu*	*Cd*	*As*
First layer	125000	BDL	83800	19.75	168.125	93.7	315	61.25	25.84	BDL	BDL
	75000	BDL	21193.75	8.25	0.5	66.7375	195.9375	30	668.85	BDL	BDL
	100000	BDL	43793.75	11.5	40.125	16.7375	116.875	43.75	28.99	BDL	BDL
	115000	BDL	28543.75	7.75	21.125	14.7	196.5625	71.25	66.81	BDL	BDL
	130000	BDL	21168.75	25.75	6.125	46.45	117.5	25	126.7	BDL	BDL
Second layer	150000		84075	15.75	120	65.125	382.8125	40	41.6	BDL	BDL
	135000	BDL	84975	41.125	120.25	72.1875	2156.25	67.5	76.27	BDL	BDL
	150000	BDL	70075	28.5	118.875	103.2375	1134.688	58.75	88.88	BDL	BDL
	145000	BDL	48612.5	6.125	44.25	88.55	353.125	40	73.12	BDL	BDL
	160000	BDL	74812.5	17.875	72.75	78.025	303.4375	38.75	69.97	BDL	BDL
Third layer	55000	10000	50587.5	7.125	4.25	0	195.3125	35	51.44	BDL	BDL
	70000	10000	60500	26.25	0	56.1375	212.8125	36.25	75.25	BDL	BDL
	30000	55000	54862.5	41.875	38.375	111.3	417.1875	32.5	68.58	BDL	BDL
	60000	20000	84287.5	40.125	29.625	34.1125	757.5	32.5	69.22	BDL	BDL
	65000	BDL	51625		0	60.775	120.625	31.25	72.43	BDL	BDL
Fourth layer	15000	50000	59975	4.625	44.75	348.1375	43.4375	147.5	55.68	BDL	BDL
	20000	65000	59037.5	10.125	56.25	316.825	116.25	67.5	80.12	BDL	BDL
	15000	60000	41137.5	44	86.125	339.175	1479.688	57.5	65.25	BDL	BDL
	15000	25000	91212.5	75.625	129	276.925	1493.75	163.75	72.12	BDL	BDL
	20000	60000	49118.75	12.75	12.75	313.8625	78.4375	106.25	75.11	BDL	BDL

Table 2 : Correlation Matrices of The Elements of Maroda Tank

	Ca	*Mg*	*Fe*	*Ni*	*Cr*	*Zn*	*Mn*	*Pb*
Ca	1							
Mg	-0.92	1						
Fe	0.33	-0.50	1					
Ni	-0.72	0.77	-0.88	1				
Cr	0.47	-0.10	-0.14	-0.21	1			
Zn	-0.69	0.90	-0.41	0.55	0.29	1		
Mn	0.20	0.16	-0.68	0.34	0.81	0.39	1	
Pb	-0.61	0.83	-0.27	0.40	0.37	0.98	0.37	1

Table 3 : Statistical Analysis of the Trace Elemental Concentration (mg/kg) of Sediment of Maroda Tank

	Ca	*Mg*	*Fe*	*Ni*	*Cr*	*Zn*	*Mn*	*Pb*
Mean	82500	17750	147495	23.68	55.66	125.13	509.35	59.31
Standard Error	28845.85	12263.6	83218.8	3.47	16.92	65.04	151.76	16.73
Median	82500	9500	66441.3	25.35	56.48	66.94	491.49	47.62
Mode	#N/A	0	#N/A	#N/A	#N/A	#N/A	#N/A	#N/A
Standard Deviation	57691.70	24527.2	166438	6.95	33.84	130.09	303.53	33.47
Sample Variance	3328333333	6E+08	2.8E+10	48.43	1145.68	16923.64	92132.7	1120.89
Kurtosis	-2.12	1.00	3.97	-1.26	0.08	3.72	-2.53	3.31
Skewness	0	1.31	1.99	-0.84	-0.13	1.92	0.23	1.74
Range	131000	52000	336904	14.82	80.77	271.32	677.69	75
Minimum	17000	0	60096.3	14.6	14.45	47.66	188.37	33.5
Maximum	148000	52000	397000	29.42	95.22	318.98	866.06	108.5
Sum	330000	71000	589979	94.73	222.64	500.52	2037.42	237.25
Count*	4	4	4	4	4	4	4	4

*Average of all the four layers of the samples of Maroda tank.

The schematic of the water distribution system in the sampled area is as given below in Fig. 1.

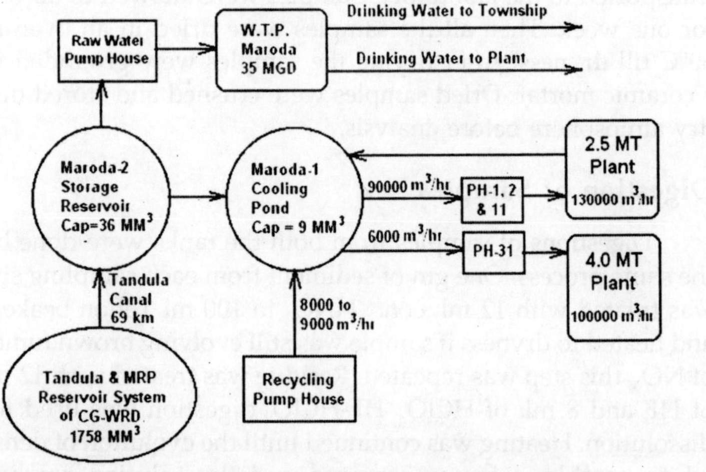

Fig 1 : Water distribution system of Bhilai city.

Sampling of Sediment from Upper Maroda Tank

Samples of sediments were collected at different depths across the cross-section of the water tank during enhancement capacity of the tank taken up by the Bhilai Steel Plant. Samples at four different depth levels (0 cm, 90cm, 180cm, 275cm) from five different locations (across the cross-section of the area taken up for tank capacity enhancement) were taken during June 08 to November 08 for analysis purpose to have a better overview of the analysis.

Sampling of Sediment from Recycle Tank

Samples of sediments were collected from the dredging machine outlet when dredging operation was in progress at various depths across the cross section of recycle tank. 18 samples were taken at six different locations across the cross section of the tank during June 08 to Nov 08. At each location the samples were taken from the silt level 0 cm, nearly 150 cm depth & lastly at nearly 305cm depth level *i.e.* top layer, middle layer & bottom layer of the Recycle Maroda Tank.

The samples were carefully collected from the freshly exposed surfaces in large plastic bags and immediately transported to the laboratory. Samples were allowed to air dry for one week. Then all the samples were dried in an oven at 60°C till dryness. After drying the samples were grounded in a ceramic mortar. Dried samples were crushed and stored in a dry atmosphere before analysis.

Digestion of Samples

Digestions of samples from both the tanks were done by the same process. One gm of sediment from each sampling site was treated with 12 ml. conc. HNO_3 in 100 ml Teflon beakers and heated to dryness if sample was still evolving brown fumes of NO_X this step was repeated. Residue was treated with 12 ml of HF and 8 ml. of $HClO_4$. HF-$HClO_4$ digestion was used for dissolution. Heating was continued until the evolution of dense white perchloric fumes occurred and the solution reached incipient dryness & than allowed to cool. In this residue 15ml of conc. HCl was added and left overnight. Then this solution was filtered using 0.45 micron filter paper and volume made up to 50 ml with deionised water (APHA, 1992).

RESULT AND DISCUSSION

This chapter presents the study of sediment contamination profile in the recycling pond and Maroda tank (Fig. 2) The various sediments reported include Ca, Mg, Fe, Ni, Cr, Zn, Mn, Pb, Cu, Cd and As. Results of recycle tank show a predominance of Ca and Fe in the major elements and Pb, Cu, Ni, Cr and Zn in the trace elements. The Cd and As levels have been found to be conspicuously absent in the studied sediments. Table 1 shows the trace elemental concentration in the Maroda tank.

The analysis was carried out at different depths to study the depositional behaviour with respect to the time. For the purpose the specified depth was top sediment (0 cm), second depth (90 cm), third depth (180 cm), fourth depth (275 cm) in maroda tank. Total five fixed locations were used to collect the samples. Similarly six locations were used to collect the samples

Fig. 2 : Picture of both the tanks showing sampling locations.

from recycle tank with the depths 0 cm, 150 cm and 305 cm. The detailed results are presented in the Table 1 and Table 4. It is very important tot note that the levels of calcium are very high in the Upper Maroda Tank followed by iron. However, this trend is reversed in the case of recycle pond.

The recycle tank receives its input water which is the waste water used in the iron and steel making process the influence of industrial activities is visible in the lower level of calcium and higher level of iron. However it is worth noting that the levels of trace elements are comparatively lower in recycle tank than the Maroda tank. The analysis of first layer of sediment of recycle tank shows mean and standard deviation values (Table 5 and 6) for Ca is 67000, Mg 1408.33, Fe 208855.4, Ni 28.16, Cr 11.97, Zn 701.38, Mn 2692.23, Pb 114.8, and Cu 54.12. The standard deviation values are in acceptable range. Looking to the fact that sediment contamination at different points in a large water body is respected to vary. The correlation matrix of recycle tank in first layer of sediment shows a significant correlation between Fe, Cr and Mn, Zn and Mn, Cr and Mn, Ni and Pb, Mn and Pb. These correlation indicates that source of these elements in the sediment is same and is probably coming from metallurgical operations.

Table 4 : Analysis of Trace Elemental Concentration (mg/kg) of Sediment of Recycle Tank (2008)

		Ca	*Mg*	*Fe*	*Ni*	*Cr*	*Zn*	*Mn*	*Pb*	*Cu*	*Cd*	*As*
Location 1	First layer	75000	BDL	63287.5	15.75	3.75	883.4	1061	85	60.12	BDL	BDL
	Second layer	25000	2000	25250	33.25	11.25	245.5	785.5	12	22.45	BDL	BDL
	Third layer	55000	BDL	39712.5	66.75	42.5	612.05	2063.5	30	20.44	BDL	BDL
Location 2	First layer	40000	5000	148237.5	16.75	23	1435.85	3378.5	140	58.55	BDL	BDL
	Second layer	22000	1100	33460	25.5	11.33	547.5	120.5	10	25.9	BDL	BDL
	Third layer	40000	BDL	66537.5	79.25	130.25	1191.5	298.5	45	25.87	BDL	BDL
Location 3	First layer	65000	BDL	221057.5	51.75	11.75	548.7	4021	215	10.08	BDL	BDL
	Second layer	35000	2100	79075	38.25	BDL	257.95	2166	101.25	31.2	BDL	BDL
	Third layer	35000	BDL	93162.5	75.5	31.75	BDL	5298.5	170	22.69	BDL	BDL
Location 4	First layer	75000	BDL	141575	23.5	BDL	772.22	333.12	65	69.97	BDL	BDL
	Second layer	33000	1200	25750	29	BDL	195.05	63.12	BDL	23.66	BDL	BDL
	Third layer	45000	2300	141375	14.5	BDL	238.95	701.25	57.5	32.14	BDL	BDL
Location 5	First layer	80000	2200	146775	33.75	BDL	348.7	460.62	67.5	64.36	BDL	BDL
	Second layer	40000	2200	27900	19.75	BDL	174.2	99.37	10	28.99	BDL	BDL
	Third layer	45000	BDL	180150	11.75	BDL	99.57	1291.87	55	35.45		
Location 6	First layer	67000	1250	532200	27.5	33.33	219.41	6899.17	116.67	61.66	10	BDL
	Second layer	45000	BDL	104700	43.33	35	304.91	1219.17	16.67	34.16	BDL	BDL
	Third layer	45000	BDL	377700	22.5	35	408.16	3931.67	83.33	70	2.5	BDL

Table 5 : Correlation Matrices of the Elements in Recycle Tank

	Ca	*Mg*	*Fe*	*Ni*	*Cr*	*Zn*	*Mn*	*Pb*	*Cu*	*Cd*
Ca	1									
Mg	–0.23	1								
Fe	0.39	–0.07	1							
Ni	–0.10	–0.40	–0.20	1						
Cr	–0.13	–0.28	0.06	0.68	1					
Zn	0.17	0.21	–0.10	0.07	0.44	1				
Mn	0.15	–0.04	0.73	0.20	0.13	–0.11	1			
Pb	0.32	0.01	0.45	0.20	0.00	0.14	0.74	1		
Cu	0.56	0.14	0.52	–0.50	–0.13	0.27	0.15	0.08	1	
Cd	0.25	–0.03	0.85	–0.15	0.11	–0.19	0.67	0.19	0.38	1

Table 6 : Statistical Analysis of the Trace Elemental Concentration (mg/kg) of Sediment (Recycle Tank)

	Ca	*Mg*	*Fe*	*Ni*	*Cr*	*Zn*	*Mn*	*Pb*	*Cu*	*Cd*
Mean	48166.67	1075	135994.7	34.90	20.49	471.31	1899.57	71.10	38.76	0.73
Standard Error	4112.98	322.32	31152.3	4.89	7.39	90.86	473.28	14.05	4.59	0.59
Median	45000	550	98931.25	28.25	11.29	326.80	1140.08	61.25	31.67	0
Mode	45000	10								
Standard Deviation	17449.93	1367.50	132168	20.77	31.35	385.51	2007.96	59.60	19.47	2.46

Cont...

	Ca	*Mg*	*Fe*	*Ni*	*Cr*	*Zn*	*Mn*	*Pb*	*Cu*	*Cd*
Sample Variance	3.05E+08	1870074	1.75E+10	431.72	983.26	148621.4	4031906	3553.28	379.30	6.06
Kurtosis	-0.75	2.53	4.30	0.21	9.11	1.19	0.78	0.56	-1.27	14.61
Skewness	0.52	1.44	2.00	1.10	2.73	1.28	1.24	0.97	0.53	3.76
Range	58000	5000	506950	67.5	130.25	1435.85	6836.05	215	59.92	10
Minimum	22000	0	25250	11.75	0	0	63.12	0	10.08	0
Maximum	80000	5000	532200	79.25	130.25	1435.85	6899.17	215	70	10
Sum	867000	19350	2447905	628.33	368.91	8483.62	34192.36	1279.92	697.69	12.5
Count	18	18	18	18	18	18	18	18	18	17

Table 7 : Mean of All The Three Layers of Recycle Tank

	Ca	*Mg*	*Fe*	*Ni*	*Cr*	*Zn*	*Mn*	*Pb*	*Cu*	*Cd*	*As*
First layer	67000	1408	208855.4	28.16	11.97	701.38	2692.24	114.86	54.12	10	0
Second layer	33333	1433	38287	31.51	9.59	287.51	742.27	24.98	27.73	0	0
Third layer	44167	383.3	149772.9	45.04	39.91	425.03	2264.22	73.47	34.43	2.5	0
Mean	48167	1075	132305.1	34.903	20.49	471.307	1899.58	69.92	38.76	0.69	0

Correlation studies provide important information about the source specific interrelationship among the studied elements. Accordingly the upper layer of sediment shows a strong negative correlation between Ca and Zn, Cr, Pb, Ni shows that they belong to two different sources. It can be easily presumed that the Ca is of crustal origin whereas the trace elements and Mg are of industrial origin. The results of the two studied water bodies were compared with the other water bodies in the vicinity. Tables 7 and 8 presents the results of both the tanks. On comparing the results (Fig. 3) it is found that the Recycle tank and Maroda tank are among the most contaminated water bodies in the Bhiai-Durg region. The level of almost all studied elements was elevated in these water bodies. It clearly suggest that the proximity to the huge integrated steel is responsible for higher elemental contamination.

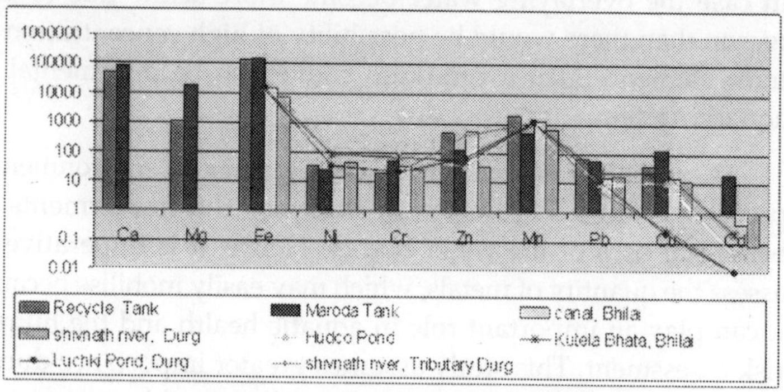

Fig. 3 : Sediment enrichment ratio.

The concept of the sediment enrichment ratio (SER) is quite important in understanding the impact and economic cost of chemical contamination.

The Sediment Enrichment Ratio (SER) is defined as :

SER = Concentration of chemical 'X' in the transported sediment / Concentration of chemical 'X' in the sediment.

If, the enrichment factors are considered then it can be seen that elemental enrichment is consistently high in all three layers of the studied sediment. From the point of elevation in the levels the uppermost layer is found maximum enriched. Enrichment in the levels of copper and lead is worth mentioning as these elements possess high toxicity and their higher presence can cause adverse impact on the aquatic life. By comparing the enrichment factor of recycle and Maroda tank (Tables 9 and 10), it is clear that recycle tank is much more enriched than Maroda tank. Among all the elements Fe, Pb, Cu and Cr are more enriched compare to other elements.

Table 11 shows elemental composition of overlaying water in the two water bodies. It can be seen that overlaying water is showing appreciable concentration of Ni, Zn, Mn and Cu. This presence can be attributed to the presence of some elements in the sediment in appreciable quantity. This composition is likely to further alter if there is a chang in the pH of overlaying water. In case the overlaying water become more acidic due to any reason than there would be possibility of high concentration of trace element. Such a condition could also be detrimental to aquatic system.

Contaminated sediment is a major cause of environmental problems and a key factor in many of the impairments to beneficial uses of the water bodies. Hence, it is imperative to assess the quantity of metals, which may easily mobilise because it can play an important role in aquatic health and the human risk assessment. This study assumes greater importance due to the reason that the Maoda Tank studied in this project is a source of drinking water to the large population of about three hundred thousand people in Bhilai. It is because of the fact that the sediment with high metals concentration can play important role to the surface as well as ground water quality because sediments usually act as a sink for pollutants. Similarly, sediments associated contaminants can influence the concentration of trace metal in both the water column and biota if they are desorbed or become available to benthic organisms. One of the major problems that heavy metals cause with respect to their effect on aquatic organisms is their long biological half-life.

Table 8 : Mean of Trace Elemental Concentration (mg/kg) of Sediment of Upper Maroda Tank (2008)

	Ca	*Mg*	*Fe*	*Ni*	*Cr*	*Zn*	*Mn*	*Pb*	*Cu*
First layer	109000	0	397000	14.6	47.2	47.66	188.37	46.25	183.4
Second layer	148000	0	72510	21.87	95.22	81.42	866.06	49	131.9
Third layer	56000	19000	60372.5	28.84	14.45	52.46	340.68	33.5	112.9
Fourth layer	17000	52000	60096.25	29.42	65.77	318.98	642.31	108.5	103.5
Mean	82500	17750	147494.7	23.683	55.66	125.13	509.355	59.313	132.9

Table 9 : Enrichment Factor of Elements of Sediment of Recycle Tank

Elements	*Upper Layer*	*Middle Layer*	*Lower Layer*
Mg	1.78	1.81	0.48
Fe	12.28	2.9	8.81
Ni	1.4	1.57	2.25
Cr	0.47	0.38	1.59
Zn	7.79	3.19	4.72
Mn	8.97	2.47	1.83
Pb	2.87	0.62	1.83
Cu	12.98	6.65	8.22
Cd	0.16	0	0.041
As	0	0	0

Table 10 : Enrichment Factor of Elements of Sediment of Maroda Tank

Elements	*First Depth*	*Second*	*Third*	*Fourth*
Ca				
Mg	0	0	24.05	65.82
Fe	23.35	4.26	3.55	3.53
Ni	0.73	1.09	1.44	1.47
Cr	1.88	3.8	0.57	2.63
Zn	0.52	0.9	0.58	3.54
Mn	0.62	2.88	1.13	2.14
Pb	1.15	1.22	0.83	2.71
Cu	7.33	5.27	4.51	4.14
Cd	0	0	0	0
As	0	0	0	0

Table 11 : Analysis of Elemental Concentration (mg/L) of Water of Recycle Tank (2008)

	Ca	*Mg*	*Fe*	*Ni*	*Cr*	*Zn*	*Mn*	*Pb*	*Cu*	*Cd*	*As*
Recycle tank	48	112	0.06	0.01	0	0.007	0.004	0	0.069	0	0
Maroda tank	16	144	0.27	]0	0	0.002	0.005	0	0.04	0	0

The sediment analysis also demonstrates a high metal loading in the surficial waters. Based on the results of sediment it is surmised that the recycling of metals from the sediments to the water column is likely to increase in response to any future decrease in metal concentrations in the overlying water. The toxic metal mobility in the sediments can be related to the bulk composition (specifically iron and manganese content) and the redox status of the sediments. As the studied sediments contain a moderate concentration of Fe (132305.1 mg in recycle and 1474945 mg in Maroda tank) with a similar median value) the metal sequestration capacity of the sediment is not very high (Table 4). It also known that the mobility of toxic metals in sediments is correlated with sediment toxicity that is the reason that sediment of studied region already appears less toxic to the aquatic life.

REFERENCES

APHA (American Public Health Association) (1992) : Standard method for the examination of water and waste water. Washington DC, U.S

Atgn, R.S.; El-Agha, O.; Zararsz, A.; Kocatas, A.; Parla, H. and Tuncel, g. (2000) : Investigation of the sediment pollution in Izmir Bay, Trace elements. *Spectrochimica Acta Part B*. 55 : 1151-1164.

Badri, M.A.; and Aston, S.R. (1983) : Observation on heavy metal geochemical association in polluted and non-polluted estuarine sediment. *Environmental Pollution, Series, B*, 6 : 181-193.

Banat, K.; Forstner, U.; Muller, G. (1972) : *Naturwissenschaften* 59: 525-528.

Chopra, A.K.; Pathak, C.; and Prasad, G. (2009): Scenario of heavy metals contamination in agricultural soil and its management. *J. Applied Nat. Sc*, 1: 99-108.

Christensen, J. M. (1995) : Exposure to toxic metals; factor influencing interpretation of biomonitoring results, *Sci. Total Environ*, 166 : 89-135.

Domingo, J.L. (1994) : Metal induced developmental toxicity in mammals; A review, *Toxicol. Environ. Health,* 42 : 123-41.

Forstner, U.; & Muller, G. (1974) : Schwermetalle in Flüssen und Seen. *Springer Verlag Heidelberg,* pp : 203.

Galloway, J.N.; and Liken, G.E. (1979): *Limnol. Oceanog.* 24, 427.

Hamilton-Taylor, J. (1979): *Environ. SCi. Technol.* 13, 693.

Jumbe, A.S. and Nandini, N. (2009) : Heavy metals assessment of wetlands around Peenya industrial area, Bangalore, India. J. Res. *Environ. Life Sci,* 2 : 25-30.

Jumbe, A.S.; and Nandini, N. (2009): Impact assessment of heavy metals pollution of Vartur Lake, Bangalore. *J. Applied Nat. Sci,* 1 : 53-61.

Kansanen, P.H.; and Jakkola, T.(1985) : *Ann. Zool. Fenn.* 22 : 13.

Kishe, M.A., and Machiwa, F.J. (2003): Distribution of heavy metals in sediments of Mwanza Gulf of lake Victoria, Tanzania. *Environmental International,* 28 : 619-625.

MacFarlane, G.R.; Burchett, M.D. (2000) : Cellular distribution of Cu, Pb and Zn in the Grey Mangrove *Avicennia marina* (Forsk) Vierh. *Aquatic Botany* 68 : 45-59.

Muller, G. (1979) : *Umschau in Wissenschaft und Technik* 79, pp : 778-783.

Muller, G. (1980) : *Naturwissenschaften* 67, pp : 308-309.

Muller, G. (1991) : In : Muller, G. (Hrsg.) : 3. Neckar-Umwelt-Symposium 7.-8. Oktober 1991. *Heidelberger Geowiss. Abh.* 48. pp : 1-15.

Pekey, H. Karakas, D.; Ayberk, S.; Tolun, L. and Bakoglu, M.(2004): Ecological risk assessment using trace elements from surface sediments of Izmit Bay (Northeastern Marmara Sea) Turkey. *Marine Pollution Bulletin,* 48 : 946-953.

Peters, J.M.; Thomas, D.; Falk, H.; Oberdorster, G. and Smith, T. J. (1986): Contribution of meals to respiratory cancer, *Environ, Health, Perspect. 70 : 71-83.*

PILZ, B. (1999), Diplomarbeit, *Heidelberg University,* pp : 118.

Shukla, R.; and Sharma, Y.K. (2009) : Heavy Metal Toxicity in Environment. In: Environmental Monitoring and Management, Trivedi, A.; Jaiswal, K.; Pandey, B.N.; and Trived, S. P.i (Eds.). *Alfa Publications,* ISBN-978-81-907843-8-2, pp : 137-162.

Smith, S.L.; MacDonald, D.D.; Keenleyside, K.A.; Ingersoll, C.G. and Field, J. (1996) : A preliminary evaluation of sediment quality assessment values for freshwater ecosystems. *J Great Lakes Res,* 22 : 624–638.

31

Studies on Phytoplankton Community of Bhalki Tank, Bhalki, Karnataka, India

Shiddamallayya, N. and Pratima M.

ABSTRACT

A study of algal taxa and abundance of various groups of algae has been carried out from February-2002 to January-2004. A total 146 algal taxa of four classes were identified. Of which Chlorophyceae is the dominant group, followed by Cyanophyceae, Bacillariophyceae and Euglenophyceae. The abundance of Cyanophyceae ranged up to 520 O/l, Chlorophyceae 2720 O/l, Euglenophyceae 340 O/l and Bacillariophyceae 2500 O/l were recorded. Bacillariophyceae appeared in most of the season and Cyanophyceae remained least in the appearance in the water body.

Keywords

Abundance, Fresh water, Limnology, Phytoplankton, Community.

INTRODUCTION

Planktonic organisms form the first trophic level of the water body. It can be used as a good index of water quality. There is a great variation in the composition of plankton, not only in different regions and different depths and also at different seasons at different hours probably even in different years. The conditions that lead to maxima and minima, as well as to minor fluctuations in abundance of phytoplankton are complex in their physical, chemical and biological aspects. A considerable amount of work has been done in different types of fresh water bodies in relation to phytoplankton by Seenayya (1971), Berman and Pollinger (1974), Hegde *et al.* (1985), Pettersson (1990), Wilk-Wozniak (1998), Calijuri *et al.* (2002), Angadi *et al.* (2005). Although several qualitative observations and reports are available on both lotic and lentic aquatic ecosystems, information on man, made aquatic bodies is limited in Karnataka, especially in Hyderabad, Karnataka region. The selected work for the present paper is about the abundance of phytoplankton in Bhalki tank, Bhalki and it also to throw light on a fresh water body from the region of Hyderabad, Karnataka.

Bhalki is one of the towns in Bidar district and lies at 18° 03′ North latitude and 77° 12′ East longitudes, located at 590 metres above mean sea level. Bhalki tank is situated at the Northwestern side of the city (Figure-1). The tank covers an area of 3.18 square kilometer. The climate of the Bhalki city is dry, where temperature fluctuates between 10° C to 45° C. The total rainfall of Bhalki was 1455.4 millimeters during the study period. The tank is perennial and fed by Monsoon rain. The water level in tank ranged from 1.2 meter in Summer to 4.35 metres during Monsoon season. The mean depth of the tank during the study period was 2.8 metres. Tank water is used for agriculture, fisheries, and other domestic activities like bathing of cattle, washing of clothes, disposing of anthropogenic wastes and cleaning of vehicles.

Material and Methods

Phytoplankton samples collected at an interval of a month from February 2002 to January 2004. Ten litres of water from ten stations were filtered through plankton net (Mesh size No. 20) to get the plankton and preserved immediately in 100 ml plastic bottle by using 4% formaline. Preserved algal samples were scanned under microscope, measured by micrometer and identified. Algal taxa were named by following the keys and literature of Desikachary (1959), Prescott (1982), Sarode and Kamat (1984) and Anand (1998). Different algal groups couted separately by using Sedgewick-rafter cell under scanning microscope and readings are represented as organisms per litre (Edler, 1979).

Results and Discussion

Algal flora, periodical abundance, seasonal average abundance and percentage composition and seasonal dominance of phytoplankton of Bhalki tank are presented in Tables 1 to 4. The study of algal flora has shown the presence of 146 taxa of 4 different classes (Table-1). Of which Chlorophyceae was dominant with 82 species followed by Cyanophyceae (28 sp.), Bacillariophyceae (19 sp.) and Euglenophyceae (17 sp.).

Periodical abundance of phytoplankton of different classes of algae has shown wide variation in the water body (Table 2). Cyanophyceae members were not found during Monsoon and Winter months of 2002 and 2003. The highest count of 520 O/l was noted during Winter month of 2003. Cyanophyceae blooms in Winter season were also recorded by Wilk-Wozniak (1998) in Dam reservoir, Poland. A complete absence of Chlorophyceae members and highest count of 2720 O/l was observed during Monsoon month of 2003. Similarly, Kirsten and Arnold (1993) and Sadguru Prakash (2001) have also noticed BGA blooms in Monsoon respectively in Hjarbeek Fjord Lake, Denmark and Nai Basti Tal, Balrampur.

A complete disappearance of Euglenophyceae species has been noticed in few months of all seasons during the study period. The highest count of 340 O/l was recorded during Summer months of 2002. Bacillariophyceae members were completely absent during Monsoon and Winter months of 2003 and Winter month of 2004. The highest count of individuals 2500 O/l was recorded during the Summer month of 2002. Maichael and Tempte (1990) have also recorded the maximum appearance of diatoms during summer in Mendota lake, Wisconsin and also by Kirsten and Arnold (1993) and Parvateesam and Maneesha (1993) have found in their studies.

The minimum count of phytoplankton, including all algal groups found in Summer season of 2003 and maximum during Summer 2002 (Table 3). Seasonal percentage composition of phytoplankton has shown summer is favorable to Euglenophyceae, Monsoon to Chlorophyceae and winter to Cyanophyceae and Bacillariophyceae.

Seasonal dominance of phytoplankton abundance has shown Bacillariophyceae is dominated Euglenophyceae in summer season and on Chlorophyceae in all the seasons (Table 4). Similarly, Chlorophyceae dominated on Bacillariophyceae in Monsoon and on Cyanophyceae in winter in the successive year of the study period. It is also noted that, Cyanophyceae and Euglenophyceae are not dominated on any group of algae in through out the study period. It is a good sign that, the water body is not polluted.

ACKNOWLEDGEMENT

The authors are indebted to late Dr. S.B. Angadi, Reader, Dept. of Botany, for his valuable guidance. We also thank the head of the Dept. of Botany, Gulbarga University, Gulbarga; Assistant Director i/c, RRI, Bangalore and Director, CCRAS for providing the facilities and encouragement.

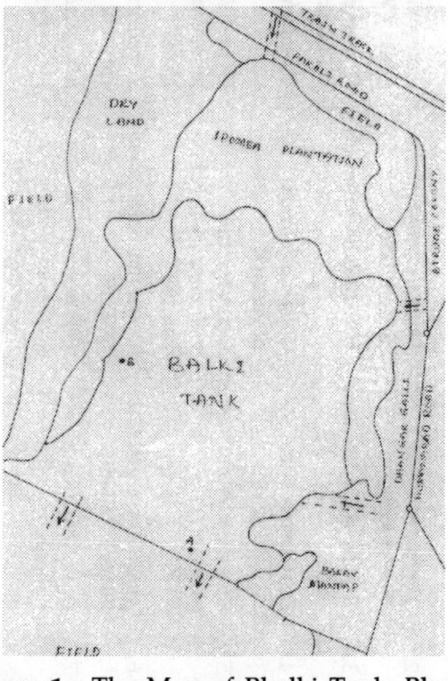

Figure 1 : The Map of Bhalki Tank, Bhalki.

Table 1 : Algal Flora of Bhalki Tank, Bhalki

S.No.	*Cyanophyceae*	*Chlorophyceae*		*Euglenophyceae*	*Bacillariophyceae*
1.	*Microcystis robusta* (Clark) Nygaard	*Chlamydomonas snowii* Printz	*Stigeoclonium nanum* Kutz.	*Trachelomonas acanthostoma* (Stokes) Deflandre	*Synedra acicularis*
2.	*Gloeocapsa atrata* (Turp.) Kutz.	*Pandorina morum* (Muell.)Bory	*Oedogonium crassum* (Hass.) Wittrock	*T. hispida* var. *coronata* Lemm.	*S. ulna* (Nitzch) Ehrenb.
3.	*G. luteo-fusca* Martens	*Volvox aureus* Ehrenb.	*O. princeps* (Hass.) Wittrock	*T. pulcherrima* Playfair	*Fragillaria brevistriata* Grun.
4.	*G. rupestris* Kutz.	*Chlorella vulgaris* Beyerinck	*O. spiralidens* Jao	*T. robusta* Swirenko Ehrenb.	*Cocconeis coliato*
5.	*Gloeothece rhodochlamys* Skuja	*Desmatractum bipyramidatum* (Chod.) Pascher	*O. undulatum* Fa. *senegalense* (Nordst.) Hirn.	*T. superba* var. *spinosa* Prescott	*C. placentula* Ehrenb. *Bulbochaete* sp.
6.	*Aphanocapsa grevillei* (Hass.) Rabenh.	*Characium rabenhorstii* DeToni	*T. volvocina* var. *punctata* Playfair	*Epithemia zebra* (Ehrenb.) Kutz.	
7.	*A. pulchra* (Kutz.) Rabenh.	*Pediastrum biradiatum* var. *emarginatum* Fa. *Convexum* Prescott	*Mougeotia scalaris* Hassal	*Euglena gracilis* Klebs.	*Surirella robusta* Ehrenb.
8.	*Aphanothece nidulans* Richter, P.	*P. duplex* Meyen	*Zygnemopsis hodgettsii* Transeau	*E. minuta Prescott*	*Cymbella affinis* Kutz.
9.	*Merismopedia tenuissima* Lem	*P. duplex* var. *gracilimum* West & West	*Spirogyra aequinoctialis* G.S. West	*E. oxyuris* Schmarda var. *minor* Prescott	*Gyrosigma acuminatum* Kutz. Rabenh

Cont...

S.No.	*Cyanophyceae*	*Chlorophyceae*		*Euglenophyceae*	*Bacillariophyceae*
10.	*Myxosarcinia spectabilis* Geitler	*P.duplex* var.*reticulatum* Lagerheim	*S. condensata* (Vauch.) Kutz.	*E. polymorpha* Dangeard	*Pleurosigma angulatum* (Quek.) Smith
11.	*Hydrococcus rivularis* Kutz.	*P. ovatum* (Ehrenb.) A. Braun	*S. marvillosa* Transeau	*Phacus acuminatus* var. *drezepolskii* Skvortzow	*Mastogloia recta* Hustest
12.	*Hyella caespitosa* Born. et Flah.	*P. simplex* (Meyen) Lemmermann	*S. maxima* (Hass.) Wittrock	*P. circumphlexis*	*M. smithii* Thwaites
13.	*Arthrospira jenneri* Stizenb. ex Gomont Rabenhorst	*P. simplex* var. *duodenarium* (Bailey)	*S. singularis* Nordstedt	*P. curvicauda* Swirenko	*M. smithii* var. *lacustris* Grun
14.	*Spriulina major* Kutz. ex Gomont	*P. tetras* (Ehrenb.) Ralfs	*Tolypella* (A.Braun) *Leonhardi*	*P. longicauda* (Ehrenb.) Dujardin	*Stauroneis phoenicenteron* Ehrenb.
15.	*Oscillatoria agardhii* Gomont	*P. tetras* (Ehrenb.) Ralfs var. *apiculatum*	*Chara vulgaris* linnaeus	*P. nordstedtii* Lemm. Fritsch	*Navicula cuspidata* Kutz.
16.	*O. curviceps* Ag. ex Gomont	*P. tetras* var. *tetraodon* (Corda) Rabenhorst	*Closterium acerosum* (Sch.) Ehrenb.	*Lepocinclis fusiformis* (Carter) Lemm.	*Pinnularia borealis* Ehrenb.
17.	*O. ornata* Kutz. ex Gomont	Sorastrum spinulosum Naegeli	*C. lanceolatum* Kg.	*L.ovum* (Ehrenb.) Lemm.	*P. stauroptera* Grun.
18.	*O. princeps* Vaucher ex Gomont	*Coelastrum microporum* Naegeli *in* A. Braun	*C. leibleinii* Kg.		*Gomphonema angustatum* Kutz.
19.	*Lyngbya aestuarii* Liebm. ex Gomont	*Scenedesmus acuminatus* var. *minor* G.M. Smith	*C. setaeum* Ehrenb.		*G. constrictum* Ehrenb.

Cont...

S.No.	*Cyanophyceae*	*Chlorophyceae*		*Euglenophyceae*	*Bacillariophyceae*
20.	*Nostoc linckia* (Roth) Bornet ex Born.et Flah.	*S. arcuatus* var. *platydisca* G.M. Smith	*C. tumidilum* Gay.		
21.	*Anabaena spiroides* Klebahn.	*S. armatus* (Chod.) G.M. Smith	*Pleurotaenium ehrenbergii* (Breb.) DeBary		
22.	*Anabaena sp.*	*S. bijuga* var. *alternans* (Reinsch) Hansgirg	*Staurastrum gutwinskii* Bern.		
23.	*Nodularia sp.*	*S. bijuga* (Turp.) Lagerheim	*S. javanicum* Nordst.		
24.	*Scytonema myochrous* (Dillw.) Ag. ex Born.	*S. denticulatus* Lagerheim	*S. manfeldtii* Delp. et Flah.		
25.	*Tolypothrix distorta* Kuetz. ex Born.	*S. dimorphus* (Turp.) kutz.	*Eustrum spinulosum* Delp. et Flah.		
26.	*Calothrix epiphytica* W. et G. S. West	*S. quadricauda* (Turp.) *de Brebisson in* de Brebisson & Godey	*Cosmarium angulosum* Breb.		
27.	*C. stagnalis* Gomont	*Crucigenia irregularis* Wille	*C. divergens* krieger.		
28.	*Rivularia aquatica* De Wilde.	*Oocystis gigas* Archer	*C. geminatum* Lund.		
29.		*O. pusilla* Hansgirg	*C. gostyniense* (Racib.) Groenbl.		
30.		*O. solitaria* var. *major* Wille	*C. granatum* Breb.		

Contd....

S.No.	Cyanophyceae	Chlorophyceae		Euglenophyceae	Bacillariophyceae
31.		*Closteriopsis longissima* Lemm.	*C. lundellii* Delp.		
32.		*Ankistrodesmus falcatus* (Corda) Ralfs	*C. margaritatum* Lund.		
33.		*A. radiatus* (Chod.) Lemm.	*C. medioscrobiculatum* West and West var. *egranulatum* Gutw.		
34.		*Quadrigula chodatii* (Tan-Ful.) G. M. Smith	*C. meneghinii* Breb.		
35.		*Actinastrum hantzschii* var. *elongatum* G.M. Smith	*C. ordinatum* Boerges		
36.		*A. hatzschii* var. *fluviatile* Schroeder	C. pseudo arctoum Nordst.		
37.		*Selenastrum westii* G.M. Smith	*C. quinarium* Lund.		
38.		*Ulothrix variabilis* Kutz.	*C. subprotumidium* Nordst.		
39.		*U. zonata* (Weber and Mohr) Kutz.	*C. subspeciosum* Nordst.		
40.		*Pithophora mooreana* Collins	*C. trachypleurum* Lund.		
41.		*Chaetophora sp.*	*Desmidium bengalicum* Turn.		

Table 2 : Periodical Abundance of Phytoplankton of Bhalki Tank, Bhalki (February, 2002 to January, 2004)

Months	*Cyanophyceae*	*Chlorophyceae*	*Euglenophyceae*	*Bacillariophyceae*
February-02	000	0340	340	0080
March	180	0100	340	0540
April	080	0100	040	0400
May	360	0560	000	2500
June	040	0480	000	1720
July	020	0560	100	0380
August	000	0340	000	0500
September	040	0420	060	0340
October	020	0160	000	0040
November	060	0160	000	0020
December	000	0080	040	0180
January-03	000	0170	000	1980
February	000	0120	020	0140
March	020	0120	000	0140
April	080	0140	060	0140
May	020	0080	000	0260
June	000	0000	000	0000
July	000	0020	000	0000
August	000	2720	080	0020
September	120	1720	100	0240
October	520	1040	240	0040
November	300	0340	000	0000
December	040	0100	040	0000
January-04	020	0100	000	0000

All the values are expressed in O/l

Table 3 : Seasonal Average Abundance and Percentage Composition of Phytoplankton of Bhalki Tank, Bhalki

Phytoplankton	*February 2002 to January 2003*			*February 2003 to January 2004*		
	Summer	*Monsoon*	*Winter*	*Summer*	*Monsoon*	*Winter*
Cyanophyceae	155(7.6)	25(2)	20(2.8)	30(9)	30(2.4)	220(31.7)
Chlorophyceae	275(13.6)	450(36)	143(19.6)	115(34.3)	1115(88.9)	395(56.8)
Euglenophyceae	720(35.5)	40(3.2)	10(1.4)	20(6)	45(3.6)	70(10.1)
Bacillariophyceae	880(43.4)	735(58.8)	555(76.2)	170(50.8)	65(5.2)	10(1.4)
Total	2030	1250	728	335	1255	695

Table 4 : Seasonal Dominance of Phytoplankton Abundance of Bhalki Tank, Bhalki (February, 2002 to January, 2004)

Season	*Phytoplankton*
Summer	Bacillariophyceae > Euglenophyceae > Chlorophyceae > Cyanophyceae
Monsoon	Bacillariophyceae > Chlorophyceae > Euglenophyceae > Cyanophyceae
Winter	Bacillariophyceae > Chlorophyceae > Cyanophyceae > Euglenophyceae
Summer	Bacillariophyceae > Chlorophyceae > Cyanophyceae > Euglenophyceae
Monsoon	Chlorophyceae > Bacillariophyceae > Euglenophyceae > Cyanophyceae
Winter	Chlorophyceae > Cyanophyceae > Euglenophyceae > Bacillariophyceae

REFERENCES

Anand, N. (1998) : 'Indian fresh water micro algae'. Center of advanced study in Botany, Madras. India.

Angadi, S.B., Shiddamallayya, N. and Patil, P.C. (2005): Limnological studies of Papnash pond Bidar (Karnataka). *J. Environ.Biol.* 26 (2): 213-216.

Berman, T. and Pollingher. U. (1974) : Annual and seasonal variations of phytoplankton chlorophyll and photosynthesis in lake kinneret. *Limnol. Oceanogr.* 19 : 31-54.

Calijuri, M.C., Dos Santos, A.C.A. and Jati, S. (2002) : Temporal changes in the phytoplankton community structure in a tropical and eutrophic reservoir (Barra Bonita, S.P.-Brazil). *J. plankt. res.* 24 (7) : 617-634.

Desikachary, T.V. (1959) : 'Cyanophyta' Indian Council of Agricultural Research New Delhi.

Edler, L. (1979) : Recommendation on methods for marine biological studies in the Baltic Sea. Phytoplankton and chlorophyll. *Baltic.* Marine *Biologists.* 5: 1-38.

Hegde, G.R., Bongale, U.D. and Bharati, S.G. (1985) : Diurnal variations of physico-chemical factors and phytoplankton in a pond and a lake near Dharwad, Karnataka state, India. *Comp. Physiol, Ecol.* 10(4) : 245-250.

Kirsten, O. and Arnold, N. (1993): Stress and disturbance in the phytoplankton community of a shallow, hypertrophic lake. *Hydrobiol.* 249 : 15-24.

Michael, J.V. and Tempte, J. (1990) : Seasonal patterns of grazing and nutrient limitation of phytoplankton in a Eutrophic lake. *Limnol. Oceanogr.* 35(3) : 697-709.

Parvateesam, M. and Maneesha, M. (1993) : Algae of Pushkar lake including pollution indicating forms. *Phykos.* 32 (1&2) : 27-39.

Pettersson, K. (1990) : The spring development of phytoplankton in lake Erken: Species composition, biomass, primary production and nutrient conditions-a review. *Hydrobiol.* 191 : 9-14.

Prescott, G.W. (1982) : Algae of the Western great lakes area. Otto Koeltz Science publishers. Germany.

Sadguru Prakash. (2001) : Seasonal dynamics of plankton in a fresh water body at Balrampur. *Geobios.* 28(1) : 29-32.

Sarode, P.T. and Kamat, N.D. (1984): 'Fresh water diatoms of Maharastra'. Saikrupa Prakashan. Aurangabad.

Seenayya, G. (1971) : Ecological studies on the phytoplankton of certain fresh waters ponds of Hyderabad. India. II. The phytoplankton. I. *Ibid.* 13 (1) : 55-88.

Wilk-Wozniak, E. (1998) : Late autumn mass development of *Woronichinia naegeliana* (Cyanophyceae) in a dam reservoir in southern Poland. *Biologia Bratislava.* 53(1) : 1-5.

32

Bioindicators

K.K. Sharma, Arti Sharma and Shvetambri

INTRODUCTION

Man's unplanned and unjustifiable manoeuvres of using natural resources as waste disposal sinks have impaired their quality so much, that they are usable in some instances neither for his own needs and purposes nor for the inhabiting micro and macro flora as well as fauna. As a result, the environment has undergone grave degradation and abuse due to toxicants released through pollution of air, water and soil. These unprecedented anthropogenic changes in the biosphere have resulted in the disruption of the delicate ecological balance.

Today the need to neutralize the impact of pollution is well documented. Among the holistic studies dedicated to pollution annihilation and sustainable development, the control of pollution through biological resources is globally accredited. Comprehending the fact that no man-made instrument can be as good as the biota itself, biological systems have proven to be the pertinent tools to discern the health of the milieu, an organism inhabits. Therefore, biological monitoring is a worthy technique employed in the conservation strategies in order to conserve the biological veracity of the ecosystems. Biomonitoring provides the direct evidences of alterations occurred in the ecosystem due to environmental pollution. Many species of microbes, plants, animals, individuals, populations and biological communities are effectively employed as biological indicators to define the environmental fluctuations. These organisms are very useful in forecasting the level and degree of pollution before the effect of effluent actually thickens.

A number of bioindicators or biomonitors are proposed based on their special behavioural responses to high bio accumulative ability, which offers appealing approaches for the bioremediation of pollution. A bioindicator can be defined as "a species or group of species that readily reflects the abiotic or biotic state of an environment, represents the impact of environmental change on a habitat, community, or ecosystem, or is indicative of the diversity, within an area". Bioindicators can be divided into environmental indicators (changes in physico-chemical environment), ecological indicators (impacts of factors on ecosystems), and biodiversity indicators (habitat assessment for conservation) (McGeoch, 1998).

Broadly speaking, bioindicators can be used to designate:

- a changing physical environment;
- a changing chemical environment, particularly with respect to various forms of pollution;
- the comparative quality or conservation value of habitat; and
- changes in the ecological status of the habitat with respect to time and place (Hodkinson and Jackson, 2005).

Biomagnification and bioaccumulation are some of relevant parameters through which resource portability for the flora and fauna can be adjudged. Lot of biomarkers at the cellular levels are developed due to the sensitive responses of the organisms, which speak about the toxic effects induced by effluence, and also explain the potential toxicological mechanisms.

General features of a good bioindicator are :

- It detects changes occurring in the natural environment;
- Monitors for the presence of pollution and its effect on the ecosystem in which it lives and;
- Monitors the progress of environmental clean up. (BioCanada, 2005).

Bioindicators are commonly grouped into accumulation indicators and response indicators.

Accumulation indicators store pollutants without any evident changes in their metabolisms.

Response indicators react with cell changes or visible symptoms of damage when taking up even small amounts of harmful substances.

Biomonitoring is divided into passive and active :

Passive biomonitoring is the use of organisms, organism associations, and parts of organisms which are a natural component of the ecosystem and appear there spontaneously.

Active biomonitoring includes all methods which insert organisms under controlled conditions into the site to be monitored.

A number of bioindicators are known to fulfil the requirements of convenience, consistency, rate, and evaluative capability and such bioindicators belong to both the plant and the animal realm. When the plants act as indicator species then they are designated as 'plant indicators' and when the animals are used as the indicators they are referred to as the 'animal indicators'. When the microbes are used in biomonitoring studies they are called as 'microbial indicators'.

Plant Indicators

Plants can effectively be used as cheap and naturally available monitoring tools to find out the type of air, soil and water pollution or any other environmental fluctuations in a particular habitat. Plants since very early times are being utilised not only for the food, fodder and shelter but also as a source to comprehend the altitude, soil condition, forest type, agricultural utility, mineral deposits and availability of the underground and surface water form an area.

Ranging from a plant as small as a phytoplankton to angiosperms all have definite potential to signify the biological integrity of an ecosystem. The dominant species are the most important as they sustain in the habitat for a longer time than the less prevailing ones. Many species of phytoplankton, mosses, lichens, ferns, and aquatic plants are very useful in pollution monitoring systems.

Plants as Indicators of Agricultural Utility

Plants that are aboriginal to an area certify about the availability of land for the agricultural use. Land having a growth of tall as well as small grasses indicates the accessibility of water and good fertility of the soil. Plants such as *Salvadora olevids, Zizyphus nuncularia, Prosopis cineraria* are the indicators of land suitable for the agriculture practice.

Plants as Indicators of Soil Characteristics

Plants such as *Citrullus colocynthis, Erianthus* sps., *Acacia nilotica, Calotropis* sps., *Agave* sps., *Opuntia* sps., *Argemone* sps., *Cassia auriculata* are all indicators of low soil moisture. Growth of plants such as *Rumex acetosa, Rhododendron* sps., *Polytrichum* sps., *Sphagnum* sps. are the indicators of saline soils. Plants like *Casuarina equisetifolia, Ipomea* sps., *Citrullus colocyntis, Arisuda* sps., *Cenchrus* sps., *Pennisetum* sps. and *Ermanthus* sps., show the sandy nature of the soil. Similarly, the progresses of plants such as *Imperata cylindrica, Vetiveria zizanioides* indicate clayey soils.

Plants as Indicators of Surface/Underground Waters

Many phreatophytes such as *Populus deltoids, Tamarix* sps., *Medicago sativa, Alhagi pseudalhagi, Isatis sabulosa* point towards the presence of underground waters. Growth of plants such as *Prosopis cineraria* indicate deep water table. Plants such as *Acacia greggi, Prosopis chilensis* point towards the presence of underground springs.

Plants as Indicators of Mineral Deposits

A number of metallicolus plants are found growing near rich deposits of many valuable minerals *viz.* gold, copper, zinc, cobalt, silver, diamond etc. These plants are often observed to be affected by the toxicity of these metals and exhibit various physiological disorders and indicate about the presence of the metal concentrations in their cells. These plants include *Equisetum arvense, Lonicera confuse, Papaver libonoticum, Vallozia candida, Erigonium ovalifolium, Stellaria setacea, Viscaria alpina* etc.

Plants as Indicators of Pollution

Some aquatic plants play a crucial role in maintaining the potable, recreational and aesthetic characteristics, as well as the ecological functioning of most waters. These plants display extreme plasticity in structure and morphology in relation to changing environmental conditions. They react slowly and progressively to changes in nutrient conditions, *i.e.* over several years. Macrophytes therefore function as integrators of environmental conditions to which they are subjected and thus can be used as long-term indicators with high spatial resolution (Melzer, 1999). Aquatic plants such as *Lemna* sps., *Wolffia* sps., *Utricularia* sps., *Chara* sps., generally grow in the nutrient rich waters and thus indicate about the shifting trophic status of the water body. Presence of *Eichhornia crassipes* indicate about the anoxic condition of a water body. Macrophytes are sensitive to changes in abiotic conditions, making them good indicators of health of the aquatic ecosystem.

Apart from the aquatic plants many phytoplanktonic species also reveal strong articulation concerning the health of the water bodies were they are abundantly found. Studies have revealed that pollutants have a profound effect on the growth rate and photosynthesis of the phytoplankton especially the algae. The growth and concentration of the nutrients are carried out by the phytoplankton in the presence of light. Sometimes when the effluents are coloured or disperses suspended solids in the aquatic medium then sufficient light may not be able to reach there and thus the growth of the algal species is observed to be directly affected. Pollutants in the form of heavy metals are sometimes seen to have a kindling effect on some phytoplankton species which eventually result in there dominance over the other phytoplanktons in a particular zone. *Example,* some concentrations of phenyl mercuric acetate stimulate the growth of *Chlorella* species. Some phytoplanktonic indicators of pollution and the eutrophic conditions are revealed as :

Bacillariophyceae : *Navicula* sps., *Merosira* sps., *Fragillaria* sps., *Synedra* sps., *Gomphonema* sps., *Diatomella* sps., *Tabellina* sps., *Pinnularia* sps. *Achnanthes* sps., *Nitszshia* sps., and Synedra sps.

Chlorophyceae : *Spirogyra* sps., *Nephrocytinum* sps., *Clostridium* sps., *Chlamydomonas* sps., *Pediastrum* sps., *Zygonema* sps., *Desmidian* sps., *Ankistrodesmus* sps., *Pandorina* sps., *Scendesmus* sps.

Cyanophyceae : *Chroococcus* sps., *Microcystis* sps., *Oscillatoria* sps., *Spirulina* sps.

Animal Indicators

A number of animal species including the individual species as well as the whole community are potential indicators of the effluence levels in an ecosystem. Bioaccumulation and biomagnification within the food chains gives an important insight into the extent of pollution load within the trophic levels. Biomonitoring studies can be initiated by selecting indicator animal species. Both invertebrates and vertebrates populations significantly depict the biological health of an ecosystem.

It is well observed that many invertebrates either individually or as a community show their responses at different levels of exposure to the changing trophic status of a habitat. Sometimes the organism's response to stress may involve changes in the morphology or the phenotype thus leading to a very intricate phenomenon called as 'phenotypic plasticity' or 'Cyclomorphosis'. This phenomenon is found conspicuously in both plants as well as the animals. Phenotypic plasticity is when an organism exhibit different phenotypes in response to its environment. It is believed that phenotypic plasticity evolved to allow organisms a greater chance of survival in their ever-changing surroundings. In animals there occurs a notable cyclic change in the morphological features. Cyclomorphosis may be controlled by environmental fluctuations and the changes in the predation regime. It is found to be well marked in many zooplankton organisms and is characterized by the formation of various structures of the lorica and carapace such as helmet, crest, hump and spine, correspondingly. These morphological intricacies exhibited by the animals signify about the levels of pollution generated stress in the aquatic ecosystems. Some pollution and the eutrophication indicating species in zooplankton are itemised:

Ascomorpha ovalis, Asplanchna herricki, Synchaeta grandis, Ploesoma hudsoni and Collothica libera (**indicators of oligotrophic waters**), *Kellicottia longispina,Keratella tropica Notholca squamula* (**indicators of oligo- mesotrophic waters**), *Anuraeopsis fissa, Keratella cochlearis, Trichocerca cylindrical, Asplanchna priodonta, Brachionus angularis, B. bidentatus, B.calyciflorus, B.caudatus, B.patulus, B.quadridentatus, Colurella sps., Euchlanis sps., Filinia longiseta., Philodina* sps., *Moin* sps., *Ceriodaphnia* sps., *Daphnia* sps., *Mesocyclops* sps., *and Cyclops sps.*(**indicators of eutrophy**), *Anuraeopsis fissa, A.navicula, Conochiloides coenobasis, Ploesoma triacanthum* (indicators of dystrophy).

Some macro benthic invertebrates also significantly point towards the levels of pollution in an aquatic ecosystem and these include *Tubifex tubifex, Limnodrilus* sps., *Chironomous* sps., *Culicoides* sps., *Pentaneura* sps., *Tabanus* sps., *Lymnea* sps., and *Physa* sps. etc. Further, it is observed that in some animals, revelation to pesticides, fertilizers, thermal pollution load and metal toxicity result in the distortion or asymmetry in the body shape and size of many invertebrates, usually measured either side of a bilateral line. This trend is referred to as FA or 'Fluctuating Asymmetry'. The index may increase with increasing stress levels and thus serve as simple way to monitor the impact of a wide range of known environmental stress factors (Hodkinson and Jackson, 2005).

Earthworms can be used as bioindicators, by measuring their biomass, or as indicators of biodiversity from the different earthworm species present, thus allowing good evaluation of soil quality (Pérès *et al.* 2006). Some species of tadpoles (*Rana temporaria*) have been observed to show morphological abnormalities directly, by producing a certain type of deformity, and exposure to environmental stress can increase its incidence. Some studies on wild seabirds suggest that they reflect the level of environmental contamination (especially of harmful metal cadmium), which should be monitored. Fishes are the vertebrates that are very strong indicators of the aquatic health. Some fish species such as *Salmo* sps., *Schizothorax* sps., *Tor sps.*, are found in oligotrophic and clean waters while *Channa* sps., *Heteropneustes fossilis, Labeo* sps.,*Garra* sps., *Trichogaster* sps., *Puntius* sps., *Mystis* sps., *Cyprinus sps. and Notopterus sps.* are the dwellers of eutrophic waters.

At the community level the responses are quite evident and these include species richness, species diversity and the presence of 'indicator' species. Sometimes a set of species or the structure and function of an entire biological community may function as a bioindicator. These indicator species specify the level of stress an ecosystem is experiencing and therefore, speaks about the health of an ecosystem. Moreover, the 'steno' species which are more sensitive are found to qualify as effective bioindicators as compared to 'eury' species.

Microbial Indicators

Micro-organisms respond rapidly to changing environmental conditions so that they are sensitive indicators of environmental health and commonly used for bio monitoring. Traditionally, indicator micro-organisms have been used to suggest the presence of pathogens (Berg, 1978). Microbial indicators are divided into the following three groups :

- **General (Process) Microbial Indicators** : It is a group of organisms that demonstrates the efficacy of a process, such as total heterotrophic bacteria or total coliforms.
- **Faecal Indicators (Such as E. coli)** : It is a group of organisms that indicates the presence of faecal contamination, such as the bacterial groups thermo tolerant coliforms or *E. coli*. Hence, they only infer that pathogens may be present

- **Index Organisms and Model Organisms** : It is a group or species indicative of pathogen presence and behaviour respectively, such as *E. coli* as an index for *Salmonella* and F-RNA, coliphages as models of human enteric viruses (Nicholas *et al.* 2001).

Bacteria should be excellent indicators of the early signs of degradation caused by human intervention because they have the highest surface area to volume ratio of all organisms. Use of bacteria, by analysing the products of their metabolism and identifying the main functional groups to which they belong, underlines the dual aspect of biodiversity and bioindication (Pompili *et al.* 2006).

Some important species of bacteria that are the persuasive indicators of the pollution in the air, water and soil are enumerated as: *Escherichia coli, Enterococcus* sps., *Streptococcus* sps., *Clostridium perfringens, Bifidobacterium asteroides, B. indicum, B. coryneforme, Bacteroides fragilis, Microviridae* sps., *Siphovirida sps., Salmonella typhimurium, Neurospora* sps., *Aspergillus* sps., *Nostoc microscopium, Haplosiphon welwitschii, Scolecobasidium* sps., *Mortierella* sps., *Pencillium* sps., *Candida quillermondii, Aureobasidium pullulans.*

The future holds numerous possibilities for the detection of indicator and pathogen index organisms. In the future, bioindicators may be used more widely in the holistic studies involving the effluence level and environmental impact assessment stage to the remediation and post-remediation monitoring stages.

REFERENCES

Bio Canada (2005) : Bioindicators. www.biobasics.gc.ca.

Berg, G. (1978) : The indicator system. In Indictors of Viruses in Water and Food (ed. G. Berg), Ann Arbor Science Publishers, Ann Arbor, MI. pp. 1–13.

Hodkinson, I. D., Jackson, J.K (2005) : Terrestrial and aquatic invertebrates as bioindicators for environmental monitoring, with particular reference to mountain ecosystems. *Environ Manage.* 35(5) : 649-66.

McGeoch, M.A. (1998) : The selection, testing and application of terrestrial insects as bioindicators. *Biol. Rev.* 73 : 181-201.

Melzer, A. (1999) : Aquatic macrophytes as tools for lake management *Hydrobiologia* 395/396 : 181–190.

Nicholas J.A., Willie O.K. Grabow and Mario S (2001) : Indicators of microbial water quality World Health Organization. Water Quality: Guidelines, Standards and Health. Edited by Lorna Fewtrell and Jamie Bartram. Published by IWA Publishing, London, UK. ISBN : 1 900222 280

Pérès, G., D. Cluzeau, C.Ferrand and D. Peron. (2006) : Earthworms used as indicators of agricultural management. In : BIO-BIO Project. Biodiversity-Bioindication to evaluate soil health. R.M. Cenci and F. Sena (Eds), EUR 22245 EN, 107-115, Office for the Official Publications of the European Communities, Luxembourg.

Pompili, L., A.S. Mellina, A Benedetti and J. Bloem. (2006) : Microbial indicators for assessing biological fertility status of soils. In : BIO-BIO Project. Biodiversity-Bioindication to evaluate soil health. R.M. Cenci and F. Sena (Eds), EUR 22245 EN, 46-54, Office for the Official Publications of the European Communities, Luxembourg.

33

Impact of Hydroelectric Project on Adjacent Flora in Ganga Valley of District Uttarkashi (Uttarakhand)

Ramdas, G.K. Dhingra, Ravindra Joshi and R.K. Gupta

ABSTRACT

Formation of a Dam always results in alteration of normal pattern of existence of flora and fauna, it also create disturbances on the normal living of plants and animals. Many aquatic animals coordinate their reproductive cycles with annual flood seasons. Every flood is valuable in that it takes nutrients from the land and deposits them in the river, providing food for the stream's residents. Hydroelectric dams affect Natural Cycles, Armoring the Riverbed, Sedimentation, Starving the River, Temperatures, Fish Migration, growth of Cultivated and Wild plants etc. The study area is concentrated around the Loharinag Pala, Hydro project and affected villages *viz.* Sukhi, Bhangeili, Sunagar, Tehar, Kunjan, Barsu, Pala Maradi, Kyarki, Salang, Bhukhi and Huri of Uttarkashi district in Uttarakhand State. The present research work deals with the vegetation structure and community patterns within and between various plant communities in the Loharinag Pala Dam sites *viz.* Mukh Disposal, Edit Point, Near the Tunnel, Dumping Zone around the Working area and Dam affected villages of this valley. The analysis was carried out taking various aspects into consideration like measurement of frequency of valuable plant species and harmful weed species, Impacts of chemicals, dust particles released by dynamite and vehicles etc. used in project construction on cultivated crops, Phyto and Zooplanktons, Loss of valuable medicinal plants found around the project area.

During the present investigations, the biomass of harmful weeds *viz. Lantana camara* (74.5%), *Eupatorium adenophorum (77.50%) and Ageratum conyzoides* (65.80%) were reported high from the entire affected zones as compared to unaffected zones. Frequency of climber plant species ranged from 7-13 respectively and especially this climber species (*Clematis montana*) was reported maximum. In all sites including dam affected villages Herbs were recorded maximum (frequency 115-195) followed by shrubs and trees. Three parasitic herbs, 03 twiners and 05 small trees were also noticed during this study. Negative impacts on the cultivated crops like loss of production, early and late flowering were also reported. A total of 14 phytoplankton and 9 species of Zooplankton's were observed at submerged zones.

Keywords

Edit point, Lantana camara, Mukh Disposal, Profound impacts, Phytoplanktons, Sedimentation, Submerged zones, Tunnel, Twiners, Zooplanktons etc.

INTRODUCTION

The district Uttarkashi is located between 38° 28´ to 31° 28´ North latitude and 77° 49´ to 79° 25´ East longitude. It covers an area of 8016 km^2. Kinnaur district of Himachal Pradesh and territory of Tibet form the Northern boundary of the district and Eastern boundary is formed by a part of Tibet and district Chamoli, while the Southern and Western boundaries are formed by the districts of Tehri and Dehradun respectively. Two major rivers of India *viz.* Bhagirathi (called Ganga beyond Deoprayag) and Yamuna have their origin in the snow covered peaks of Gaumukh and Bandarpunch respectively in Uttarkashi. These peaks and rivers are of high reverence to people all over India and are holy centers of pilgrimage since time immemorial. The sustenance of these mountain ranges depends on the surrounding forests (Awasthi *et al.* 2001).

The widely varying climate and topography produce a wide range of vegetation and serve as habitats to diverse species of wild life. Forests occupy a place of pride in the environment of the district not only for the sheer bulk of the area they occupy but also for the richness of variety of vegetation. As much as 88 per cent of the total area of the district is administered by the forest department. Pine forests occur between the altitudes of 900-2000 meters and Deodar forests between 2000-3000 meters. Fir and Spruce forests are found over 3000 meters and Kharshu, Birch and Junipers forests upto the height of 4000 meters. Above the fir and Spruce forest zone, alpine pastures are found throughout the district between the heights of 3500 meters to 4877 meters above Sea level. Rich varieties of grass, shrubs and herbs come up during June–September while during the remaining part of the year these areas remains covered with snow. A large number of medicinal plants of great commercial value grow spontaneously in the forests.

The Ganga (Ganges) basin extends over more than 1 million square kilometers and encompasses parts of India (about 80% of the total basin area), Nepal, China and Bangladesh. The length of the main channel is some 2,525km, while altitude ranges from 8,848m in the high Himalayas, to sea level in the coastal deltas of India and Bangladesh. The Ganga valley captures the entire gradient of Himalayan habitats ranging from subtropical Sal to Alpine meadows within a distance of about a hundred km of the river from its confluence to source. A broad based landscape ecological approach was adopted for biodiversity characterization in this valley. The waters of the Ganga carry one of the highest sediment loads anywhere in the world, with a mean annual total of 1.6 billion tones, compared to 0.4 billion tones for the Amazon. The Ganga basin is one of the most populous regions on Earth, home to 450 million people at an average density of over 550 individuals per square kilometer. Fisheries along the river are of considerable economic value and their output makes a major contribution to regional nutritional needs. The global herbal trade is of the order of $15-20 billion. The medicinal plant not only cures the ailment but also provides employment opportunities to the rural women via cultivation and marketing but also attempt to forge a linkage between people, forests and health sector (Maikhuri *et al.* 2002).

MATERIALS AND METHODS

The study area is concentrated around the Loharinag Pala Hydro project area and affected villages *viz.* Sukhi, Bhangeili, Sunagar, Tehar, Kunjan, Barsu, Pala Maradi, Kyarki, Salang, Bhukhi and Huri of Uttarkashi district in Uttarakhand state. The present research work deals with the vegetation structure and community patterns within and between various plant communities in the Loharinag Pala Dam sites (Mukh Disposal, Edit point, near to tunnel and around the working zone) and Dam affected villages of this valley. Wild flora and cultivated crops related data were studied within and around the affected villages and around the dam sites. The sampling was done within 1km on either side of the river bed and in dam affected villages.

The project site was located in the highland forest zone (1000-3000 m). Systematic sample plots were used to quantify various parameters of vegetation. Considering the terrain and topography of the area to be covered, quadrat method was used for sampling of the vegetation. Taking into consideration, the size of the vegetation patches, ten random quadrats of 10 × 10 m size were laid to study the trees and shrubs and ten random 1 x 1 m quadrats were laid to study the herbaceous component at each site. During the survey, number of plants of different species in each quadrat was identified and counted. Based on the data, frequency, density and cover (Basal area) of each species were calculated. During the vegetation survey, herbaria were prepared for the plants those had flowers. The Red Data Book of India and other available literature, flora and herbarium pertaining to the rare/endangered species of western Himalayas were referred to identify the endemic, rare and other threatened categories of plants. In Dam affected villages cultivated crops related data were reported with the help of local peoples. Some major impacts on their cultivated crops were also noticed during field survey.

For studying the diversity of phytoplankton and zooplankton, bulk water samples were collected in polythene jars for obtaining net plankton from water sample, 150ml of water was filtered through a 50 µm net and was centrifuged at 1500 rpm for 10 minutes. The sediment of the centrifuge tubes was made to volume of 5ml. An aliquot of 0.5 ml of this concentrate was used for identifying phytoplankton and zooplankton population. A plankton chamber of 0.5 capacities was used for identification of plankton under a light microscope and studied its impact on biodiversity. Earlier the data was analyzed for density, frequency, species richness and species diversity. During the present research work following parameters were studied :

- Frequency of valuable plant species and harmful weed species.
- Impacts of chemical, dust particles, released by dynamite and vehicles etc. used in project construction on cultivated crops, Phyto and Zooplanktons.
- Loss of valuable medicinal plants found around the project area.

RESULTS AND DISCUSSIONS

The biomass of *Lantana camara* (74.5%), *Eupatorium adenophorum* (77.50%) and *Ageratum conyzoides* (65.80%) were high for the entire affected zones as compared to unaffected zones. Frequency of climber plant species ranged from 7-13. In all sites including dam affected villages, herbs were recorded in maximum frequency (168-375) as compared to shrubs and trees. Three parasitic herbs (*Cuscuta europea, C. reflexa and Orobanche epithymum*), 03 twiners (*Dioscorea belophylla, D. bulbifera and D. deltoidea*) and 05 small trees (*Aegle marmelos, Eriobotrya japonica, Lindera pulcherrima, Litsea monopetala* and *Pyrus malus*) were also noticed during field study. Shrubs were found in frequency of 75-120 while trees were ranged from 25-35 respectively. *Lonicera* sp., *Rosa* sp. *Rubus* sp. *Juniperus* sp. *Indigofera* sp. and *Berberis* sp. were reported in maximum frequency while *Adhatoda* sp., *Buddleja asiatica, Callicarpa macrophylla, Caryopteris odorata, Cassiope fastigiata,* and *Coriaria nepalensis* were noticed in minimum frequency respectively. Negative impacts on cultivated crops like loss of production, early and late flowering were also reported by local peoples. A total of 14 phytoplankton and 9 species of Zooplanktons were observed at submerged zones. Maximum frequencies of Phytoplanktons were recorded to class Chlorophyta and minimum in class Dinophyta while in Zooplanktons maximum frequency were observed in class Rotifera (99%) and minimum in class Rhizopoda respectively. Highly medicinal plant species *viz. Aconitum heterophyllum* (Atis), *Bacopa monarii* (Brahmi), *Allium strychi* (Faran), *Orchis mascula* (Salam mishri), *Emblica officinalis* (Amla), *Withania somnifera* (Aswagandha), *Swertia chiraita* (Chiraita), *Commiphora wightii* (Guggul), *Solanum nigrum* (Makoi), *Berberis aristata* (Kingore) and *Podophyllum hexandrum* (Bankakari) were directly affected by the cementing of dam sites.

Due to maximum frequency of *Lantana, Eupatorium and Ageratum,* an increase in temperature was recorded around the construction area. In affected villages, a loss of fertility in agricultural land was reported and growing seasons were changed which ultimately led to low crop production. Loss of biodiversity was resulted in disruption of ecosystem services important for agriculture, such as pollination by bees and impact on fisheries. *Pinus wallichiana* (Kail) and *Alnus nepalensis* (Utis) were the dominant tree species observed in all dam affected villages. *Quercus semicarpofolia* was the other species. *Debergeasia longifolia* and *Desmodia tilaefolium* were the major shrubs observed in the area. *Geranium nepalensis, oxalix coniculata, Artemisia nilagirica* etc. were the dominant herb species at these dam affected zones while around the working sites (mukh diposal, barrage site, edit point) *Eupatorium adenophorum, Lantana camara, Ageratum conyzoides* weeds were dominant plant species. *Cyanthula tomentosa* and *Rubus ellipticus* were the major shrubs observed at Aungi, Huri, Bhukhi and Gangnani zones.

The growth of *Lantana* was highest in its own associations in the Loharinag Pala dumping zone. Cultivated crops viz.

Eleusine coracana (Mandua), *Fagophyrum esculantum* (Ogal), *Fagophyrum tataricum* (Fafra), *Phaseolus* spp., *Hordeum vulgare* (Jau), *Glycine* spp., T*riticum aestivum* (wheat), *Panicum miliaceum* (Cheena), *Hordeum luinalayans* (Uana), *Amaranthus blitum* (Chaulai), *Echinochloa frumentacea* (Jhangora) and all vegetables were indirectly affected by the dam construction. Negative impacts on these cultivated crops like loss of production, early and late flowering were also reported by locally affected inhabitants. This study was also supported by the studies of Caballero *et al.* (2007), De Witt *et al.* (2007), Hattermann *et al.* (2008), Minville *et al.* (2007), and Bronstert *et al.* (2007).

Rivers tend to be fairly homogenous in temperature. Reservoirs, on the other hand, are layered. They are warm at the top and cold at the bottom. Fish passage is a concern with dams. Many fishes must move upstream and downstream to complete their lifecycles. Dams are often built without fish ladders. When fish ladders are provided, they seldom work as needed. If enough adult fishes do manage to climb above a dam, there remains the issue of their young that how will they get back downstream. Many are killed by predators while they wander, lost, in the reservoir above the dam. Many are killed in their fall downward through the dam to the river below. Most of the rivers originating in the high Himalayas are nourished by snow and glaciers and flow to the south to provide water to millions. They also threaten the livelihood of people of the Trans-Himalayan region through floods, erosion, landslides, water-borne epidemics, etc. Water resources of Nepal can be profitably utilized for power generation, irrigation, domestic uses, aquaculture and recreational fisheries. Cold water fish of Bhagirathi are facing problems due to an increasing number of hydropower projects. Once abundant indigenous fish stocks have been declining due to overfishing, harmful fishing practices (electro fishing, dynamiting, use of chemicals), pollution and developmental works. Developmental works such as river damming have a major impact on river ecology, aquatic flora and fauna, including fish. Cold water fish of Nepal are affected by the increasing number of hydropower dams in the country.

Figure 1 : Hydro project construction depleting important vegetation.

Figure 2 : Cemented water spilling into River Bhagirathi.

Figure 3 : Slide showing vegetation loss through project debris.

Figure 4 : Destruction of herbs and Shrubs due to Dam.

CONCLUSION

We concluded by this study that hydroelectric projects are threatening wild floral vegetation and cultivated plants of the region. Hydroelectric dams affect Natural Cycles, Armoring the Riverbed, Removing Sediment, Starving the River, Changing Temperatures and Stopping Fish Migration and check growth of wild and cultivated plant species.

ACKNOWLEDGEMENT

The authors are grateful to Forest Department of Uttarkashi and Irrigation department, Uttarkashi for their supports in the form of providing data and relevant literature. Thanks are also due to local villagers who gave invaluable time, cooperation and hospitality during field visits. RKG and GKD are thankful to UGC, New Delhi and UCOST, Dehradun, Uttarakhand for financial assistance.

REFERENCES

Awasthi A. (2001) : Ecological impacts of anthropogenic pressures on high altitude forests along, the upper Catchment of river Bhagirathi. Ph. D. Thesis submitted to the Deemed University, Forest Research Institute Dehradun.

Bronstert, A., Kolokotronis, V., Schwandt, D. and Straub, H. (2007): Comparison and evaluation of regional climate impact analysis: General scheme and application example, *Int. J. Climatol.* **27** : 1579–1594.

Caballero, Y., Voirin-Morel, S., Habets, F., Noilhan, J., LeMoigne, P., Lehenaff, A., and Boone, A. (2007): Hydrological sensitivity of the Adour-Garonne river basin to climate change, *Water Resour. Res.* **43**, W07448. Doi : 10.1029/2005WR004192.

De Witt, M.J.M., Van den Hurk, B., Warmerdam, P. M. M., Torfs, P.J.J.F., Roulin, E., and van Deursen, W. P. A. (2007) : Impact of climate change on low-flows in the river Meuse, Climatic Change. **82** : 351–372.

Hattermann, F., Post, J., Krysanova, V., Conradt, T., and Wechsung F. (2008) : Assessment of Water Availability in a Central-European, River Basin (Elbe) under Climate Change, Adv. Clim. Change Res. Suppl. **4** : 42–50.

Maikhuri, R.K., Nautiyal, S. and Rao, K.S. (2002) : Nanda Devi Jaivmandel Arakshit Kshetra Mein Aushadhiya Padapon Ki Paramparagat Krishi Takneak, Sustainable Development of Rural Ecosystem programme, G.B. Pant Institute of Himalayan Environment and Development, Kosi-katarmal, Almora.

Minville, M., Brisette, F., and Leconte, R. (2008) : Uncertainty of the impact of climate change on the hydrology of a Nordic watershed, *J. Hydrol.* **358** : 70- 83.

34

Bioremediation

K.K. Sharma, Arti Sharma and Shvetambri

INTRODUCTION

Bioremediation refers to the productive use of biodegrdative processes to remove or detoxify pollutants that have found their way into the environment and threaten public health, usually as contaminants of soil, water, or sediments (Crawford and Crawford, 1996). It seems to be a good alternative to the conventional clean up technology and has been used worldwide with varying degree of success. In this process, the organic wastes are biologically degraded under controlled conditions to an innocuous state (Mueller *et al.*, 1996).

Bioremediation has originated from the combination of two words; 'bio' (biological) and 'remediation' (remedy). It is an option that offers the possibility to destroy or render harmless contaminants using natural biological activity, often be carried out on site, using low cost, low technology techniques, and generally having high public acceptance (Vidali, 2001).

Bioremediation can occur on its own (natural) or can be spurred via addition of fertilizers so as to increase the bioavailability within the medium. Recent advancements have been proven to be successful via addition of microbial strains to the medium so that the resident microbe population ability is enhanced to break down the contaminants. Microbes may be indigenous to the contaminated area or they may be isolated from elsewhere and brought to the contaminated site.

Biodegradation of the compound is often the result of action of multiple organisms which enzymatically attack the pollutant and convert them to harmless product.

Bioremediation can be broadly classified as :

- *In-situ* Bioremediation
- *Ex-situ* Bioremediation

In-situ Bioremediation

In this, Bioremediation can be used at the site of contamination. These techniques is generally more desirable due to less cost & less disturbance and provides treatment in place avoiding excavation and transport of contaminants.

Main types of bioremediation are (BioCanada, 2005) :

1. Biostimulation
2. Bioaugmentation
3. Intrinsic Bioremediation
4. Biosparging

Biostimulation

Bioremediation can occur on its own (natural) or can be spurred via addition of fertilizers/nutrients/oxygen in a liquid or gas form to the soil to encourage the growth of bacterial activity within the medium. The disappearance of contaminants is monitored to ensure that remediation occurs.

Bioaugmentation

Micro-organisms that clean up a particular contaminant are added to the contaminated soil or water. It is more commonly and successfully used on contaminants removed from the original site, such as in municipal waste water treatment facilities. This method has not been very successful when done at the site of contamination because it is difficult to control site conditions for optimal growth of micro-organisms added.

Intrinsic Bioremediation

Also known as 'Natural attenuation', occurs naturally in contaminated soil or water. This natural bioremediation is the exertion of micro-organisms and is seen in petroleum contamination sites such as old gas stations with leaky underground tanks.

Biosparging

It involves injection of air under pressure below water table to increase ground water oxygen concentrations and enhance the rate of biological degradation of contaminants by naturally occurring bacteria. It increases mixing of saturated zone by increasing contact between soil and ground water.

Ex-situ Bioremediation

This technique involves the excavation or removal of contaminated soil from ground. The most common methods are: **Land farming, Composting, Biopiles, Bioreactors.**

Land farming

It is a special technique in which contaminated soil is excavated and spread over a prepared bed and periodically tilled until pollutants are degraded. The goal is to stimulate indigenous biodegradable micro-organisms and facilitate the aerobic degradation of contaminants. It has received much attention due to its less monitoring and maintenance costs.

Composting

It is a technique that involves combining contaminated soil with non hazardous organic amendants such as manure or agricultural wastes. The presence of these organic materials supports the development of a rich microbial population and elevated temperature characteristic of composting.

Biopiles

They are a hybrid of land farming and composting.

Essentially, engineered cells are constructed as aerated composted piles (Raskin and Ensley, 2000). Typically, used for treatment of surface contamination with petroleum hydrocarbons. Biopiles, provide favourable environment for indigenous aerobic and anaerobic micro-organisms.

Bioreactors

Bioremediation in reactors involves the processing of contaminated solid material (soil, sediments, and sludge) or water through an engineered contaminant system. The rate and extent of biodegradation is greater in bioreactor because the contained environment is more manageable, controllable and predictable. However, the contaminated soil requires pre-treatment or alternatively the contaminant can be stripped from the soil via washing or physical extraction before being placed in a bioreactor.

USAGE OF BIOREMEDIATION

Air Pollution Control

Biological Gas Treatment System (Biofilters, Biofilms, Bioscrubbers, Biotrickling Filters) : In bioscrubbers and biotrickling filters, multiple microbial communities are grown on solid surface area to produce multi layered complexes called 'biofilms'. When gas streams containing organic pollutants are passed through these systems, the pollutants are degraded. Degradation by these systems is more than 95 per cent of the organic compounds in the gases. Some fungi like *Candida tropicalis* are able to assimilate styrene, toluene, xylenes and propene used chiefly in the making of synthetic rubber, resins and plastics. This fungus is grown on a ceramic solid support. The mycelium of the fungi gives the biofilters a large surface area and greater capacity to eliminate pollutants.

Replacement for Petrochemicals (Hydrogen Fuels)

Biotechnology research for producing hydrogen for use as automotive fuel that will not contribute to global climate change is of particular interest. Major problem of air pollution is because of combustion of gases and productivity of carbon dioxide and other harmful gases. Certain algae are also found to be good sources of this clean fuel.

Biodegradable Plastics

Biodegradable plastics made of microbially produced polyhyroxybutyrate have been recently developed and are being marketed. A strain of bacterium *Alcaligenes eutrophus* has been isolated. This bacterium produces copolyester of 3- and 4-hydroxybutyrate. The cells of this bacterium produces 80 percent of their biomass as this polymer, when grown on 1, 4 butanediol.

Biodesulphurisation

The BIO-SR is microbial system devised to remove inorganic sulphur from industrial air emissions. Industrial exhaust gas containing hydrogen sulphide passes through an acidic ferric sulphate solution which oxidizes the sulphide to elemental sulphur while reducing the ferric ions to ferrous ions.

An iron and sulphur oxidizing bacterium *Thiobacillus ferroxidans* is grown in a bioreactor at a pH between 2 and 3 which regenerates the ferric sulpahte solution. Bacterium *Rhodococcus sps.* biodegrades dibenzothiophenes and other organosulphur compounds in the fuel. Biodesulphurisation produces a cleaner fuel that does not produce SO_2 emission when it burns.

Reversal of Global Warming

Besides looking for hydrogen and other fuels that would not contribute to global climate change there are also simultaneous efforts to develop systems to remove carbon dioxide already emitted into the atmosphere from the burning of carbon based fossil fuels. Various microbes, algae and marine organisms are used to remove this green house gas and convert carbon dioxide to various organic compounds.

The alga, *Chlorococcum littorale* has been growing in the bio reactors. This alga tolerates high levels of carbon dioxide and converts it to polysaccharides. *Prasinococcus capsulatus* alga coverts CO_2 into extracellular mucilaginous polysaccharide. Marine foraminifera are also well documented in converting carbon from CO_2 into calcium carbonate.

Intensive Plantations

Planting all the available lands from village fields, to community land, to road/rail sides to counteract the pollution is globally well acknowledged. Indigenous and exotic species both can be used for plantation of the land.

- **Afforestation :** It involves the plantation of more and more trees on the barren lands.
- **Agroforestry :** It is the most successful scheme wherein farmers establish plantations of Eucalyptus, Casuarina, Poplar, Teak etc., on their agricultural lands.
- **Strip Plantation :** Saplings of tall growing tree species are established in multiple lines along the railway tracks, canal bunds and roadsides.
- **Homostead Agroforestry :** In this method trees, crops, vegetables are grown in a small area around home to help control pollution.
- **Reforestation :** It involves plantation of trees on those lands where trees have already been grown earlier.

Water Pollution Control

- **Removal of Organic Wastes :** Some waste water system, originally designed only to remove organic compounds aerobically to reduce BOD, have been modified to include anaerobic zone to remove inorganic compounds as well. The reactor contains the methylotrophic bacteria such as *Methylophilus methylotrophus* that carry out denitrification.

Methanol is first added to bioreactor to support the growth of methylotrophs and later removed by trickling filter and granular activated carbon filters before the water is discharged for use. The bacteria converts nitrate to nitrite and then to molecular nitrogen, which is released into the atmosphere.

• **Removal of Toxic Chemicals from Industrial Waste Water** : New waste treatment systems can also maintain populations of desired specific micro-organisms that can biodegrade toxic compounds like hydrocarbons and chlorinated solvent found in industrial plant waste water. Bacteria like *Pseudomonas cepacia* are able to biodegrade chlorinated hydrocarbon present in the heptachlor, chlordane etc.

• **Biogas (Energy) from Solid Wastes and Refuse** : In this process both anaerobic and aerobic microbes are used. The first phase of treatment, bio waste is manually sorted and treated in hammer mills then anaerobic digestion takes place in a larger tank lasting for about three weeks.

Radioactive Pollution Control

Microbes like algae, bacteria and fungi as well as higher plants have capabilities to uptake these pollutants. After uptake these either accumulate or are assimilated by them. Accumulated heavy metals are recovered for recycling or disposal.

- *Zooglea ramigera* absorb copper and cadmium.
- *Rhizopus arrhizus* uranium and thorium
- *Pencillium chrysogenum* absorb radium.
- Bacteria *Thiobacillium thiooxidans* brings about bio leaching of Zinc, Cobalt and Nickel.
- Fungi *Trichoderma* sps., *Aspergillus* sps., *Ophiostoma* sps. and *Rhodotorula* sps. plays important role in detoxification of the industrial effluents and heavy metals.
- Water Hyacinth (*Eichhornia crassipes*) has come into prominence for purifying industrial and radioactive waste. Plant regenerates rapidly and has tremendous capacities to accumulate heavy and radioactive metals. The polluted water fed into reservoirs or lagoons with

Water Hyacinth becomes markedly clean and free from 70-90% of pollutants.

Noise Pollution Control

Noise pollution means the unwanted sound dumped into the atmosphere leading to health hazards. Noise pollution is because of industries, heavy traffic, and electric equipments. All these have added to the noise nuisance in the environment. The WHO has fixed 45db as the safe noise level for a city.

Biological control of noise pollution involves the creation of vegetation cover along the highways, streets and other places. Trees absorb and dissipate sound energy and thus act as buffers zones. Trees that operate as good buffers include Ashoka, Neem, and Tamarind etc.

Soil Pollution Control

- Use of organic manure and prevention of synthetic fertilizers and pesticide is one of the biological methods to control soil pollution.
- Biofertilization is one of the biological methods to control soil pollution in which certain microorganisms are used to increase the fertility of the soil.
- Reuse of materials such as plastics, glass, cloth, paper, etc. This can be effectively reused at domestic levels and solid wastes soil pollution can be greatly abridged.
- Recycling and recovery of materials.
- Reforesting.
- Aerobic or anaerobic decomposition of biodegradable domestic and municipal wastes is one of the biological methods to control soil pollution that can be followed in treating wastes.

Bioremediation vis-a-vis Biotechnology

Nature has mechanisms for self renewal. The role of

biotechnology in bioremediation is to efficiently apply these existing mechanisms to clean up environmental contamination. Various microorganisms are studied to see if they remediate various chemicals present at contaminated industrial stations. Also, scientists are currently looking into certain genetically engineered microorganisms to increase their ability to metabolize specific chemicals such as hydrocarbons in contaminated sites.

The elimination of a wide range of pollutants and wastes from the environment is an absolute requirement to promote a sustainable development of our society with low environmental impact. New methodological breakthroughs in sequencing, genomics, proteomics, bioinformatics and imaging are producing vast amounts of information. Functional genomic and metagenomic approaches are increasing the understanding of the relative importance of different pathways and regulatory networks to carbon flux in particular environments and for particular compounds and they will certainly accelerate the development of bioremediation technologies and biotransformation processes (Diaz, 2008).

Bioremediation vis-a-vis Natural Resources

It is very frequently used in natural resources industry like forestry, mining & energy.

For example :

Bioremediating Oil and Gas Industry : The oil industry uses bacteria to clean up pollution created by spills and underground leaks and to clean up waste production from oil production.

Bioremediating Contaminated Sediment : Organic wastes have contaminated many of world's harbours, lakes, rivers. They affect water quality and aquatic life sometimes combining with other naturally existing compounds in water to produce foul smelling toxic gases. The most common way to deal with sediment contamination is to scoop it and move it away for long term storage. But these are expensive and disrupt the natural habitat.

Bioremediation and Sustainable Development : Bioremediation uses the resources available in nature to clean up contamination and also less disturbing to the environment. Using biological processes means lower costs compared to chemical treatment processes which are expensive.

ADVANTAGES

There are number of cost effective advantages of bioremediation which may be employed in areas that are otherwise inaccessible without excavation e.g. Hydrocarbon spills (specifically petrol spills) or certain chlorinated solvents may contaminate groundwater.

Bioremediation is a natural process and is therefore, perceived by the public as an acceptable waste treatment process for contaminated material such as soil. Microbes or 'microscopic bugs' that live in soil underground water like to eat certain harmful chemical substances such as those found in gasoline and oil spills. When microbes completely degrade these chemicals, they change them into water and harmless gases like carbon dioxide.

BIOREMEDIATION

Constraints

- Bioremediation is limited to those compounds that are biodegradable. Not all compounds are susceptible to rapid and complete degradation.
- There are some concerns that the products of biodegradation may be more persistent or toxic than the parent compound.
- Biological processes are often highly specific. Important site factors required for success includes the presence of metabolically capable microbial populations, suitable environmental growth conditions and appropriate levels of nutrients and contaminants. It is difficult to extrapolate from bench and pilot scale studies to full-scale filed operations.

- Research is needed to develop and engineer bioremediation technologies that are appropriate for sites with complex mixtures of contaminants that are not evenly dispersed in the environment. Contaminates may be present as solids, liquids and gases.
- Bioremediation often takes longer than other treatment options such as excavation and removal of soil or incineration.

BIOREMEDIATION

Future

Even though bioremediation holds great promise for dealing with intractable environmental problems, it is important to comprehend that how microorganisms interact in different hydrologic environments.

Bioremediation integrates the tools of many disciplines. As each of the disciplines advances and as new cleanup needs arise, opportunities for new bioremediation techniques will emerge. As these new techniques are brought into commercial practice, the importance of sound methods for evaluating bioremediation will increase. However, further research and better education of those involved in bioremediation will improve the ability to apply the strategy and understanding of the fundamentals behind bioremediation (National Academy of Sciences, 1993).

REFERENCES

BioCanada (2005) : Bioremediation. www.biobasics.gc.ca.

Crawford, R and Crawford, D.L. (1996) : Bioremediation: Principles and applications. Cambridge University Press. Published by The Press Syndicate of University of Cambridge.

Diaz, E. (2008) : Microbial Biodegradation: Genomics and Molecular Biology (1st ed.). Caister Academic Press. ISBN 978-1-904455-17-2. ISBN 1904455174.

Mueller, J.G., Cerniglia, C.E. and Pritchard, P. H. (1996) : Bioremediation of Environments Contaminated by Polycyclic Aromatic

Hydrocarbons. In Bioremediation : Principles and Applications, Cambridge University Press, Cambridge : 125–194.

National Academy of Sciences (1993) : *In Situ* Bioremediation: When Does it Work? Committee on In Situ Bioremediation, National Research Council ISBN: 978-0-309-04896-5, 224 pages, 6 × 9, hardback

Raskin, I. and Ensley, B.D. (2000) : Phytoremediation of Toxic Metals: Using Plants to Clean Up the Environment, Wiley, New York.

Vidali, M (2001) : Bioremediation. An Overview. *Pure and Appl. Chem.* 73(7) : 1163-1172.

35

Histological Alterations in *Oreochromis Mossambicus* (Peters) Due to Retting Activities in Kadinamkulam Estuary, South India

R. Sreejai and D.S. Jaya

ABSTRACT

Studies on water physico-chemical characteristics and assessment of histopathologial changes in fishes were carried out in the selected stations of the Kadinamkulam estuary in Kerala, South India during different seasons. The results showed increased biochemical oxygen demand, hydrogen sulphide and organic pollutants in the retting areas of the estuary. Histopathological studies were conducted in *Oreochromis mossambicus* (Peters), one of the abundant species in the estuary. The microscopic observations showed distinct histological alterations in liver such as appearance of blood streaks, vacuolation of hepatic cells and necrosis. In gills, vacuolar degradation, clubbing of secondary gill lamellae and hyper chromatic cells were seen. Inflammatory cells, vacuolation and degradation of tubules, swollen renal tubules and necrosis were seen in the kidney of fishes collected from the retting areas of the estuary in different seasons. From these observations it is obvious that the fishes were subjected to toxicity due to the different gaseous and organic pollutants produced during the coconut husk retting activity in the Kadinamkulam estuary.

Keywords

Kadinamkulam, Estuary, Oreochromis mossambicus, Retting, Histopathology, Physico-chemical characteristics.

INTRODUCTION

The backwaters of Kerala provide waterfront for several major and small-scale industries, amongst which coir-retting industry ranks first. This industry provides employment to a large number of people, especially women folk. The raw material for the industry is obtained by immersing the coconut husk in water for eight to nine months. Retting of coconut husk presents unique and extremely serious problems along the coastal belt, changing the hydro-ecology of the water body (Nandan, 2002). The present investigation was carried out in the Kadinamkulam estuary (latitude 8° 35' to 8° 40' N and longitude 76° 45' to 76° 52' E), the largest among the estuaries in Thiruvananthapuram district, Kerala, South India. The average depth of the Kadinamkulam estuary is 8 m. This estuary is polluted mainly by coconut husk retting activities.

Retting of coconut husk is basically a biological process involving the release of a variety of biochemical compounds. The conventional coconut husk retting causes extreme pollution due to the production of organic wastes and the liberation of hydrogen sulfide, which is a potent inhibitor of aerobic respiration. Sulphide is an important environmental factor and toxicant for aquatic plants, invertebrates and fishes (Bagarinao, and Vetter, 1992). The cell membranes are highly permeable to sulphide and it is therefore, very likely that sulphide quickly penetrates the gill epithelium of fish. Sulphide entering the blood may interact with haemoglobin (Hb) by forming sulphaemoglobin (SHb). The major biochemical lesion caused by sulphide appears to be the inhibition of cytochrome 'c'oxidase, the terminal enzyme in the electron transport chain of mitochondria.

Fishes are relatively sensitive to changes in their surrounding environment including an increase in pollution. Fish health may thus reflect, and give a good indication of the health status of a specific aquatic ecosystem. Early toxic effects of pollution may, however, only be evident on cellular or tissue level before significant changes can be identified in fish behavior or external appearance. One of the great advantages of using histopathological biomarkers in environmental monitoring is that this category of biomarkers allows examining specific target organs, including gills, kidney and liver that are responsible for vital functions such as respiration, excretion and the accumulation and biotransformation of xenobiotics in fishes.

Histological analysis appears to be a very sensitive parameter and is crucial in determining cellular changes that may occur in target organs, such as the gills, liver and gonads (Dutta, 1996). The histological changes occur earlier than reproductive changes and are more sensitive than growth or reproductive parameters and, as an integrative parameter, provide a better evaluation of organism's health than a single biochemical parameter. Therefore, the major objective of the research study was to find out the histopathological changes in the tissues of fishes in the Kadinamkulam estuary due to anthropogenic activities and the physico-chemical characteristics of the water in the study stations of the estuary.

MATERIALS AND METHODS

Four different stations (SI, SII, SIII and SIV) in the Kadinamkulam estuary were selected for the study. The station I-Kotrakiri area in Kadinamkulam estuary is one of the interior most inlets of the estuary and is fully used for retting purpose. The station II- Kadinamkulam, represents one of the main arms of the estuary. Retting, sand mining, clay mining and land filling activities are still going on, navigation and such other estuary based activities are high. The station III-Madanvila area is a typical retting pit (a semi enclosed retting yard) situated on the banks of the estuary. The station IV- Perumathura is situated at the marine zone of the estuary close to the Lakshadweep Sea and with no retting activity.

The fish *Oreochromis mossambicus* (Peters) is one of the abundant fish species in the Kadinamkulam estuary and was selected to study the histopathological changes in important organs due to aquatic pollution. Healthy female fishes (n=10) of body weight (30 ± 5 gm) and body length (12 ± 1.4 cm) were captured using a fisherman's net from four selected stations of Kadinamkulam estuary in pre monsoon, monsoon and post monsoon seasons during the period 2004 - 2005. The surface and bottom water samples from the four stations during the different seasons were also collected in plastic bottles using a water sampler. Through trans-spinal dissection liver, kidney and gill tissues were removed, gently rinsed with the physiological saline solution (0.9% NaCl) and immersed in fixative composed of glacial acetic acid, formaldehyde and ethanol (1 : 3 : 7). The fixed tissues were passed through alcohol series 50 to 100% for dehydration, and finally cleared in chloroform. Tissues were then impregnated with paraffin wax (58°C) and were embedded. After embedding, the sections were cut in 6 micron thickness using a rotary microtome, and then stained with haematoxylin and eosin (H&E). The stained slides were rendered transparent with xylene and mounted in DPX (Gurr, 1962). The stained sections were observed under a research microscope attached with digital camera (Leica, Germany). Appropriate sections and portions of tissues were photographed for the purpose of description and comparison using the Leica image analyzer software. The various physico-chemical parameters of water samples were analyzed following the procedures described in APHA (1995) and by Grasshoff (1983).

RESULTS AND DISCUSSION

Physico-Chemical Characteristics of Water

The results of the physico-chemical characteristics of bottom and surface water samples of the four different stations of Kadinamkulam estuary is depicted in Table 1 to 7. The water quality parameters analysed indicate that the estuary is moderately to grossly polluted in different seasons. The intense colour and odour of the water samples from the retting zones (SI, SII, and SIII) and the extremely high values obtained for the concentration of hydrogen sulphide, BOD, phenols, total hardness and low dissolved oxygen content in different study stations of the Kadinamkulam estuary indicate the degradation of the water quality in the estuary. Most of the physico-chemical characteristics (pH, sulphates, Total Nitrogen, H_2S, Turbidity, Total Suspended Solids, Total Solids, BOD and Total Hardness) of the water in the retting areas (SI, SII and SIII) of the estuary showed values above the permissible limits of Indian Environmental Standards for the discharge of effluents into inland surface water (Raman and Devotta, 2006). In station IV, the hydrogen sulphide concentration is below detectable limit and BOD values are less in all the seasons studied, and these results show that this station is less polluted compared to other stations.

Histopathological Studies

The histopathological changes in the liver, gills and kidney of the fishes collected from different stations during different seasons are given in Plates 1, 2, 3, 4, 5 & 6.

Changes in Liver : The fish liver appears, as does the liver of other vertebrates, as a key organ which controls many life functions and plays a prominent role in fish physiology, both in anabolism and catabolism (Brusle *et al.* 1996). Liver, the first organ to face any foreign molecule through portal circulation is subjected to more damage. The parenchymatous hepatic tissue in teleosts has many important physiological functions and also detoxification of endogenous waste products as well as externally derived toxins, drugs, heavy metals and pesticides (Roberts and Rodger, 2001).

The microscopic observations of the hepatic tissues of fishes in the different study stations of the Kadinamkulam estuary showed structural alterations. In station IV fishes, the histopathological alterations include appearance of blood streaks (ABS), glycogen granules (GG) visible nucleus (VN) and degradation of hepatic cytoplasm (DHC) in premonsoon, monsoon and post monsoon seasons. The liver samples of fishes collected in premonsoon season from station I showed vacuolation of hepatic cells (VHC), appearance of blood streaks and hyperchromatic nucleus in fishes. The liver tissues of fishes

(SI) collected during monsoon showed necrosis, vacuolation of hepatic cells and appearance of blood streaks. The hepatic tissues of fishes (SI) collected during post monsoon season showed appearance of blood streaks and degradation of hepatic cytoplasm.

The liver tissues of fishes collected from station II, exhibited appearance of blood streaks, vacuolation of hepatic cells. In monsoon season, the SII fish tissue showed degradation of hepatic cytoplasm, visible nucleus and in post monsoon season visible nucleus, appearance of blood streaks and degradation of hepatic cytoplasm. In station III during pre monsoon season the liver tissue exhibited vacuolar degradation (VD) and necrosis (NE), while vacuolation and nuclear hypertrophy were observed in monsoon liver samples. Necrosis, lymphocyte infiltration and degraded cells are detected in hepatic tissues of fishes (SIII) collected during post monsoon season.

Therefore, liver of *Oreochromis mossambicus* in the retting areas of the estuary showed marked histopathological changes. Degeneration and necrosis of the hepatocytes in the fishes may be due to the cumulative effect of dissolved pollutants and the increase in their concentrations in the liver (Chang *et al* 1998). The cellular degeneration in the liver may be due to oxygen deficiency as a result of gill degeneration and/or to the vascular dilation and intravascular haemolysis (Mohamed, 2001). Pacheco and Santos (2002) described increased vacuolisation of the hepatocytes as a signal of degenerative process that suggests metabolic damage, possibly related to exposure to contaminated water.

Changes in Gills : The gill is the main organ for different functions, such as gas exchange, ion regulation and excretion of metabolic waste products (Wood, 1991). Its complexity and constant contact with the external environment make the gill the first target to waterborne pollutants (Perry and Laurent, 1993; Fracacio *et al.* 2003). Gill histology of fishes in the study stations in different seasons are briefly illustrated in Plate 3 and 4.

The histological analysis of Station IV gill tissues showed clubbing of secondary gill lamellae (CLB), hyperchromatic nuclei and curling of secondary gill lamellae (CSGL). The Station I tissue showed clubbing of secondary gill lamellae, filling up of space between secondary gill lamellae (FSGL) and curling of secondary gill lamellae (CSGL) in pre monsoon season, clubbing of secondary gill lamellae and curling of secondary gill lamellae in monsoon season. In post monsoon season the tissue samples of station I exhibited curling of secondary gill lamellae, clubbing of secondary gill lamellae and hyperchromatic nucleus. The Station II gill tissues showed clubbing of secondary gill lamellae and dilation of primary gill lamellae (DPGL) in pre monsoon, clubbing of secondary gill lamellae and DPGL in monsoon season. In post monsoon season the gill tissues (SII) showed clubbing of secondary gill lamellae, curling of secondary gill lamellae and filling up of space between secondary gill lamellae (FUS). In SIII the tissues exhibited clubbing of secondary gill lamellae, filling up of space between secondary gill lamellae and curling of secondary gill lamellae in pre monsoon season. The tissue showed filling up of space between secondary gill lameilae and clubbing of secondary gill lamellae in monsoon, and in post monsoon season the tissue exhibited curling and clubbing of secondary gill lamellae, filling up of space between secondary gill lamellae.

The histopathological changes observed in the gills of *Oreochromis mossambicus* in the estuary may be an adaptive mechanism to prevent the entry of the pollutants through gill surface and the findings of this study are in good agreement with the reports of Rao *et al.* (2003). Dhanapakiam *et al.* (2002) states that the tannery effluent and the associated pollution of hydrogen sulphide can cause severe damage in the surface structure of gills in *Labeo rohita*. Poleksic and Mitrovic-Tutundzic (1994) in their studies found anomalies such as clubbing of secondary gill lamellae, filling up of space between secondary gill lamellae and curling of secondary gill lamellae, hyperplasia, dilation of the marginal channel and aneurysms in Neotropical fish, *Astyanax altiparanae*, collected from Cambe stream, which corroborates the hypothesis that the water of the Kadinamkulam estuary is really contaminated and that exposure to this water causes structural damage to fish gills.

Changes in Kidney : The kidneys of *Oreochromis mossambicus* are paired elongated structures above the gut and below the vertebral column. Each kidney opens through a duct which may fuse to form a single duct known as urino-genital sinus. Kidney histology of fishes in the selected stations of the estuary is briefly illustrated in Plate 5 and 6.

Histopathological analysis of the kidney tissues of retting zones of Kadinamkulam estuary showed no gross morphological changes but progressive degenerative changes in kidney histology were noticed. The kidney tissue of fishes from the retting zones (station I) of Kadinamkulam estuary exhibited inflammatory cells, swollen renal tubules, and decrease in size of epithelial cells (DEC) in pre monsoon tissues, increased tubular lumen (ITL) in monsoon tissues samples, vacuolation and degradation of tubules (VDT), swollen renal tubules and glomerulus were found in post monsoon tissue samples. And SII tissue samples showed swollen renal tubules, formation of vacuoles in the renal interstitial tissue (FVRT) and decrease in size of the epithelial cells in pre monsoon tissues. In monsoon samples decrease in size of the epithelial cells, formation of vacuoles in the renal interstitial tissue and swollen renal tubules are seen but in post monsoon season there is increased tubular lumen, swollen renal tubules and vacuolation and degradation of tubules were observed.

In Station III samples swollen renal tubules, necrosis (NE) and heavy infiltration of inflammatory cells with some debris (HCS) were observed. The pre monsoon season tissue sample showed vacuolation and degradation of tubules, while increased tubular lumen and swollen renal tubules were seen in monsoon. Decrease in size of the epithelial cells and swollen renal tubules are observed in the fish kidney in post monsoon season. In the kidney of fishes in Station IV, only changes like formation of vacuoles in the renal interstitial tissue and hyper trophiod convoluted segments were observed in premonsoon, monsoon and postmonsoon seasons.

The teleostean kidney is one of the first organs affected by contaminants in the water (Thophon *et al.* 2003). In the present study the common alterations found in the kidney of fishes exposed to water contamination are vacuolation, degradation of tubules, swollen renal tubules and necrosis. Handy and Penrice (1993) found swollen renal tubules, inflammatory cells and melanomacrophages in the kidney of trout (*Salmo trutta*) and tilapia (*Oreochromis mossambicus*) exposed to mercuric chloride. Similar histological alterations were found in fishes exposed to organic contaminants (Veiga *et al.* 2002) and mixed environmental contaminants (Schwaiger *et al.* 1997; Pacheco and Santos, 2002). The presence of inflammatory cells, swollen renal tubules, decrease in size of epithelial cells, vacuolation and degradation of tubules coupled with necrosis in kidney of fishes in the retting areas (SI, SII and SIII) indicate that the kidney suffered damage due to the exposure to gaseous (H_2S) and organic pollutants in Kadinamkulam estuary.

CONCLUSIONS

The study on the physico-chemical characteristics of estuary water showed that the Kadinamkulam estuary is facing serious water quality deterioration and pollution stress due to human interventions like coconut husk retting, sand mining etc. The below detectable levels of dissolved oxygen and high amount of hydrogen sulphide makes this water body as a puddle of toxic pollutants. The histopathological investigations in the liver, gills, and kidney of *Oreochromis mossambicus* in the estuary indicate that the fishes were responding to the direct effects of the contaminants as much as to the secondary effects caused by stress.

The histological alterations like appearance of blood streaks, vacuolation and necrosis of hepatic cells, curling and clubbing of secondary gill lamellae, dilation of primary gill lamellae, and the vacuolation and degradation of tubules, swollen renal tubules and necrosis in kidney authenticate the intensity of coconut husk retting effluents integrate with cell function of the different organs of this fish, even though it is a tolerant species in such contaminated environmental conditions.

Table 1 : Seasonal Variations in the Physical Characteristics of Surface and Bottom Water Samples

Stations	Temperature						pH						Electrical Conductivity						Turbidity					
	PrM		M		PoM		PrM		M		PoM		PrM		M		PoM		PrM		M		PoM	
	SW	BW	SW	BW	SW	BW	SW	BW	SW	BW	SW	BW	SW	BW	SW	BW	SW	BW	SW	BW	SW	BW	SW	BW
SI	28	30	24	27	34	34	6.82	6.97	6.7	6.6	7.3	7.11	29.7	29.6	32.4	30.9	30.1	28.6	24.6	26.5	36.5	33.1	22.4	20.9
SII	28	31	25	26	31	32	7.54	7.57	6.5	6.4	7.53	7.53	32.4	32.1	36	35	29.2	30.6	25.4	19.6	35	32	10.8	14.6
SIII	29	32	26	27	31	31	7.62	7.68	6.3	6.3	7.52	7.54	29.7	29.6	32.4	30.9	30.1	28.6	27.6	20.8	34.8	32.4	21.3	22.8
SIV	27	31	26	27	33	33	6.48	6.31	6.8	6.8	7.62	7.62	32.0	29.8	30.6	31.5	27.8	27.6	16.7	13.3	21.4	21.8	13.2	12
Seas. Av.	28	31	25.3	26.7	32.5	32.7	7.12	7.13	6.58	6.53	7.49	7.45	30.95	30.28	32.85	32.08	29.3	28.85	23.58	20.05	31.93	29.83	16.93	17.58

Seas. Av. = Seasonal Average; PrM = Pre monsoon; M = Monsoon; PoM = Post monsoon; SW = Surface water; BW = Bottom water

Table 2 : Seasonal Variations in the Chemical Characteristics of Surface and Bottom Water Samples

Stations	Total Dissolved Solids (mg/L)						Total Suspended Solids (mg/L)						Total Solids (mg/L)					
	PrM		M		PoM		PrM		M		PoM		PrM		M		PoM	
	SW	BW	SW	BW	SW	BW	SW	BW	SW	BW	SW	BW	SW	BW	SW	BW	SW	BW
SI	90.2	90.4	60.4	60.2	60.8	60.4	82	78	114	98	88	84	172.2	168.4	174.4	158.2	134.8	117.4
SII	180.2	180.2	124.3	124.4	120.8	120.3	78	70	96	88	66	62	258.2	250.2	219.3	212.4	186.8	182.3
SIII	360	360.2	240	241.2	780	781	62	54	86	74	78	60	422	414.2	326	315.2	858	841
SIV	460.4	461	660	660.5	660	659.4	66	55	112	101	66	68	526.4	516	772	761.5	748	727.4
Seas. Av.	272.7	272.95	271.18	271.58	405.4	405.28	72	64.25	102	90.25	74.5	68.5	344.7	337.2	372.93	361.83	481.9	467.03

Seas. Av.=Seasonal Average; PrM = Pre monsoon; M = Monsoon; PoM = Post monsoon; SW = Surface water; BW=Bottom water

Table 3 : Seasonal Variations in the Chemical Characteristics of Surface and Bottom Water Samples

Stations	Total Dissolved Solids (mg/L)						Total Suspended Solids (mg/L)						Total Solids (mg/L)					
	PrM		M		PoM		PrM		M		PoM		PrM		M		PoM	
	SW	BW	SW	BW	SW	BW	SW	BW	SW	BW	SW	BW	SW	BW	SW	BW	SW	BW
SI	1.45	1.42	1.13	1.12	0.4	0.4	18.64	19.78	15.02	15.2	20.23	21	16	14	12.8	12.4	16	16
SII	1.73	1.72	1.14	1.14	0.4	0.3	17.64	18.3	20.14	20.26	21.3	21.3	16	14	12.42	12.08	14.95	14
SIII	1.2	1.2	1.12	1.02	0.3	0.2	18.5	17.48	18.93	18.68	26.4	26.8	17.7	16.4	14.2	14.2	17.9	17.2
SIV	5.9	5.8	5.74	5.74	6.12	6.02	6.14	6.2	6.18	6.02	4.3	3.8	BDL	BDL	BDL	BDL	BDL	BDL
Seas. Av.	2.57	2.54	2.28	2.26	1.81	1.73	15.23	15.44	15.07	15.04	18.06	18.23	12.43	11.1	9.86	9.67	12.21	11.8

Seas. Av. = Seasonal Average; PrM = Pre monsoon; M = Monsoon; PoM = Post monsoon; SW = Surface water; BW = Bottom water

Table 4 : Seasonal Variations in the Chemical Characteristics of Surface and Bottom Water Samples

Stations	Total Dissolved Solids (mg/L)						Total Suspended Solids (mg/L)						Total Solids (mg/L)					
	PrM		M		PoM		PrM		M		PoM		PrM		M		PoM	
	SW	BW	SW	BW	SW	BW	SW	BW	SW	BW	SW	BW	SW	BW	SW	BW	SW	BW
SI	48.66	48.68	38.46	38.52	48.22	48.26	121.6	122.1	120	120	122.6	122.4	10.04	10.03	16.68	16.67	18.44	18.27
SII	32.86	32.74	32.64	32.68	38.4	38.46	122.9	122.6	126	126.2	120	120	16.07	16.07	18.69	18.68	16.41	16.40
SIII	39.62	39.66	34.78	34.82	36.32	36.4	121.4	121	122.2	122.4	126	126	15.70	15.70	17.40	17.40	16.40	16.42
SIV	32.26	32.34	34.46	34.48	30.28	30.24	125	125	135	135	131	131	22.50	22.50	21.60	21.60	25.44	25.44
Seas. Av.	38.35	38.35	35.09	35.13	38.31	38.34	122.72	122.67	125.8	125.9	124.9	124.85	16.07	16.07	18.59	18.58	19.17	19.13

Seas. Av. = Seasonal Average; PrM = Pre monsoon; M = Monsoon; PoM = Post monsoon; SW = Surface water; BW = Bottom water

Table 5 : Seasonal Variations in the Chemical Characteristics of Surface and Bottom Water Samples

Stations	Nitrite-Nitrogen (mg/L)						Nitrite-Nitrogen (mg/L)						Total Nitrogen (mg/L)					
	PrM		M		PoM		PrM		M		PoM		PrM		M		PoM	
	SW	BW	SW	BW	SW	BW	SW	BW	SW	BW	SW	BW	SW	BW	SW	BW	SW	BW
SI	0.89	1	0.46	0.68	0.95	1	1.82	1.94	0.98	1.2	2.97	3.21	3.51	3.48	3.85	3.85	3.8	3.81
SII	1.94	1.82	0.5	0.7	1.15	1.16	2.10	2.04	0.74	1.21	1.74	1.96	3. 83	3. 88	3. 84	3. 84	3.82	3.81
SIII	1.93	1.96	0.69	0.76	1.16	1.17	1.96	1.42	1.21	1.28	2.10	3.10	3. 92	3. 94	3. 95	3. 96	3.89	3.9
SIV	0.14	0.15	0.12	0.18	0.17	0.25	1.62	1.86	0.24	0.26	1.28	1.29	2.39	2.3	2.44	2.44	2.26	2.21
Seas. Av.	1.25	1.25	0.45	0.58	0.85	0.85	1.88	2.32	0.79	0.99	2.15	2.69	2.95	2.89	3.14	3.14	3.44	3.43

Seas. Av.=Seasonal Average; PrM=Pre monsoon; M=Monsoon; PoM=Post monsoon; SW=Surface water; BW=Bottom water

Table 6 : Seasonal Variations in the Chemical Characteristics of Surface and Bottom Water Samples

Stations	Silicates (mg/L)						Phosphate-Phosphorus (mg/L)						Sulphates (mg/L)					
	PrM		M		PoM		PrM		M		PoM		PrM		M		PoM	
	SW	BW	SW	BW	SW	BW	SW	BW	SW	BW	SW	BW	SW	BW	SW	BW	SW	BW
SI	28.6	28.8	70.6	60.2	56.4	56.8	0.82	0.9	0.24	0.26	0.74	0.78	0.61	0.59	0.58	0.54	0.74	0.72
SII	57.3	57.2	94.8	87.6	49.6	49.4	0.97	0.98	0.42	0.42	0.7	0.81	0.69	0.66	0.64	0.66	0.7	0.71
SIII	62.8	63	96.0	82.58	59.6	59.6	0.96	1.04	0.62	0.68	0.89	0.94	0.72	0.78	0.62	0.61	0.81	0.83
SIV	23.3	23.4	116.3	114.4	49.4	49.2	0.5	0.62	0.3	0.41	0.56	0.61	0.23	0.22	0.13	0.11	0.22	0.23
Seas. Av.	43.0	43.1	94.43	86.20	53.7	53.7	0.81	0.89	0.40	0.44	0.72	0.79	0.56	0.56	0.49	0.48	0.62	0.62

Seas. Av.=Seasonal Average; PrM=Pre monsoon; M=Monsoon; PoM=Post monsoon; SW=Surface water; BW=Bottom water

Table 7 : Seasonal Variations in the Chemical Characteristics of Surface and Bottom Water Samples

Stations	Sodium (mg/L)						Phosphate (mg/L)						Total Phenols (mg/L)					
	PrM		M		PoM		PrM		M		PoM		PrM		M		PoM	
	SW	BW	SW	BW	SW	BW	SW	BW	SW	BW	SW	BW	SW	BW	SW	BW	SW	BW
SI	18.7	18.5	19	19	18.2	18.4	13.7	13.6	15.8	15.7	13.6	13.5	0.11	0.11	0.09	0.08	0.12	0.12
SII	16.4	16.7	12.8	12.7	16.7	16.9	12.3	12.2	12.6	12.7	12.8	12.6	0.12	0.12	0.08	0.07	0.11	0.11
SIII	21.3	21.6	20.4	20.3	22.1	22.3	16.3	16.3	16.3	16.4	17.6	17.4	0.14	0.14	0.11	0.11	0.16	0.15
SIV	32.6	32.4	32	32	38	38	22.8	22.6	23.4	23.6	24.1	24.2	BDL	BDL	BDL	BDL	BDL	BDL
Seas. Av.	22.25	22.3	21.0	21	23.7	23.9	16.27	16.17	17.0	17.1	17.02	16.9	0.1233	0.1233	0.0933	0.0867	0.13	0.1267

Seas. Av.=Seasonal Average; PrM=Pre monsoon; M=Monsoon; PoM=Post monsoon; SW=Surface water; BW=Bottom water

The analysis of the seasonal variations in the histological parameters of fish tissues leads to the conclusion that the changes observed in the liver, gills and kidney were apparently related to the concentration of pollutants in the Kadinamkulam estuary. The study also confirms that histological alterations in fishes are good biomarkers for field assessment of pollution load, especially in tropical areas that are naturally subjected to a multiplicity of environmental variations.

ACKNOWLEDGEMENTS

The authors gratefully thank Dr.Sobhana, CMFRI, Kochi, Kerala for her help and constructive suggestions in histological analysis.

REFERENCES

APHA (1995) : Standard methods for the examination of water and wastewater, 15th Edn., American Public Health Association-AWWA-WEF. Washington.

Bagarinao, T. and Vetter, R.D. (1992): Sulfide-hemoglobin interactions in the sulfide - tolerant salt marsh resident, the California killifish *Fundulus parvipinnis. J. Comp. Physiol.* B : Biochemical,systemic and Environmental Physiology, 162 (7) : 614–624.

Brusle, J. Gonzàlez I, and Anadon, G. (1996) : The structure and function of fish liver. In: *Fish Morphology* (Munshi, J.S.D. and Dutta. H.M.,ed.) Science Publishers Inc., Newyork, p: 458.

Chang, S., Zdanowicz, V.S. and Murchelano, R.A. (1998) : Associations between liver lesions in winter flounder (*Pleuronectes americanus*) and sediment chemical contaminants from north-east United States estuaries. *Journal of Marine Sciences.* 55 : 954-969.

Dhanapakiam, P., Sampoorani, V., Kavitha, M., Ramasamy, V.K., Chandrakala, A. and Aruna, K.C. (2002) : Gill lesions in the major carp, *labeo rohita* exposed to sublethal concentrations of tannery effluent *J. of. Experimental biology.* 25 (3) : 333-336.

Dutta, H. M. (1996): A composite approach for evaluation of the effects of pesticides on fish. In: *Fish morphology* (Munshi J.S.D. and Dutta H.M.,ed.) Science Publishers Inc. p : 334.

Fracacio, R., Verani, N.F., Espíndola, E.L., Rocha, O., Rigolin-Sa, O. and Andrade, C. A. (2003). Alterations on growth and gill morphology of *Danio rerio* (Pisces, Ciprinidae) exposed to the toxic sediments. *Braz. Arch. Biol. Technol.* 46(4) : 685-695.

Grasshoff, K. (1983). Methods of Sea water analyses. Verlag. Chemic. Gymb weinheim. New York. 317 pp.

Gurr, E. (1962). Staining animal tissues: practical and theoretical, Leonard Hills Ltd, London. pp : 631.

Handy, R.D. and Penrice, W.S. (1993) : The influence of high oral doses of mercuric chloride on organ toxicant concentrations and histopathology in rainbow trout, *Oncorhynchus mykiss. Comparative Biochemistry and Physiology.* (C) 106 : 717-724.

Mohamed, F.A. (2001) : Impacts of environmental pollution in the southern region of Lake Manzalah, Egypt, on the histological structures of the liver and intestine of *Oreochromis niloticus* and *Tilapia zillii. Journal of Egyptian Academic Society for Environmental Development.* 2 : 25-42.

Nandan, S. (2002) : Impact of retting activity in the wetland ecosystems of Kerala-critical analysis and management strategies, Proceedings of the Fourteenth Kerala Science Congress, 452-454.

Pacheco, M. and Santos M.A. (2002) : Biotransformation, genotoxic and histopathological effects of environmental contaminants in European eel (*Anguilla anguilla* L.). *Ecotoxicology and Environmental Safety.* 53. 331-347.

Perry, S.F. and Laurent, P. (1993) : Environmental effects on fish gill structure and function. (J.C. Rankin and F.B. Jensen, ed.) *Fish Ecophysiology.* Chapman and Hall, London. 231-264.

Polcksic, V. and Mitrovic-Tutundzic, V. (1994): Fish gills as a monitor of sublethal and chronic effects of pollution. (Müller, R. and Lloyd, R. ed.) Sublethal and Chronic effects of pollutants on freshwater fish. Oxford, Fishing News Books. 339-352.

Rao, J.V., Rani. C.H.S., Kavitha, P., Rao, R.N. and Madhavendra, S.S, (2003) : Toxicity of chlorpyrifos to the fish *Oreochromis mossambicus, Bull. Environ. Contam. Toxicol.* 70. 985-992.

Raman. N.S. and Devotta. S (2006) : Indian Environmental Standards for Industrial and sewage effluents discharge. Handbook on Indian Environmental Standards, Chapter 3. NEERI, Nagpur. pp : 154.

Roberts, R.J. and Rodger, H.D. (2001): The pathophysiology and systematic pathology. (Ronald J. and Robert, S. ed.) In : *Fish pathology,* 3rd Edn. Harcourt Publishers Ltd. 88 p.

Schwaiger, J.R., Wanke, S., Adam, M., Pawert, W., Honnen,W. and. Triebskorn, R. (1997) : The use of histopathological indicators to evaluate contaminant-related stress in fish. *Journal of Aquatic Ecosystem, Stress and Recovery.* 6 : 75-86.

Thophon, S., Kruatrachue, M., Upathan, E.S., Pokethitiyook, P., Sahaphong, S. and Jarikhuan, S. (2003) : Histopathological alterations of white seabass, *Lates calcarifer* in acute and subchronic cadmium exposure. *Environmental Pollution*. 121 : 307-320.

Veiga, M.L., Rodrigues, E.L., Pacheco, F.J. and Ranzani-Paiva, M.J.T. (2002) : Histopathologic changes in the kidney tissue of *Prochilodus lineatus*, 1836 (Characiformes, Prochilodontidae) induced by sublethal concentration of Trichlorfon exposure. *Brazilian Archives of Biology and Technology*. 45 (2) : 171-175.

Wood, C.M. (1991) : Branchial ion and acid-base transfer in freshwater teleost fish: Environmental hyperoxia as a probe. *Physiol. Zool*. 64(1): 68-102.

Address of Authors

Ashish Kr. Dwivedi
Facility for Ecological and Analytical Testing (FEAT),
Indian Institute of Technology,
Kanpur (208016), India.

Anand K. Dubey
Department of Environmental Sciences,
Dr. R.M.L Avadh University,
Faizabad-224001 (U.P.), India.

A.K. Singh
National Bureau of Fish Genetic Resources
Canal Ring Road P. O. Dilkusha,
Lucknow-226002 (U.P.) India.

Amit Dubey
Lake Conservation Authority of MP
E-5 Arera Colony, Bhopal- 246016, India.

A. Arunachalam
Division of Natural Resource Management,
Indian Council of Agricultural Research,
KAB-II, Pusa, New Delhi-110012, India.

A.K. Das
Department of Botany, Rajiv Gandhi University,
Rono Hills, Itanagar, Arunachal Pradesh-791 112, India.

Altaf H. Ganai
Limnology Research Laboratory, Department of Zoology,
Aligarh Muslim University, Aligarh- 202002 (U.P.), India.

Abubakar Ansari

National Bureau of Fish Genetic Resources
Canal Ring Road P.O. Dilkusha, Lucknow-226002.

Asif Ali Khan
Limnology Research Laboratory, Department of Zoology,
Aligarh Muslim University, Aligarh- 202002 (U.P.), India.

Anita Rana, Director
Janhit foundation, Shashtrinagar, Meerut (U.P.).

Arti Sharma
Department of Zoology, University of Jammu, Jammu
(180006), India.

Anjali Goel
Department of Chemistry,
Kanya Gurukula Mahavidyalaya,
Haridwar (U.K.).

Akansha Bhatia
Department of Environmental Science,
Kanya Gurukul Mahavidyalaya,
Gurukul Kangri University,
Haridwar (U.K.). India.

Anamika Srivastava
2/393-Viram Khand, Gomti Nagar,
Lucknow-226010 (India).

B.B. Sandwar
Department of Chemistry, Samastipur College,
Samastipur (Bihar).

B.R. Bamniya
Department of Environmental Sciences,
Mohan Lal Sukhadia University,
Udaipur- 313002, India.

Bharti Tyagi
Department of Zoology and Environmental Science
Kanya Gurukula, Gurukul Kangri University,
Haridwar (U.K.).

C.S. Kapoor
Department of Environmental Sciences,
Mohan Lal Sukhadia University,
Udaipur- 313002, India.

D.K. Sinha
Department of Chemistry, K.G.K. (P.G.) College,
Moradabad-244001, India.

D.S. Jaya
Department of Environmental Sciences,
University of Kerala. Kariavattom, P.O,
Kerala, PIN 695 581. India

D. Vyas
Dept. of Botany, Dr. H.S. Gaur University, Sagar,
M.P., India.

D.S. Malik
Department of Zoology & Environmental Sciences,
Gurukul Kangri University, Haridwar (U.K.).

D. Balasubramanian
Restoration Ecology Lab, Department of Forestry,
North Eastern Regional Institute of Science & Technology,
Nirjuli, Arunachal Pradesh-791 109, India.

D.R. Khanna
Department of Zoology and Environmental Science
Gurukula Kangri University, Haridwar (U.K.).

G. Matta
Department of Zoology and Environmental Science
Gurukula Kangri University, Haridwar (U.K.).

Geetika
Rural Institute for Higher Studies,
Birauli, Samastipur, (Bihar).

Gaurav
Shriram Institute for Industrial Research,
Delhi-7.

G.S. Mahecha
Department of Environmental Sciences,
Mohan Lal Sukhadia University,
Udaipur-313002,
India.

G.K. Dhingra
Dept. of Botany, Govt. P.G. College,
Uttarkashi, (U.K)., India.

Gaurav Kumar Rastogi
Department of Applied Sciences & Humanities,
Sunderdeep Engineering College, Ghaziabad-201001, India.

G.K. Dhingra
Department of Botany R. C. U. Govt. P. G. College
Uttarkashi- 249 193 (Uttarakhand)

Habeeba A. Kabir
Limnology Research Laboratory, Department of Zoology,
Aligarh Muslim University, Aligarh- 202002(U.P.), India.

Jaswant Singh
Department of Environmental Sciences,
Dr. R.M.L Avadh University,
Faizabad-224001 (U.P.), India.

Kishore Boodhoo
Faculty of Science, University of Mauritius,
Reduit, (Mauritius).

K.P. Singh
Membrane Biophysics & Nanobiosensor Research Laboratory,
CBSH G.B. Pant University of Agricultural & Technology,
Pantnagar-263145, (Uttarakhand).

K. Arunachalam
School of Environment & Natural Resources,
Doon University, Dehradun, Uttarakhand-248001, India.

K. Boodhoo
Faculty of Science, University of Mauritius,
Reduit (Mauritius).

Krishna Gopal
Indian Institute of Toxicology Research,
Lucknow, India.

K.K. Sharma
Department of Zoology, University of Jammu Jammu
(180006), India

Mukesh Dixit
Lake Conservation Authority of MP, E-5,
Arera Colony, Bhopal- 246016, India.

M.K. Sharma
National Institute of Hydrology,
Roorkee.(U.K.) India

Mukesh Kumar
Dept. of Botany, Sahu Jain P.G. College,
Najibabad, (U.P)., India.

Manjul Mungali
Membrane Biophysics & Nanobiosensor Research
Laboratory, CBSH G. B. Pant University of Agricultural
& Technology, Pantnagar-263145, (Uttarakhand).

Madhurima Pandey
Centre for Environmental Science & Engineering,
Department of Engineering Chemistry,
Bhilai Institute of Technology, Durg, (C.G.), India

Nitin Kamboj
Department of Zoology & Environmental Sciences,
Gurukula Kangri University, Haridwar (India).

Namita Agrawal
J-336, Shivalik nagar, Haridwar, Uttarakhand.

Padma S. Vankar
Facility for Ecological and Analytical Testing (FEAT),
Indian Institute of Technology, Kanpur (208016), India.

P. Unauth, Faculty of Science, University of Mauritius,
Reduit (Mauritius).

Piyush Kant Pandey
Bhilai Institute of Technology, Raipur, New Raipur,
(CG), India

P.K. Bharti
Shriram Institute for Industrial Research,
Delhi-7.

Preetika Shukla
Department of Zoology & Environmental Science,
Gurukula Kangri University, Hardwar (Uttarakhand).

Prashant Tyagi
Department of Zoology and Environmental Science
Gurukula Kangri University, Haridwar (U.K.).

Pawan Kumar
Department of Zoology and Environmental science,
Gurukula Kangri University, Haridwar (U.K.).

Praveen Chauhan
GRD, IMT, Engineering College Dehradun (U.K.).

Pratima M.
Dept. of P.G. Studies and Research in Botany,
Gulbarga University, Gulbarga (585 104), India.

Ruchi Singh
National Bureau of Fish Genetic Resources
Canal Ring Road P. O. Dilkusha, Lucknow-226002
(U.P.) India.

Rashmi Yadav
Sector-8, Faridabad, Haryana.

R. Bhutiani
Department of Zoology and Environmental Science
Gurukula Kangri University, Haridwar (U.K.).

R.S. Meena
Department of Environmental Sciences,
Mohan Lal Sukhadia University,
Udaipur- 313002, India.

Rajan K. Gupta
Dept. of Botany, Pt. L.M.S. Govt. P.G. College,
Rishikesh, Dehradun, (U.K),
India.

Ritesh Saxena
Department of Chemistry, K.G.K. (P.G.) College,
Moradabad-244001, India.

Ramdas
Department of Botany R.C.U. Govt. P.G. College Uttarkashi-249 193 (Uttarakhand)

Ravindra Joshi
G.B.P.H.*I.E.*D., Kosi-katarmal, Almora (U.K.), India.

R.K. Gupta
Department of Botany Pt. L.M.S. Govt. P.G. Autonomous College, Rishikesh (Dehradun) Uttarakhand.

R. Sreejai
Department of Environmental Sciences,
University of Kerala. Kariavattom.P.O,
Kerala, India PIN 695 581.

Roshan T Ramessur
Faculty of Science, University of Mauritius,
Reduit, (Mauritius).

Sunil Kumar
Department of Zoology, D.A.V. (P.G.) College,
Dehradun (U.K.).

Swati Nagar
Department of Zoology & Environmental Sciences,
Gurukul Kangri University, Hardwar (U.K.).

Subrata Pani
Lake Conservation Authority of MP E-5,
Arera Colony, Bhopal- 246016, India.

Sanjay Kumar Gupta
Environmental Engineering,
Department of Civil Engineering Indian Institute of Technology Delhi, Hauzkhas, New Delhi – 110016.

Sunita Gupta
Department of Engineering Chemistry,
Bhilai Institute of Technology, Durg, (CG), India

Saltanat Parveen
Limnology Research Laboratory, Department of Zoology,
Aligarh Muslim University, Aligarh- 202002 (U.P.), India.

Shazia Ansar
Limnology Research Laboratory, Department of Zoology,
Aligarh Muslim University, Aligarh- 202002 (U.P.), India.

Sangeeta Madan
Department of Environmental Science,
Kanya Gurukul Mahavidyalaya, Gurukul Kangri University,
Haridwar (U.K.).

Shvetambri
Department of Zoology, University of Jammu, Jammu
(180006), India

Shalini T. Ramessur
Faculty of Social Studies and Humanities,
University of Mauritius, Reduit, (Mauritius).

Shiddamallayya N.
Regional Research Institute (Ay.), Ashoka Pillar,
Jayanagar, Bangalore (560 011), India.

Umesh Bharti
Department of Zoology & Environmental Sciences,
Gurukula Kangri University, Haridwar (U.K.) India.

Uzma Ahmad
Limnology Research Laboratory, Department of Zoology,
Aligarh Muslim University, Aligarh- 202002 (U.P.), India.

Virendra Mishra
Indian Institute of Toxicology Research, Lucknow, India.

V. Prasad
Dept. of Botany, TRM Campus, Tribhuvan University,
Birganj, Nepal.

Vidhya Kapasya
Department of Environmental Sciences,
Mohan Lal Sukhadia University, Udaipur- 313002, India.

Vijender Singh
Shriram Institute for Industrial Research, Delhi-7.

Index